Lecture Notes in Civil Engineering

Volume 257

Series Editors

Marco di Prisco, Politecnico di Milano, Milano, Italy

Sheng-Hong Chen, School of Water Resources and Hydropower Engineering, Wuhan University, Wuhan, China

Ioannis Vayas, Institute of Steel Structures, National Technical University of Athens, Athens, Greece

Sanjay Kumar Shukla, School of Engineering, Edith Cowan University, Joondalup, WA, Australia

Anuj Sharma, Iowa State University, Ames, IA, USA

Nagesh Kumar, Department of Civil Engineering, Indian Institute of Science Bangalore, Bengaluru, Karnataka, India

Chien Ming Wang, School of Civil Engineering, The University of Queensland, Brisbane, QLD, Australia

Lecture Notes in Civil Engineering (LNCE) publishes the latest developments in Civil Engineering - quickly, informally and in top quality. Though original research reported in proceedings and post-proceedings represents the core of LNCE, edited volumes of exceptionally high quality and interest may also be considered for publication. Volumes published in LNCE embrace all aspects and subfields of, as well as new challenges in, Civil Engineering. Topics in the series include:

- Construction and Structural Mechanics
- Building Materials
- Concrete, Steel and Timber Structures
- Geotechnical Engineering
- Earthquake Engineering
- Coastal Engineering
- Ocean and Offshore Engineering; Ships and Floating Structures
- Hydraulics, Hydrology and Water Resources Engineering
- Environmental Engineering and Sustainability
- Structural Health and Monitoring
- Surveying and Geographical Information Systems
- Indoor Environments
- Transportation and Traffic
- Risk Analysis
- Safety and Security

To submit a proposal or request further information, please contact the appropriate Springer Editor:

- Pierpaolo Riva at pierpaolo.riva@springer.com (Europe and Americas);
- Swati Meherishi at swati.meherishi@springer.com (Asia - except China, and Australia, New Zealand);
- Wayne Hu at wayne.hu@springer.com (China).

All books in the series now indexed by Scopus and EI Compendex database!

More information about this series at https://link.springer.com/bookseries/15087

Dmitry Ivanov · Aleksandr Panin · Inna Sukhanova
Editors

Proceedings of ECSF 2021

Engineering, Construction, and Infrastructure
Solutions for Innovative Medicine Facilities

 Springer

Editors
Dmitry Ivanov
Department of Forensics
SPbGASU
St. Petersburg, Russia

Aleksandr Panin
SPbGASU
St. Petersburg, Russia

Inna Sukhanova
SPbGASU
St. Petersburg, Russia

ISSN 2366-2557   ISSN 2366-2565 (electronic)
Lecture Notes in Civil Engineering
ISBN 978-3-030-99879-0   ISBN 978-3-030-99877-6 (eBook)
https://doi.org/10.1007/978-3-030-99877-6

This Springer imprint is published by the registered company Springer Nature Switzerland AG
The registered company address is: Gewerbestrasse 11, 6330 Cham, Switzerland

Contents

Research on the Community Renewal of "Active Adaptation to the Elderly" in the Unit Compound

Linfei Han and Bo Wang

Abstract In aging society, the unit compound, a representative residential form in Chinese cities, is facing a trend of adaptation to the aging. Based on the spatial characteristics of the community and neighborhood interaction, this article guides the renewal of the "actively suitable for the elderly" community in the unit compound. We should make use of the existing resources of the compound for the elderly, improve the spatial organization and utilization of the compound and the communication with the community, and increase and optimize the layout of elderly care facilities in a targeted manner. Under the background that the characteristics of the unit compound are becoming increasingly equivalent to the urban community, the introduction of the excellent experience of typical foreign "actively suitable for the elderly" community construction, the reconstruction of the unit compound's elderly care space system, and the establishment of an integrated elderly care community information system, which serves as a strategy for the community renewal of the unit compound.

1 Introduction

1.1 Background

1.1.1 Population Aging Society and "Active Aging" Theory

The aging of the population is a phenomenon of the natural development of society under the rapid economic development. Since the beginning of the twenty-first century, the structure of the elderly population has increased sharply, and the problem of population aging has gradually become prominent and has risen to China's national strategic issues [6]. In recent years, the issue of population aging has attracted the attention of people from all walks of life. Through construction and measures such as controlling the composition of the population, improving the pension mechanism,

L. Han (✉) · B. Wang
Beijing Jiaotong University, Beijing, China
e-mail: hanlf@bjtu.edu.cn

and creating a new type of elderly community, we will focus on the problem of population aging and improve the quality of life of the elderly in urban communities.

Since the late 1980s, with the continuous progress of social development, the aging theory has undergone the development from successful aging, healthy aging, productive aging, and then to "active aging" [2]. The "active aging" theory focuses on the self-development of the elderly and is committed to enabling the elderly in the twenty-first century to realize their physical health and personal value [15]. According to the status of the elderly in China, the "active aging" theory is more suitable to guide the community renewal of the unit compound.

1.1.2   Main Features of Chinese Unit Compound

In the era of China's planned economic system, the unit compound was one of the main residential models built by major state-owned enterprises. The long history of construction shows that the base and proportion of the elderly in the compound greatly exceeds that of other residential areas. Therefore, the unit compound should be the focus of research and attention to the problem of population aging. In the early days of the founding of the People's Republic of China, major units represented by factories, hospitals, and party and government agencies, to facilitate the work and life of their employees, usually select locations near their worksites according to units and departments. The state allocates land, and the construction is only for the employees of the units to live in. And the residential area used [4, 9], this type of residential area with Chinese characteristics is called the unit compound.

1.1.3   An Overview of the Problem of Aging Communities in Chinese Unit Courtyards

As the times change, the unit compound faces the problem of aging communities. The service facilities, community management, and internal environment in the compound all show the problem of single function and aging, which reflects many phenomena of discomfort. The original service facilities such as water supply, power supply, management, and sanitation are far from being able to meet the diversified functional needs of the elderly for business, culture, medical care, education, sports, and entertainment; the closed community self-management model of the compound, Compared with the humanized and systematic property management in modern communities, there are disadvantages such as rigid management and low efficiency; the internal environmental functions of the compound are relatively simple, and there is a lack of thinking about the planning of places for the elderly to travel, services for the elderly, and community activities.

With the continuous loss of the young population, the proportion of elderly people in the unit compound is rising, and the pressure on the community to provide for the elderly is increasing [11]. Due to the limited living space in the compound, most of the children of each family choose to move out of the compound and live separately from their elders due to factors such as family needs and home habits. The phenomenon of family separation between generations has exacerbated the aging trend of the unit compound, making the limited community pension resources even more scarce. The old-age management models of the unit compound are uneven, and there is a lack of unified and efficient management for the elderly. Affected by national conditions and traditional culture, family and social pensions that target the main body of subsidy, due to changes in the demographic structure and the continuous evolution of the pension system, the pension model has gradually changed to home pension and institutional pension [20]. Due to differences in funding and implementation efforts, the pension model that focuses on home-based care for the elderly has completely different pension policies in different community courtyards. The service facilities of each unit's compound cannot realize the unification of functions and the coordination of management, and it is difficult to reflect the reasonable allocation of community pension resources.

1.1.4 The Need for Community Renewal in the Compound from the Perspective of "Actively Adapting to the Elderly"

As of the end of 2019, the proportion of the elderly population aged 60 and over in China is 18.1%. The age structure of the elderly population continues to tilt towards an aging population, and the pension burden will become the first social burden after 2022 [12, 17]. The residential areas of the unit compound nature within the Fifth Ring Road in Beijing account for about 44.2% of all residential areas in the city. The large scale of the unit compound is of representative significance for my country's "actively adapting to the elderly" community renewal.

Building an "actively suitable for the elderly" compound community can effectively alleviate the physical and mental health problems of the elderly. Home quarantine in special periods such as urban epidemics [3] brings safety hazards to empty-nest elderly; most of the elderly over 80 have problems that cannot take care of themselves [13]; affected by factors such as illness, family changes, retirement, and making friends, The elderly often have a sense of loneliness and loss psychologically [18]. The concept of "active aging" focuses on the protection of the health of the elderly and provides the elderly with opportunities for self-learning and further improvement.

At present, the process of community renewal in the process of urban renewal is in the stage of "utilizing the stock". The primary goal of community renewal is to realize the micro-renewal of different community spaces through a diversified strategy of government-led, public participation, and community autonomy [8]. Based on my country's national conditions, to solve the problem of the elderly living alone and empty nests, the "active aging" renewal of the unit compound community is conducive to quickly alleviate the phenomenon of "the old have nothing to support". This in-situ renewal method has low cost, Easy to implement, quick to take effect, and easy for the public to accept.

2 Materials and Methods

2.1 Analysis of Active Aging Theory and Summary of Main Points

In 2002, the World Health Organization formally proposed that a policy framework for active aging should be established, with the goal of "health, safety, and security", and based on ensuring the health of the elderly, enhancing their value and social identity outside the elderly [14]. During the 13th Five-Year Plan period, my country took "actively responding to population aging" as an important national strategy and incorporated it into the party's development strategy, which also demonstrated my country's determination and active preparation to respond to the aging social transformation. Taking the compound community as a pilot and carrier for the development of the "active aging" theory is of positive significance for accelerating the healthy development and transformation of the community and creating a community suitable for the elderly.

2.1.1 A New Understanding of the Theory of Active Aging

The emergence of the theory of "active aging" started from a macro perspective at the national and social levels and is of great significance to the practical role of my country's urban planning system. The old urban community represented by the unit compound is the focus of future research and community renewal. Combining with the theory of active aging, the method of spatial organization and community environment construction from the perspective of urban planning should be used to establish a widely used and easy-to-expandable unit compound community renewal model.

2.1.2 Exploration of Community Renewal Under the Concept of Active Aging

The community is the most basic social management unit in a modern city [19]. In the era of market economy dominance, a unit compound not only tends to be integrated into the urban structure in space, but also gradually presents the characteristics of community management in management. The commercialization of housing and the openness of settlements have changed the traditional internal management structure of the unit compound. Social management agencies have replaced the self-management of enterprises, which has injected new vitality into the unit compound. The community management agency integrates more urban infrastructure, embeds diversified community services and urban functions in the compound, which virtually eliminates the external "soft boundary" of the unit compound, and better caters to social development. At the same time, the urban community renewal work is also actively advancing, and the pace of urban renewal and old community renewal is in the ascendant, which has brought a new source of power to the aging and renewal of the unit compound.

The fundamental purpose of building an "active and age-appropriate" community is to provide security for the elderly. The renewal of the community "actively adapting to the elderly" should mainly be innovated in the creation of community space and community management. Through the reconstruction of the community's space, the problems of inconvenience and walking difficulties for the elderly have been solved; the improvement of the community's management model has improved the quality of life and ensured physical and mental health, enabling the elderly to participate more in community activities and gradually adapt to the new social rhythm:

1. "Age-friendly" is the main concern for the future of the elderly community

The community's care for people of all ages, especially the attention and care of the elderly who are prone to illness and poor mobility, is an important part of the concept of "active aging". The construction and reform experience of foreign communities has been integrated into the interconnected relationship between the elderly and community institutions, solved the pain points of the elderly living in the geographical dimension, and reflected the characteristics of the elderly-friendly community. The future community for the elderly must be built around the physical and mental health of the elderly, and the elderly-friendly ideas should be reflected based on environmental friendliness. For communities of different sizes, especially for the unit compound in our country, a grid-based and node-based, top-down community service system should be established to provide all the elderly with accessible service paths.

2. Old-fashioned space network and information network

The systematic setting of resources for the elderly in the community helps to allocate resources rationally and comprehensively care for and manage the elderly. In terms of spatial organization, set up primary and secondary tiered community service centers and community service nodes, covering basic health care facilities for the elderly, and set up elderly care facilities combining medical care and rehabilitation and healing at the core. Construct an information network from within the community to between the communities as an invisible framework for the aging and renewal of the community. The use of scientific and technological means to monitor the activities and physical conditions of the elderly reflects the advanced nature of the Internet of Things and the Internet in the elderly care mechanism.

2.2 Unit Compound Community Renewal and "Active Aging"

It is necessary to comprehensively analyze the space and neighborhood character-istics of the unit compound, rationally use the resources of the compound that is suitable for the elderly, pay attention to the current situation of the elderly of all ages, and improve the problems and contradictions that are not beneficial to the community. The well-distributed spatial scale of the compound and the long-lasting neighborhood relationship structure are strong support for promoting the "actively adapted to the elderly" community renewal of the unit compound. Factors such as motor vehicle traffic interference and fragile neighborhood structure are the main disadvantages of elderly care in the compound. Multi-level improvements should be made in terms of space reengineering and management innovation, and a new space organization and management system for all-age elderly care should be created.

2.2.1   The Spatial Characteristics of the Unit Compound and "Actively Adapting to the Elderly"

The initial form of the unit compound is close to square and integrated. The enclosed courtyard wall encloses the living space, and the courtyard space becomes the only carrier for the internal residents to work, live and communicate. The living space of the unit compound is mainly composed of houses for the family members of employees, and the management department is gradually transferred to the living area; commercial, medical, education and other public service facilities are also increasing; to meet the higher requirements for the ageing of the space environment, the compound Old-age facilities such as landscaping, and greening should be added (Fig. 1).

The unit compound has many problems in terms of spatial organization. The unit compound is facing the renewal trend of urban renewal and community openness, but the unorganized spatial evolution will bring the disadvantages of not being old. The community renewal model under the urban renewal thinking emphasizes the flexible self-regeneration of the community, and mainly focuses on the suitability renewal of the community street micro-space [1, 7]. With the increase in residential buildings, the internal space of the unit compound presents the spatial characteristics of primary

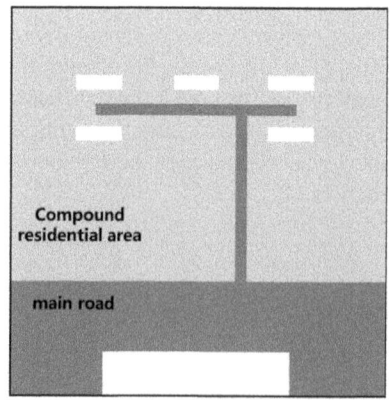 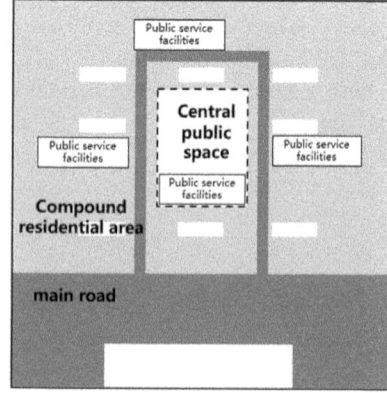

Fig. 1 The internal space characteristics and evolution of the unit compound

and secondary levels, and there is a lack of connection between scattered public service facilities. In the unit compound where people and vehicles are mixed, the circulation of the elderly is easily disturbed by motor vehicles, causing inconvenience and traffic safety problems for the elderly; the increasing number of motor vehicles leads to tight parking, which gradually reduces the elderly's mobility. Activity space: due to the early construction period, the unit compound lacks a scientific plan for the elderly walking system, which brings safety hazards to the elderly's activities.

2.2.2 The Original Neighborhood Characteristics of the Unit Compound and "Active Aging"

The independent construction and closedness of the unit compound determine the unique neighborhood relationship in terms of space. The residential area of the compound extends the industry management system of the unit system, and the crowd structure also revolves around "unit-compound-home". The industry relationship of the unit compound and the clustering characteristics of the population under the distribution mechanism naturally gather people from the same income class. The industry relationship of the unit and the geographic relationship of the compound space has become the main driving force for neighborhood communication [5].

The role of neighborhood characteristics for the elderly in the unit compound is invisible and indispensable. The crowd relationship network of the unit compound determines the harmony and stability of the neighbor relationship. The residents in the compound have a high degree of overlap in their lifestyles, and neighborhood communication activities are more frequent under the tribal life characteristics, which makes it easier for the residents of the compound to have a centripetal neighbor relationship. Enhancing the collective identity of the elderly in the compound has an obvious promotion effect on the development of the elderly care work, and it should be used protectively in the process of renewing the unit compound.

Over time, the neighborhood network has changed due to the influence of external factors such as the movement of people. Some families have gradually moved away from the compound and gradually alienated from their original colleagues and neighbors. The original neighbor relationship in the compound will gradually weaken; as some foreign tenants gradually flood into the unit compound, the originally stable neighbor relationship network will be new families, and populations are added to form new neighborhood characteristics [16]. The constantly updated neighborhood characteristics have broken the social habits of the elderly in the unit compound and brought challenges to the elderly. The elderly in the compound often finds it difficult to adapt to changes in intergenerational communication patterns. Due to the weakened physical function and the more sensitive psychology, the elderly group's activities and social circle in life are more limited, and they rely more on the stable and simple neighborhood communication mode in the unit compound. The original neighborhood relationship is an important factor in maintaining the community communication of the elderly. Using the original neighborhood relationship to cultivate and establish a new community-appropriate communication model for the elderly is an important goal for the renewal of the neighborhood communication model of the unit compound.

3 Results

The unit compound should absorb the advanced experience of the international community for the elderly, reconstruct the community space system. It should reflect the construction of a community digital management system that is "active for the elderly". In the compound, in addition to satisfying basic medical services and community operating facilities, new technologies and new technologies should also be combined to realize the data statistics of the old community and the interconnection of information inside and outside the community.

3.1 Reconstruct the Spatial System of the Old Community

The construction unit's courtyard is suitable for the old space system, and the traffic flow lines for the separation of people and vehicles are planned to reduce the interference of motor vehicles and parking spaces on the activities of the elderly, which reflects the renewal concept of priority. Through the core-corridor-node three-level community-appropriate service space system for the elderly, supplemented by barrier-free pedestrian corridors for the elderly distributed throughout the compound to connect the pedestrian circulation in series (Fig. 2).

The centralized community service center, with the elderly complex as the development goal, should be placed in the center of the community's 15-min life circle to improve the radiation scope and service efficiency of community elderly care services. If the size of the unit compound exceeds the reach of a 15-min walking circle (800 m), one should consider setting up two service centers at a time. The service center is highly integrated with various community elderly care service functions, and theoretically should include all the functions of modern community life. The service center should include elderly service facilities that integrate activities for the elderly, employment for the elderly, nursing care for the elderly, medical treatment for the elderly, and healing for the elderly.

Renew the public space in the house into a dedicated corridor for the elderly in the community for the elderly to rest, entertain, communicate, and walk. The scale of the corridor is suitable to meet the width of two wheelchairs facing each other, not less than 1.8 m; in areas such as corridors and residential unit entrances, road crossings, idle buildings, and other areas, focus on expanding public space and reflect the age-appropriateness Features. Set up seats and wind and rain corridors to meet the short-term rest of the elderly; set up a full-time monitoring system in the corridors, interconnect with the cloud management system of the community service center

Fig. 2 Schematic diagram of the community space network update of the unit compound

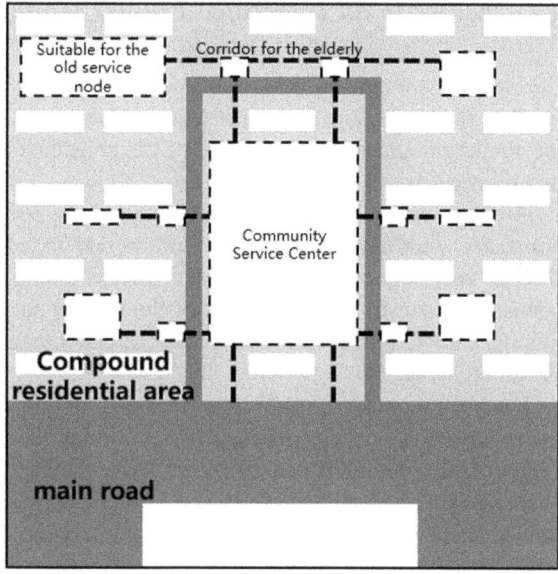

and provide real-time feedback on the behavior trajectory and health status of the elderly (Fig. 3).

Attention should be paid to the establishment of small-scale service nodes suitable for the elderly in the unit compound. Focus on using unit entrances and residential entrances to set up small nodes for elderly care services. Basic facilities such as monitoring devices for the elderly, health alarms for the elderly, and leisure seats should be added to meet the daily small-scale communication activities and self-health assistance of the elderly. Each service node suitable for the elderly shall cover all households of at least one residential unit.

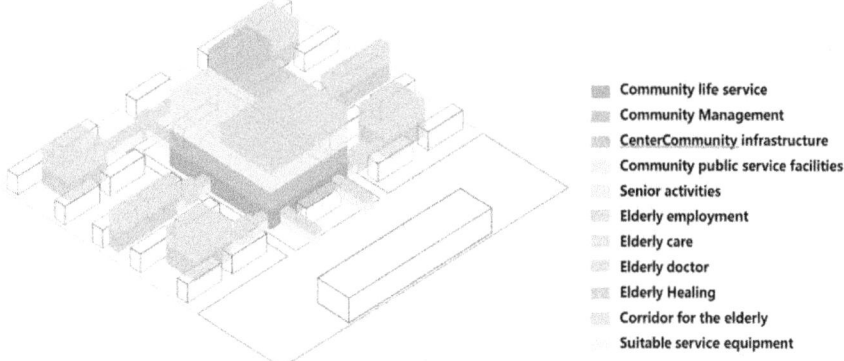

Fig. 3 Schematic diagram of the updated function distribution of the unit compound community

3.2 Establish an Integrated Elderly Management System

Improve the community pension agreement between the unit compound, the social medical institution, the government, and every family with the elderly, and improve the funding sources and feasibility of the pension system using pension subsidies and property-based pensions.

Archives should be established for all elderly people, and resources for elderly assistance within and outside the community should be allocated reasonably to provide real-time protection for the elderly at home. According to age and health status, distinguish the information of the elderly such as the youngest-the elderly, nursing care-self-care, etc., and provide targeted management and care services for the elderly (Fig. 4).

Establish the Internet of Things mechanism in the community and the cloud data interconnection mechanism between the communities. The elderly information cloud center of the community can realize real-time monitoring and management and build an urban elderly health network with other communities, an elderly information management node. It is mainly used to process the elderly information of several small residential units in the community and use products such as health bracelets and health cards to provide health information updates and real-time positioning services for each elderly person, especially to solve the home isolation period during the epidemic The health protection of the elderly.

For the people aged 60–70, certain community jobs are provided according to their hobbies and expertise. Let some elderly people become a member of the internal management system of the unit compound, give play to the strong self-organization ability of the unit compound, and re-strengthen the fragile neighbor relationship in

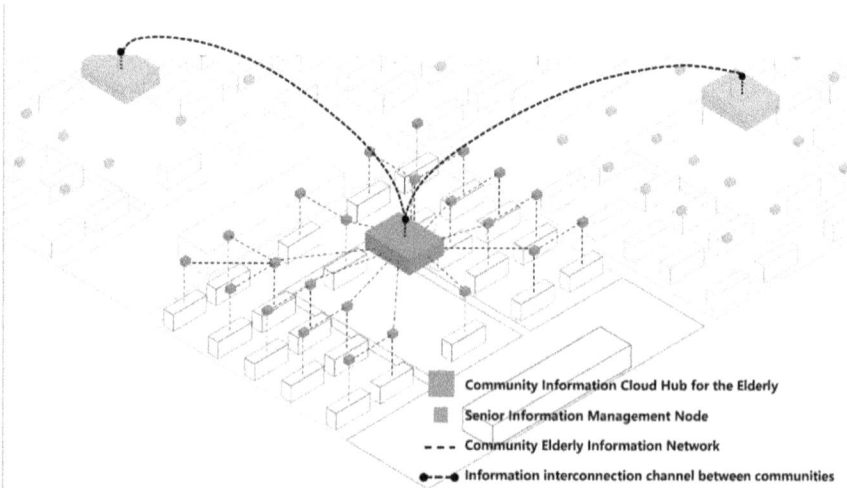

Fig. 4 Schematic diagram of community information network after unit compound community update

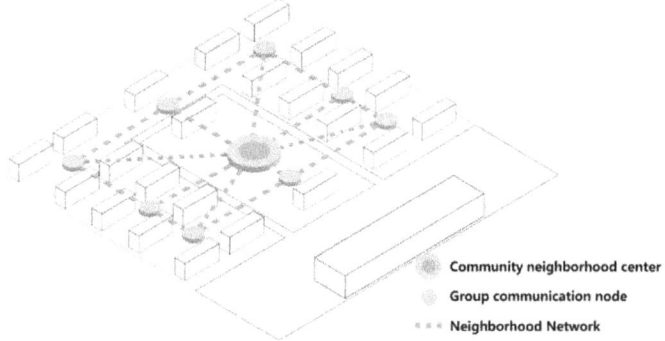

Community neighborhood center

Group communication node

Neighborhood Network

Fig. 5 Schematic diagram of community neighborhood network after unit compound community update

the unit compound. Rebuilding the neighborhood network of the unit's compound encourages the elderly to participate in community care work, psychologically not only bringing a positive attitude towards life to themselves but also having a positive impact on the elderly neighbors around them (Fig. 5).

4 Discussion

Some foreign age-appropriate communities have established a new type of elderly care service system, which has important reference significance for the "active age-appropriate" community renewal of the unit compound. Singapore's naval complex and Japan's Toyoshikitai community mainly reconstructed the community's elderly care space, integrated the community's elderly care service facilities, and created a small-scale elderly care neighborhood unit. Its successful experience is the community's "active aging" renewal thinking, which mainly covers the concept of friendship for the elderly and the construction of a community network. For the unit compound in China, it is necessary to build a community space network and an age-appropriate management network based on the current state of care issues and learn from foreign advanced experience.

4.1 Foreign Experience of "Actively Adapting to the Elderly" Community Construction

Some countries in the world have entered the stage of population aging earlier, such as Singapore and Japan. Thanks to advanced design ideas and special social driving methods, the above-mentioned countries have made rapid progress in building communities suitable for the elderly. Community renewal in foreign countries uses the creation of a multi-functional elderly community and the renewal and transformation of the community suitable for aging to create a comprehensive elderly care building suitable for the characteristics of the country and a comprehensively managed community elderly care center.

4.2 Singapore Kampung Admiralty

As an elderly care community building complex integrating elderly care services, medical care, elderly entertainment, and elderly apartments, the Singapore Kampung Admiralty represents the future development direction of urban elderly care building space intensification. In terms of the design concept, the Kampung Admiralty retains the concept of "village". By connecting different community spaces and using vertical traffic and vertical functional partitions as the connection, the old people's impression and use of traditional nursing homes are maintained. habit.

The design and construction of the Kampung Admiralty embody the concept of active aging. The Kampung Admiralty represents the development direction of the community elderly care complex on intensive urban land. Its successful experience lies in: reflecting the spatial classification in the vertical space, with the central elderly care service facility as the core, connecting the service nodes of other floors. To form an integrated vertical community network for the elderly. For the unit compound, it should focus on the integration of the internal space of the community and allocate the elderly service facilities in the compound in a targeted manner to create a hierarchical activity space for the elderly.

4.3 Toyoshikidai Community, Japan

Toyoshikidai community is a representative old residential area in Kashiwa City. The migration of young people decreases with the aging of residential facilities, and the aging rate of the community exceeds 40% [10], forcing the community to rebuild in situ. The Toyoshikidai community has re-planned the elderly service facilities in the key reconstruction areas and created a management mechanism for the elderly suitable for the elderly based on the tenet of "the original site is suitable for the elderly". The community has reconstructed some old spaces and vacant buildings, placing community medical care, community dining, reemployment of the elderly, parenting homes, special care centers, and other institutions in the center of the 30-min living circle. The old-fashioned buildings with full-time care provide different service modes for self-care and nursing care of the elderly. The medical and nursing system for the elderly at home, the establishment of a data-sharing system, the establishment of files for the elderly, and the comprehensive monitoring of the health of the elderly.

The Toyoshikidai community is a representative case of community renewal in Japan. It mainly exerts the advanced nature of community elderly care management, carries out community renewal on the original site, and centralizes the elderly care functions such as medical care, healing, and special care. The unit compound in our country should learn from the updated management experience of the community, build the service radius of community elderly care management, and create an integrated service facility for all-aged seniors; combine big data and information interconnection to build archives and share resources for the elderly.

5 Conclusions

The unit compound should closely follow the development of the times and carry out the "active aging" community renewal from a systematic perspective. The community elderly care space system reorganized the elderly care service facilities in the unit compound, emphasized the transportation planning based on the concept of pedestrian and vehicle branches and the elderly walking network, and rationally planned the level of functional space suitable for the elderly, and solved the limited walking accessibility of the elderly and elderly care services. The problem of insufficient radiation coverage of the facility. The integrated community elderly management system improves the original unit compound courtyard management model, protects the physical and mental health of the elderly, and continues the original neighborhood emotions of the compound. It is also necessary to combine the geographical characteristics of each unit compound, optimize community management according to local conditions, improve the management level of the compound community, strengthen the quality of age-appropriate management, and provide new and effective solutions for the unit compound's "active aging".

References

1. Baoxing C (2009) Complex science and urban planning changes. Urban Plann 33(04):11–26
2. Chunfen T, Jiatong L (2017) Progress and prospects of active aging research. Aging Sci Res 5(09):69–78
3. Dong L (2020) Community protection strategies for novel coronavirus pneumonia. Med Herald 39(03):315–318
4. Fan Z (2006) The decomposition of the unit compound (part one). Beijing Plann Constr 02:67–70
5. Fengfeng X (2004) Research on neighborhood communication in urban communities. Archit J 04:26–28
6. Guangzong M, Tuan Z (2011) The development trend of China's population aging and its strategic response. J Central China Normal Univ (Humanities Soc Sci) 50(05):29–36
7. Hao L (2018) Disappearance and awakening of publicity—the value paradigm of inner city street community renewal governance based on space justice. Planner 34(02):25–30

8. Jingjing L, Yuelai L, Xiao L (2020) Exploration of the realization ways of public participation in the micro-renewal of old communities: taking the community of Lane 580, Zhengli Road, Yangpu Chuangzhi District Shanghai as an example. Landscape Archit 27(10):92–98
9. Le W (2010) The evolution mode of unit compound and its influence on urban space. Dalian University of Technology
10. Lianghua D, Dian Z, Jing H et al (2018) A study on the aging of community based on comprehensive regional care—taking Toyoshikidai in Kashiwa City Japan as an example. Archit J S1:45–49
11. Meimei W, Yongchun Y, Yiming T et al (2015) Intergenerational separation/cohabitation behaviors of Chinese urban residents and their influencing factors: taking Chengdu as an example. Acta Geogr Sin 70(08):1296–1312
12. National Bureau of Statistics (2019) The National Economic and Social Development Statistical Bulletin of the People's Republic of China in 2019. People's Daily, 2020–02–29(005). http://www.stats.gov.cn/english/PressRelease/202002/t20200228_1728917.html. Accessed 21 May 2021
13. Peng D (2013) An analysis on the heath status of the older persons in China. Population Econ 6:3–9
14. Rudnicka E, Napierała P, Podfigurna A et al (2020) The World Health Organization (WHO) approach to healthy ageing. Maturitas 139:6–11
15. Xin Y, Jin Niu Y (2020) Population urbanization, "active aging" and urban planning suitable for aging. Urban Rural Plann (03):25–30+36
16. Zengyuan L (2013) Separation and integration: transforming the flow of farmers and community integration in society. Central China Normal University
17. Zhai Z, Chen J, Li L (2016) Population aging in China: megatrends, new features and corresponding pension policies. J Shandong Univ (Philos Soc Sci) 1:27–35
18. Zhenyun W (2003) The connotation, evaluation and research overview of the mental health of the elderly. Chin J Gerontol 12:799–801
19. Zhuo Z (1997) The changes of China's grass-roots community organizations. Soc Res 04:15–25
20. Zongchuan Y (2000) Home-based care and China's old-age care model. Econ Rev J (03):59–60+68

Application of BIM Technology in Hospital Engineering Project

Zhaodong Wang

Abstract On account of the outpatient medical technology building's characteristics of multi-subjects of the mechanical and electrical system, large and difficult of construction coordination, and high demands of post-operation and maintenance management in the removal project of one county People's Hospital, the BIM technology application will optimize the profession and process of the hospital project, simulate the construction process and auxiliary coordinated management, finally summarize the advantage of BIM technology application in the medical building projects.

1 Introduction

BIM (Building Information Modeling) is a technology that describes the physical and functional characteristics of Building facilities based on three-dimensional geometric models. With the help of the ability of BIM's three-dimensional visualization and building information integration, it has become a common practice for major projects to guide the design and construction process [1]. It takes the BIM technology application of the outpatient medical technology building project of the removal project in one County People's Hospital as an example, researching BIM how to solve the problems that exist in the professional integrated application during the hospital construction project, such as the architectural engineering project construction, structure, mechanical and electrical, hardcover and medical special projects, and promoting the elaborating management level about project progress, quality, safety and cost, further exploring the integration of operation and maintenance data and the interactive operation of operation and maintenance model based on BIM technology.

Z. Wang (✉)
Henan Urban Construction University, Pingdingshan City, Henan Province, China
e-mail: intj@hncj.edu.cn

© The Author(s), under exclusive license to Springer Nature Switzerland AG 2023
D. Ivanov et al. (eds.), *Proceedings of ECSF 2021*, Lecture Notes in Civil
Engineering 257, https://doi.org/10.1007/978-3-030-99877-6_2

2   Materials and Methods

Project Profile. The removal project of outpatient and emergency medical technology building of a county People's Hospital is located at the intersection of Jingsi Road and South Second Ring Road of a county in Henan Province. The building has one floor underground and five floors above ground, and its total length is 155.5 m and the width is 104.5 m, and the building height is 22.35 m, and the total construction area is 76182.7 m². It is raft foundation and frame-shear wall structure.

The basement on the first floor has a single-storey area of 19,774.11 m², with up to 6 post-cast belts. The concrete consumption of the basement is about 24,600 m³, and the steel reinforcement is about 2400 t, and the formwork is about 55,000 m². There are 22 wards in the project. After completion, it can accept more than 2500 hospitalized patients and elderly rehabilitation personnel at the same time, which will play an active role in the development of local social security. The rendering of the completed project is shown in Fig. 1.

Key and Difficult Points of the Project. The scale of the hospital project is huge. It has a complex process design, and many technical difficulties and professional docking work. What's more, the time limit is tight. In the hospital, patients, family members, visitors, medical workers, administrative staffs, and other personnel are dense and multifarious, so the operation and maintenance service are difficult. Different areas of the hospital, such as public area, special area and special area, have different requirements for temperature, room pressure, exhaust, sewage discharge and ventilation. And the forms of energy (including electricity, water, gas and steam) are various, with high requirements for quality assurance of operation and maintenance.

At the same time, the hospital project has many specialized systems and complex medical equipment. The construction among various specialties has high complexity, so it involves numerous specialties and units. The participating units include civil engineering, mechanical and electrical, steel structure, curtain wall, weak current

Fig. 1 Project renderings of the outpatient and emergency medical technology building of a county people's hospital relocation project

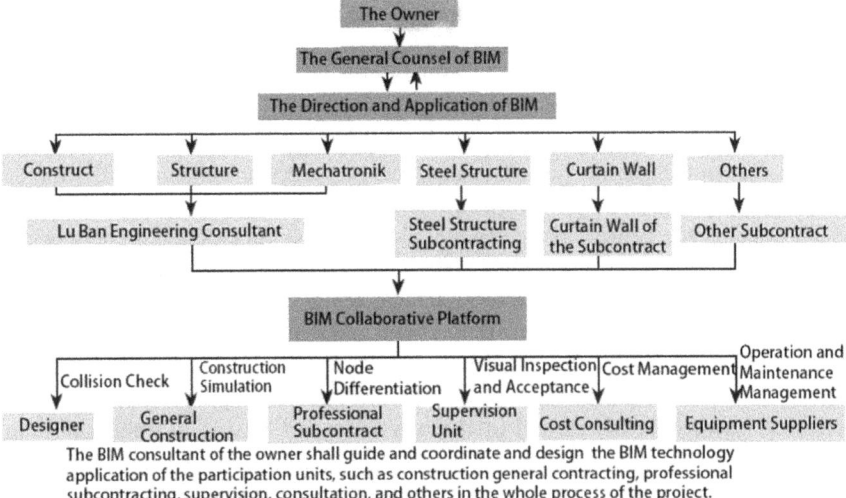

The BIM consultant of the owner shall guide and coordinate and design the BIM technology application of the participation units, such as construction general contracting, professional subcontracting, supervision, consultation, and others in the whole process of the project.

Fig. 2 Data integration management platform of BIM technology of this project satisfying the whole process

intelligence, elevator equipment, purification, and medical treatment, etc., so it is difficult to share information and coordinate on-site.

Application of BIM Technology. The BIM technology application of the hospital project aims to use BIM technology to review the design model and design drawings and deepen and coordinate the collision problems of various professions. Visualization technology is used to simulate and review complex construction processes [2]. At the same time, the use of BIM integrated management platform can carry out the fine management of project quality, schedule, safety, and cost. Also, to ensure the smooth progress of fieldwork, a whole-process data integration management platform for this project is built based on the whole-process collaborative management pattern of BIM, as shown in Fig. 2.

3 Results

Create BIM Model and Review Drawings. In the early stage of the project, a special BIM technical team was set up. Adopting the Luban series modelling software to create the model, the BIM technical team created all professional models of civil engineering, steel bar and installation combined with the project design drawings. During the modelling process of BIM, we found problems, such as the deficiency of rebar of civil engineering, the lack of appliances of water supply and drainage system in installing, the missing of one electrical system diagram, and the pipeline of equipment connection, and loop, etc. Through the model was established, the drawings problems can be found before pre-construction, which largely avoid the design changes during the construction process and even rework, and others.

Pipeline Layout and Optimal Arrangement. The mechanical and electrical major of this project involves many majors and needs to consider structural avoidance or reserved holes, etc. When using the traditional operation mode, the collision among electromechanical specialties and between electromechanics and structure occurring in the process of the project, often needs to be dismantled and reworked, which affected the construction schedule [3]. Besides, the pipeline arrangement is unreasonable, which will affect the indoor clearance height. Through the application of BIM technology to establish professional models, problems of unreasonable space layout and professional collision will be found and solved. Converting the model to format, Navisworks software is used in the software to find out the contradictions and components that do not meet the requirements through roaming function and collision detection function. With the aid of the ways, we can solve the design problems in architectural space in-depth. Finally, through the characteristics of visualization and coordination of BIM, we point out problems and communicate and discuss for them based on the BIM model, making the participants have a deep understanding of the design intention and improving the efficiency of communication, which is conducive to reducing design changes during construction and reducing construction risks, thus improving construction quality, shortening construction period, reducing construction cost and reducing resource waste.

Collision Check. The pipeline arrangement of this project is intricate. Especially near the equipment room and the walkway, the collision is easy to happen between the air duct, bridge frame and water supply pipe. The BIM team views the collision points through the BIM multi-specialty integration platform, carries out the check of each specialty and space collision in the BW system, and outputs detailed collision information and the report of collision point to facilitate the collaborative work among various departments, posts, and specialties. For these problems of various collision situations, the collision avoidance scheme standard was formulated. Through collision inspection, 374 effective collisions in the five floors above ground and 156 effective collisions on the first floor below ground were found and solved. Collision check provides the final basis for the project to reserve holes [4].

Check the Clear Height and Teserve Hole Positioning. Use the BIM model to check the clearance height of each area in advance. Problems are found in advance, and the clearance is optimized in advance. The clearance analysis report is issued, and changes in the later stage are reduced. Reducing engineering costs and improving the comfort of later use. At the same time, the BIM model after the comprehensive pipeline layout is used to check the accuracy of the reserved openings in the construction drawings and issues the analysis report of the reserved openings to guide the reserved openings accurately and improving the accuracy of the reserved holes.

According to the collision inspection results, the pipeline should be reasonably optimized and improving the overall quality under the condition that the design requirements are met. As shown in Fig. 3, it is the collision point example of a B2 refrigerating room. The safety issues of the design of structures of optimization results will be reported to the supervision unit through the general contractor, and

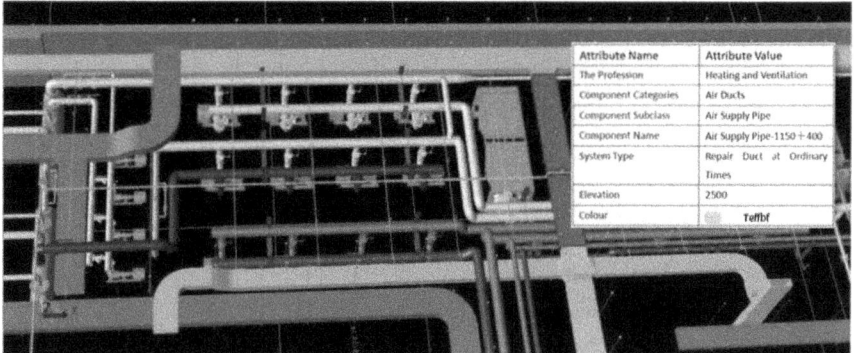

Attribute Name	Attribute Value
The Profession	Heating and Ventilation
Component Categories	Air Ducts
Component Subclass	Air Supply Pipe
Component Name	Air Supply Pipe-1150 + 400
System Type	Repair Duct at Ordinary Times
Elevation	2500
Colour	Teffbf

Fig. 3 B2 Example of collision point in refrigeration room

Fig. 4 The comprehensive layout of a certain part of the road after optimization

the final optimization scheme will be determined by the design unit, which will be changed after approval by the owner unit. According to the collision results, pipeline layout is optimized in combination with the corresponding pipeline layout principles. As shown in Fig. 4, it is the comprehensive layout of a certain part of the corridor after optimization.

Optimization of Construction Site Layout. In this project, we use BIM technology to arrange the site of construction and build a three-dimensional field model. As shown in Fig. 5, it changed the defect that the traditional ground floor plan can't intuitively show each professional information, so BIM technology is used to simulate the on-site construction, and reasonably arrange various components on-site, which reflects the scene layout problems in advance and adjust in time. Determine and optimize the position of tower cranes to further optimize the area of steel and wood storage and processing and living and office areas of project personnel, which achieves the goal of rational distribution, ensuring the on-site hydropower road meet the requirements, reducing the secondary handling distance of on-site materials, facilitating the work and life of project personnel, improving the space utilization, and reducing the production cost. An example of synchronous site layout is shown in Figs. 6 and 7.

Fig. 5 Three-dimensional construction site layout drawing

Fig. 6 The example 1 venue layout synchronization

Virtual Simulation of The Construction Process. The use of BIM technology can simulate the construction of the stage of this project and realize the functions of parametric design, virtual reality, structural simulation, computer-aided design and others, and have the realistic three-dimensional simulation of the flow of human, financial, and material information during construction, achieving the virtual simulation of the construction process on the computer to find the existing or possible problems in the project as early as possible. Figure 8 shows the virtual layout of the refrigerating machine room. This virtual simulation can provide each participant with a controllable, non-destructive, low-cost, low-risk and repeatable test method.

Fig. 7 The example 2 of venue layout synchronization

Fig. 8 The virtual layout of the refrigeration room

4 Discussion

The established BIM model is used to simulate the construction process, and the construction plan of the project can be compared directly and quickly with the actual progress anytime and anywhere. As shown in Fig. 9, the simulation of the construction schedule enables the construction party, the supervisor, and even the owner leaders who are not from the engineering industry to know all the problems and situations of the project like the palm of their hands [5]. Through the model, technicians and

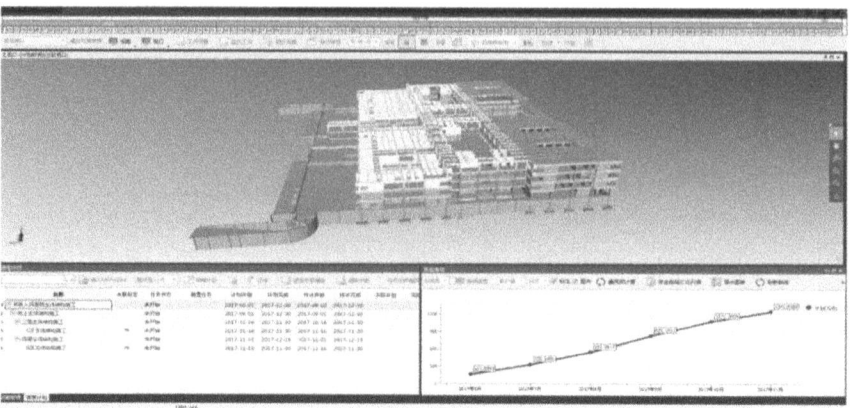

Fig. 9 Construction schedule simulation

construction personnel can predict the construction difficulties ahead of schedule, thereby improving the collaborative efficiency and speeding up the construction progress.

We divided the model into manageable working faces by flowing water segment methods and helped line managers arrange the production plan reasonably and avoided working face conflicts in advance, monitoring project progress intuitively. In the case of dust control and other construction encountered, we can adjust the planned duration to ensure the continued progress of the latter construction process can be controlled.

Delicacy Management of Material. Using BIM technology, and combing with the construction organization design, we carry on the fine management of the project material and use the function such as box selection to accurately calculate the material requirements of different parts and different time nodes, and reasonably arrange the entry of materials. Otherwise, we reasonably arrange the secondary masonry in combination with other materials, through the exporting CAD layout of bricks with one click in the automatic layout of bricks. As shown in Fig. 10, it is convenient for

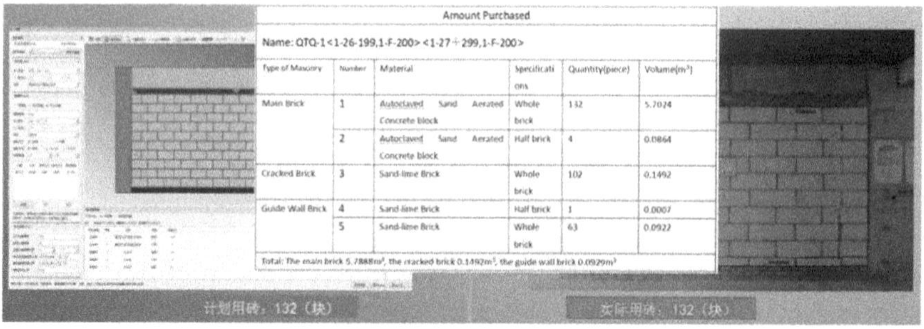

Fig. 10 Effect drawing of a secondary structure arrangement

workers to construct according to the drawing, effectively reducing the number of broken bricks, and ensuring the aesthetic quality of the wall. On this basis, combined with the field layout scheme to stack materials, and reduce the secondary handling or secondary handling distance, which is convenient for allocation of material resources [6].

Coordinating and Sharing of Information. The project uses mobile terminals (smartphones, tablet computers) to collect site data, and immediately connect with the BIM model, and establish cloud shared data database on-site quality defects, safety risks, civilized construction, and others, which can facilitate the statistical management of construction quality defects and other data during the construction process and after completion. Its advantages are mainly highlighted in the following aspects:

Visualization of Defects. Mobile terminal applications such as mobile phones are used to obtain data information like photos. For example, on-the-spot defects can be recorded by taking pictures briefly. What's more, the use of the BIM model can locate defects makes managers accurately control the location of defects.

Effective Sharing of Information. According to the privilege of all parties involved, we can check our problems, and relevant parties of project construction can make statistic and analyze all kinds of problems and implement rectification. The manager can grasp the quality defect and safety risk factors on the site at any time in the office and improve the efficiency of communication of all parties.

Multi-device Support and Simple to Use. It supports iPhone, iPad, Android and other smart devices. The construction site can record problems anytime and anywhere. As shown in Fig. 11, the construction personnel guide the rectification of construction problems with mobile devices. The operation is simple and convenient, and the implementation cycle is short, giving full play to the convenience of handheld devices.

Cloud-based management system with fast running speed being able to query all kinds of engineering-related data. The collaborative management platform based on BIM technology, combined with the Internet of things, big data, mobile communications, and others, encrypting the data of the project, and managing various professional BIM model unified and coordinately. Managing and controlling the privilege of different application personnel to safeguard the safety of data. At the same time, it realizes data sharing and collaboration and improves management level and efficiency of communication.

Fig. 11 Construction personnel hold mobile equipment to guide the rectification of construction problems

This hospital project eliminates the drawing problems by adopting BIM technology to optimize the scheme and deepen the design. The on-site paperless office, cloud data transmission and other means accelerate the efficiency of on-site drawing review and construction, significantly improving the management level, and comprehensively coordinating the on-site construction process, making the project go smoothly. By applying BIM results and orderly with visualization technology guide field each special construction, simulating the construction process, and analyzing the difficulty in the construction, it effectively avoids the problems in the construction of the rectification, greatly improving the construction efficiency and accuracy, effectively reducing rework by about 60%, saving time limit 48 days, and saving 5% than traditional period. And the use of BIM technology optimizes the technology, reasonably arranges materials according to the schedule of the project flow into the site and eliminates the backlog of funds and material misreporting. The consumption saving rate of reinforcement and concrete in the project material budget is 7%, which saves about 3 million yuan of construction cost and reduces the project cost.

5 Conclusions

Based on the outpatient and emergency medical technology building project of a county people's hospital relocation project, this paper studies the full-professional application and full-process implementation of the use of BIM technology in the hospital project. Through 3D virtual and collision inspection, we can quickly foresee the problems between civil engineering and electromechanical professions in advance, and communicate and solve problems in time, which is conducive to the overall control of project implementation risks. The unified information sharing platforms for all parties involved in the project and project implementation database, both based on BIM technology, can effectively cooperate with each other, which effectively improve the quality, progress, and the management level of information in the process of hospital engineering project construction, realizing comprehensive benefits such as improving delivery capacity, shortening the construction period and saving construction cost.

References

1. Chen S (2019) The application of BIM technology in construction project management. Dev Orientation Build Mater J 017(006):173
2. Guo J (2019) Research on cost control of hospital construction projects based on BIM. Beijing University of Architecture and Architecture
3. Guo J, Xie Y, Sun C, Song J, Liu Q, Liang X (2019) Research on the application of BIM technology in the whole life cycle of hospital engineering projects. J Hebei Univ Geosci 42(03):77–81

4. Ji X (2019) Application of BIM technology in tertiary hospital projects. Archit Technol J 50(595(07)):62–65
5. Xu W (2019) Application of BIM technology in medical construction projects. Eng Technol Res J 004(013):103–104
6. Xu X (2018) The application of BIM technology in the construction of engineering projects. China New Technol Prod J 000(009):119–120

Features of the Civil Liability Associated with Artificial Intelligence Technologies in Healthcare Services Sector

Elena N. Abramova and Elena V. Starikova

Abstract The paper discusses the points of view existing with regard to assigning the liability for actions of artificial intelligence. The approach is proposed towards effective legal regulation and attachment of the civil responsibility, that classifies the harmful actions by artificial intelligence as ultrahazardous activity.

Keywords Artificial intelligence · Medical services · Civil liability

1 Introduction

Artificial intelligence is used in medicine to help doctors diagnose and treat patients, with machine learning tools widely available to the majority of lead scientists [6]. The legal framework for the use of AI technologies is not without imperfections. Further, AI technologies are proposed to be used remotely for patients to be able to consult doctors from home [9]. Medical device manufacturers are increasingly using machine learning to innovate their products, provide more effective assistance to healthcare providers and improve health outcomes [2].

Our study involved reviewing of the articles by international researchers. It was essential to understand the place artificial intelligence occupies in the legislation of Russia and that of other countries. The application of AI in medicine is challenged by:

- large sets of healthcare data being stored in separate cloud storages, which slows down the process of their absorbing by artificial intelligence and can lead to its incorrect machine learning [11];
- statistic distortion risk. Statistic distortion risk can be caused by too much of adjustment in response to algorithms that are based on extensive datasets with limited applicability to rarer cases. However, the main problem relates to data security, which is a particularly sensitive field of healthcare [8];

E. N. Abramova (✉) · E. V. Starikova
St. Petersburg State University of Economics, Saint-Petersburg, Russia
e-mail: enastar@mail.ru

- in the event the data selected for artificial intelligence training is incomplete, poorly labeled, or biased, then the final conclusion to be made by software is likely to contain flaws [7];
- artificial intelligence, which is currently used in medicine, does not operate any clearly defined levels of confidentiality and related principles of professional medical ethics [1]. This problem is relevant in the light of the hacked databases managed by Google, Facebook and other organizations [10];
- the risk of patient management decision-making becoming more autonomous. It is necessary to develop a clear legislative framework regarding ethical and legal responsibility for cases when decision-making is autonomous or subject to an algorithm [4];
- cases when the artificial intelligence training using data from a particular geographical region may be inaccurate in patients with different ethnic or geographical backgrounds (for example, in dentistry) (Fahad et al. 2021).

These and other issues remain unresolved within the framework of Russian legislation.

2 Materials and Methods

This study made use of the following methods of scientific cognition: theoretical and regulatory source studies; modeling; comparison; description, analysis and synthesis.

3 Results

Any analysis of the AI technologies in medical services would be incomplete without addressing the ethical issues associated with artificial intelligence. A single, commonly shared universal code of ethics seem impossible from historical and geographical perspectives, although a number of generally adopted principles exist [3].

The process of providing proper care to patients involves various stages:

- obtaining patient information about through surveys and tests;
- results processing and analysis;
- source studies for correct diagnosis;
- selecting treatment method;
- implementation of selected treatment method;
- patient follow-up.

Some of these stages could be automated. It is assumed that their automation can help doctor achieve their tasks in shorted timeframes. Doctors are known to have to spend a lot of time dealing with administrative procedures, while automation and artificial intelligence are designed to partially relieve them of this burden to allow more time for patient care and health problems.

This approach to introducing artificial intelligence in the healthcare sector seems to be expedient. Once the elementary technical functions are delegated to artificial intelligence, doctors will be able to concentrate on more careful examining of their patients to provide them with high-quality medical care. At the same time, the duties being delegated to AI go beyond the elementary technical functions. Artificial intelligence is used in making diagnoses based on the declared symptoms and in performing surgeries that require high level of accuracy or may not be conducted by human hands, among other cases.

Unfortunately, not all examples of the artificial intelligence application in the field of healthcare are successful. A certain degree of risk arises, fueled by the problems and injuries received by patients as a result of system errors [13]. In his paper titled "Artificial intelligence and criminal liability: The challenges associated with emergence of the new type of perpetrator", I. N. Mosechkin cites cases when the use of artificial intelligence in exercising various medical functions can inflict harm and even death. One example is the tragic death of one-year-old child, caused by actions of the artificial intelligence. After the child was admitted to the hospital, the AI diagnosed him with flu. Appropriate treatment was prescribed, which the medical staff carried out thoroughly, but the child died a few hours later. Autopsy showed that death occurred as a result of bacterial infection, meaning that the diagnosis made by artificial intelligence was incorrect.

Sometimes doctors do manage to prevent the negative consequences of an incorrect diagnosis by AI through verification procedure, which was the case with testing of IBM Watson for Oncology system. The situation is described in the article by C. Ross, titled "IBM's Watson supercomputer recommended 'unsafe and incorrect' cancer treatments, internal documents show". This system is not defined as an alternative attending physician, but can be used for third opinion when there two human doctors have two contradictory opinions. The case was a 65-year-old man with lung cancer, to whom IBM Watson for Oncology system had prescribed a set of pharmaceuticals, one of which became the reason for severe bleeding. Due to the fact that artificial intelligence was running in a test mode, the doctors were able to establish which if the prescribed drugs was worsening the bleeding and was inappropriate for that patient.

Among the reasons cited as likely to lead to errors is, for example, the AI preliminary training that uses solely simulated data on fictional patients and can therefore look confusing to AI when dealing with the real data and the real patients. Notably, IBM Watson for Oncology system is constantly being improved using doctors' comments.

4   Discussion

Artificial intelligence algorithms operate based on the principle of internal evaluation and feedback from data they receive. Thus, the evolution of an algorithm that is relied upon to make clinical decisions appears uncontrolled by humans. The lack of understanding as to how the decision-making happens can create an obstacle to the adoption of the appropriate technology [12].

The artificial intelligence technologies are being put to test as actively as they are being developed. Consequently, it appears relevant to raise the issue of legal regulation of the public relations involving artificial intelligence. The Russian legislation in this area is represented by one Federal Law N 123-ФЗ dated 24.04.20 on conducting the experiments for introducing artificial intelligence technologies in Moscow; and the Strategy for the Development of Artificial Intelligence 2030, duly endorsed by Presidential Decree on Artificial Intelligence N 490 dated October 10, 2019. These two documents contain the definition of artificial intelligence and provisions establishing the goals, objectives, basic operating principles and prospects of AI.

Given the significant disruptions that are expected to be caused by the advances in artificial intelligence sector, it is not surprising that countries are currently considering the practical aspects of its current and future developments [5]. The issues of assigning and distributing responsibility for harm as a result of using AI remain unregulated by the Russian legislation. The same is true public relations involving artificial intelligence, which are associated with the provision of medical services. But, there is some progress with this issue. The Government of the Russian Federation has issued Decree N 2129-p dd. August 19, 2020 approving the Artificial Intelligence and Robotics Technologies Development Concept 2024 (hereinafter referred to as 'the Concept'). As stated in its second section (paragraph 1), which highlights the industry-wide tasks of using artificial intelligence technologies, priority is given to creating the implementation mechanisms for products using artificial intelligence and robotics technologies. The same paragraph establishes that opportunities be created for testing and further implementation of artificial intelligence technologies in a wide range of industries and further introduction into legislation of the norms resulting from the testing and experimental evidence. The Concept defines the artificial intelligence systems as highly sought-after by healthcare sector and medical robotics.

For the development and implementation of artificial intelligence algorithms to be effective in medical practice, an interdisciplinary approach is required that would involve all potential stakeholders, including in risk management and healthcare administration [14]. The issue of liability for damage caused by artificial intelligence and robotics is covered in paragraph 2 of Section II of the Concept and is defined as "the most vital". It is particularly highlighted that the current level of development of artificial intelligence may not be deemed basis for making fundamental changes to the institution of civil liability, the latter expected to receive improvements in the future. Special attention is paid to the liability for harm caused by artificial intelligence with high degree of autonomy. Possible improvements of the legislation include such options as no-fault liability, compulsory insurance, and establishment of special compensation funds.

These and other options for compensating for the damage caused as a result of the actions of artificial intelligence can be effective only in some cases. Before their implementation takes place, it is necessary to figure out which of the civil relations subjects will be held responsible for possible harm by artificial intelligence. This necessity is pointed out also in paragraph 2 of Section II of the Concept, but the emphasis is made on the division of responsibility between the subjects as a way to guarantee fairness.

Potentially, there can be several parties to bear the responsibility for damages caused by artificial intelligence—artificial intelligence itself; creator of artificial intelligence; user of artificial intelligence.

It is also possible for the participants of civil legal relations to bear this responsibility collectively or mutually. For more accurate and fair distribution of the civil liability arising in the case of harm caused by artificial intelligence when used in the provision of medical services, it is necessary to analyze each options separately in more detail and identify the most appropriate one.

It should be borne in mind when considering this issue that the current legislation defines artificial intelligence as titled property, as can be concluded from the analysis of the definition given to artificial intelligence in paragraph 2 of Article 2 of the Federal Law on conducting the experiments for introducing artificial intelligence technologies in Moscow. Thus, assigning the responsibility for any harm inflicted on the patient to the artificial intelligence itself seems impossible within the framework of current Russian legislation. In this sense, artificial intelligence can only be made liable for its harmful actions if it classifies as a body subject to law. One example could be an electronic entity established in the form of legal entity, meaning that the artificial intelligence will, in this case, have property and be liable to the extent of its property. However, given the current level of the AI technology development in medicine, the electronic entity, as a possible solution, sounds too premature and unlikely to be effective.

At first glance, the option where the damage compensation is imposed on the user of artificial intelligence seems to be more reasonable. At the present stage of development, artificial intelligence acts as a programmable and human-controlled technical means. Assigning the liability to the user, i.e. a medical facility as a participant in the provision of medical services, in the meaning assigned to it in sub-para 11, para 1, Article 2 of Federal Law N 323-ФЗ dd. November 21, 2011, sounds more reasonable, provided that the damage has nothing to do with artificial intelligence failures resulting from manufacturer's fault. According to paragraph 3 of Article 98 of the Federal Law on fundamental healthcare principles, the liability for causing harm to a citizen's life or health shall be borne by medical facility.

Similarly, in the event artificial intelligence is recognized as ultrahazardous activity, the responsibility for the harm caused will be assigned to the medical facility in accordance with Article 1079 of the Civil Code of the Russian Federation. The term "ultrahazardous activity" is not defined in the Russian legislation. In his research, A.A. Antonov refers artificial intelligence to sources of increased danger, defining the latter as objects of material world that, due to physical or technical deficiencies occurring in the process of their use, defy continuous and comprehensive control by humans and can cause harm to personal or property rights.

Another possible way to achieve assigning of responsibility for harm caused by artificial intelligence to its user, is by assigning artificial intelligence the legal status of an animal. In practice, animals are classified as a source of increased danger, which guarantees further imposition of the responsibility for damage on their owners, who are often found to exercise improper care and poor supervision of their animals (Article 210 of the Civil Code of the Russian Federation).

As to the option, where the liability is imposed on manufacturer of artificial intelligence-assisted medical equipment, whose use results in medical facility becoming liable for damage to a patient, it seems justified in the presence of manufacturer's guilt, one example being the delivery to medical facility of poor quality devices (Article 518 of the Civil Code of the Russian Federation).

What participants are there in the legal relations arising from the provision of artificial intelligence-assisted medical services, that can be held mutually liable for damage compensation? They are the medical facility (artificial intelligence user) and the artificial intelligence manufacturer. One prominent point of view is presented by E. Vasin in his paper "What is to be done?" AI "Who is to blame?": where a medical facility and a manufacturer (or seller) conclude a license agreement to provide the medical facility with artificial intelligence technology, the manufacturer will not be responsible for any low-quality items provided. This is due to the fact that the rights transferred under the license agreement cannot by definition be "of poor quality". It should be noted that conclusion of a license agreement for sale and purchase of artificial intelligence-assisted technical device seems impractical and contradicting the legal nature of the transaction being concluded. A system (technical device) installed with artificial intelligence is acquired by a medical facility for making use of its useful properties. In this kind of a legal relationship, a technical device installed with artificial intelligence is not considered as an object of intellectual property, but as a thing (or quasi-thing). To give one example from everyday life, when a consumer buys a computer, he or she enters into a retail sale agreement with the seller (or directly with the manufacturer acing as the seller) (Article 492 of the Civil Code of the Russian Federation), whereby the consumer intends to use the computer as a thing, not as an object of intellectual property, the latter status not being the case in seller-consumer relations. In the case in question, what the seller transfers to the buyer is the proprietary rights to technical device installed with artificial intelligence, not the exclusive rights to it.

The same is true about selling a technical device installed with artificial intelligence to a healthcare provider. Any medical facility can, of course, opt to engage in legal relations relating to intellectual property, but we assume in our analysis that the medical facility acquires artificial intelligence to use it as a means for providing medical services, i.e. as a thing. License agreement (Article 1235 of the Civil Code of the Russian Federation) will be concluded for the purpose of transferring the right to use the results of intellectual activity. The very wording "use of the results of intellectual activity" sounds misleading. When buying, for example, MRI machine, we do, indeed, intend to "use" it, but the key word here is the nature of use. Article 1227 of the Civil Code of the Russian Federation states that intellectual property rights are not related to the ownership of a thing (a material carrier of an intellectual property object). The transfer of the ownership of a thing does not entail any transfer of intellectual rights to the result of intellectual activity, as a general rule. If we are going to purchase an MRI machine in order to "plug it into the socket" and use it for its intended purpose, then it is impractical for us to conclude a license agreement with the owner of exclusive rights. Instead, we should enter into sale and purchase agreement with the seller. Any purchase by a medical organization of a technical device installed with artificial intelligence will be regulated by contract for the supply of goods (Article 506 of the Civil Code of the Russian Federation).

An additional mention of the importance of establishing the liability for harm that may be caused to a patient when providing healthcare services assisted by artificial intelligence technologies, is provided in paragraph 1 of Section III of the Concept. The paragraph states that the lack of clear understanding of how this kind of responsibility should be distributed is the main factor preventing wider use of artificial intelligence technologies in healthcare services sector. This statement sounds only logical, and so does the introduction in the legislation of the legally established list of situations allowing the use of artificial intelligence technologies in medical decision-making.

At the same time, attributing the lengthy procedure for registering artificial intelligence systems and medical robots to the "key barriers to the introduction of artificial intelligence" seems controversial. Instead, passing by AI technologies of relevant verification procedures and registration should be seen as a guarantee of artificial intelligence being able to cause only minor damage to patients during its operation.

5 Conclusion

We conclude that the most reasonable design for imposing the liability for damages inflicted during the operation of an artificial intelligence-assisted medical equipment is where this liability is imposed on healthcare provider under Article 1079 of the Civil Code of the Russian Federation and can subsequently be shifted on to the manufacturer of the artificial intelligence in question, provided there is evidence of manufacturer's fault.

References

1. Anom BY (2020) Ethics of Big Data and artificial intelligence in medicine. Ethics Med Public Health 15:1–11

2. Beckersa R, Kwadeb Z, Zanca F (2021) The EU medical device regulation: implications for artificial intelligence-based medical device software in medical physics. Physica Med 83:1–8
3. Carrillo MR (2020) Artificial intelligence: from ethics to law. Telecommun Policy 44(6). https://doi.org/10.1016/j.telpol.2020.101937
4. Currie G, Hawk E (2021) Ethical and legal challenges of artificial intelligence in nuclear medicine. Semin Nucl Med 51(2):120–125
5. Fatima S, Desouza KC, Dawson GS (2020) National strategic artificial intelligence plans: a multi-dimensional analysis. Econ Anal Policy 67:178–194
6. Galimova RM et al (2019) Artificial intelligence—developments in medicine in the last two years. Chronic Dis Translational Med 5(1):64–68
7. Greenhill AT, Edmunds BR (2020) A primer of artificial intelligence in medicine. Tech Innov Gastrointestinal Endoscopy 22(2):85–89
8. Kolanska K, Chabbert-Buffet N et al (2021) Artificial intelligence in medicine: a matter of joy or concern? J Gynecol Obstetrics Human Reprod 50(1). https://doi.org/10.1016/j.jogoh.2020.101962
9. Li, J-PO, Liu H, Ting DSJ et al (2020) Digital technology, tele-medicine and artificial intelligence in ophthalmology: A global perspective. Progress in Retinal and Eye Research:1–32.
10. Loughlin KR (2020) The world is only ten years old: the dawn of artificial intelligence in urologic oncology. Urologic Oncol: Seminars Original Investigations 38(8):646–649
11. Mann Douglas L (2021) Will artificial intelligence transform translational medicine: (not so) elementary, My Dear Watson. JACC: Basic Translational Sci 6(4):400–401
12. Somashekhar SP, Sepúlveda M-J, Puglielli S et al (2018) Watson for oncology and breast cancer treatment recommendations: agreement with an expert multidisciplinary tumor board. Ann Oncol 29(2):418–423
13. Sunarti S, Rahman FF, Naufal M et al (2021) Artificial intelligence in healthcare: opportunities and risk for future. Gac Sanit 35:67–70
14. Van Assen M, Lee SJ, De Cecco CN (2020) Artificial intelligence from A to Z: from neural network to legal framework. Eur J Radiol 129. https://doi.org/10.1016/j.ejrad.2020.109083

The Challenges of Preserving Infectious Diseases Hospitals as Architectural Monuments

Elena N. Baulina

Abstract In modern Russia, the use of architectural monuments that previously housed infectious diseases hospitals represent a huge challenge. These monuments face a triple circle of problems that relate to modern regulatory requirements to space-planning design; legal regulation of monument protection; negative perception of the adapted public spaces that formerly housed hospitals.

Keywords Infectious diseases hospital · Tuberculosis dispensary · Architectural monument · Infection transmission pathways · Final disinfection · Revitalization of monuments · Hospital renovation · Adaptation

1 Introduction

The history of medical facilities construction dates back to centuries ago. Many studies have been conducted domestically and internationally to explore typology of the design solutions for hospital buildings and complexes. Every period in this history is valuable from the perspective of historical and cultural heritage [1–3]. The Russian experience of hospital construction falls mainly on the eighteenth–early twentieth century, marked by three basic hospital design solutions that appeared one after the other.

Initially, the hospitals were of centralized design, when all departments, with exception of infectious diseases and anatomic pathology, were housed in one building. Then, the pavilion (barrack) design appeared, that had departments housed in separate buildings that formed hospital complexes with their own planning structure. The latest and current design is a mixed or block-type solution, where all somatic departments are in the main building and infectious diseases, maternity, children's diseases, polyclinic, and administrative services occupy separate buildings (Fig. 1) [4].

E. N. Baulina (✉)
Saint Petersburg State University of Architecture and Civil Engineering, Saint Petersburg, Russia
e-mail: arhi-lena@mail.ru

© The Author(s), under exclusive license to Springer Nature Switzerland AG 2023   35
D. Ivanov et al. (eds.), *Proceedings of ECSF 2021*, Lecture Notes in Civil
Engineering 257, https://doi.org/10.1007/978-3-030-99877-6_4

Types of building of hospital complexes

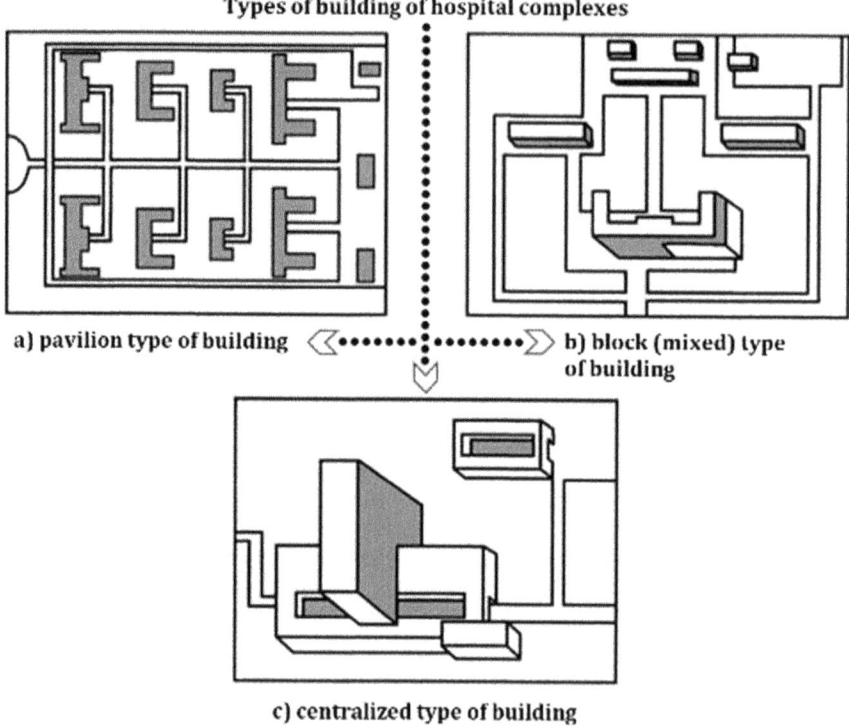

a) pavilion type of building ⟪••••••⫶••••••••⟫ b) block (mixed) type
                                              of building

c) centralized type of building

Fig. 1 Hospital design solutions

A special type of medical facilities is represented by the buildings whose original purpose was different from that of medical and which were converted into hospitals later in their lives. The strategy of converting valuable architectural sites for urban development purposes was characteristic of the young Soviet state: tuberculosis hospitals and dispensaries would be often housed in architecturally expressive, valuable mansions and estates. Such a strategy was common across the country. Examples of merchant estates converted into hospitals include Kushelev-Bezborodko's estate in St. Petersburg, which, however, was converted even before the revolution; Demidov's estate in Gatchina District of Leningrad Region; and many other premises that were once adornments of their home towns.

These medical facilities are often found in historical parts of small towns and metropolitan cities of Russia or outside the built-up areas. Functionally obsolete, abandoned, underused, dilapidated, collapsing and unrestored cultural heritage sites, these buildings are a burden. Strange as it may sound, but with the trend towards adjusting them to other purposes, the sites the used to house infectious diseases hospitals, tuberculosis dispensaries and sanatoria are not welcome for conversion in Russia, enjoying negative popularity fueled by the alarming articles in mass media about the threat of infection.

In the absence of any substantial stance of the government and local administration regarding such monuments, the fear of the threat of infection remains the main reason for not taking any measures to preserve this kind of monuments. At the same time, final disinfection measures are being safely conducted in all healthcare facilities, ensuring due level of public safety.

2 Materials and Methods

Despite the obvious urgency of infectious diseases hospitals renovation issues, there is actually no published research in this field in Russia. The scope of research in this field is limited to a handful of papers by students and faculty of Saint Petersburg State University of Architecture and Civil Engineering [4, 5], Tyumen Industrial University [6] and Tomsk State University of Architecture and Civil Engineering [7].

Our review of the international publications has shown that adapting former infectious diseases facilities for other uses has never been a problem in the developed countries. The reason why the idea of retrofitting and converting former hospitals abroad has always been welcomed by the communities abroad [8] lies in the disinfection system that inspires trust and belief that the formerly infected facilities are now safe.

Our study made use of the following methods:

- object classification;
- abstraction technique (for alternative restoration design options);
- comparative analysis (monument management practices in Russia and abroad);
- synthesis technique (for proposals for the multifaceted strategy for healthcare facilities and complexes).

3 Results

For better understanding and appreciation of the architectural value of hospital complexes, it is important to know their history and place in typological classification [1–3].

Based on the previous studies and initial design, all infectious diseases facilities can be divided into two groups: (1) facilities designed as hospitals; (2) facilities adapted for use as hospitals (sanatoria).

In terms of architectural value, of great importance for the facilities in the first group is the integrity of their original planning design. One example of group one facilities is Alexander (Botkin) City Barrack Hospital in St. Petersburg (Fig. 2).

The facilities of the second group can be divided into three categories: (1) facilities that were converted to hospitals but have preserved their architectural value; (2) facilities that had their architecturally valuable characteristics laid in their original

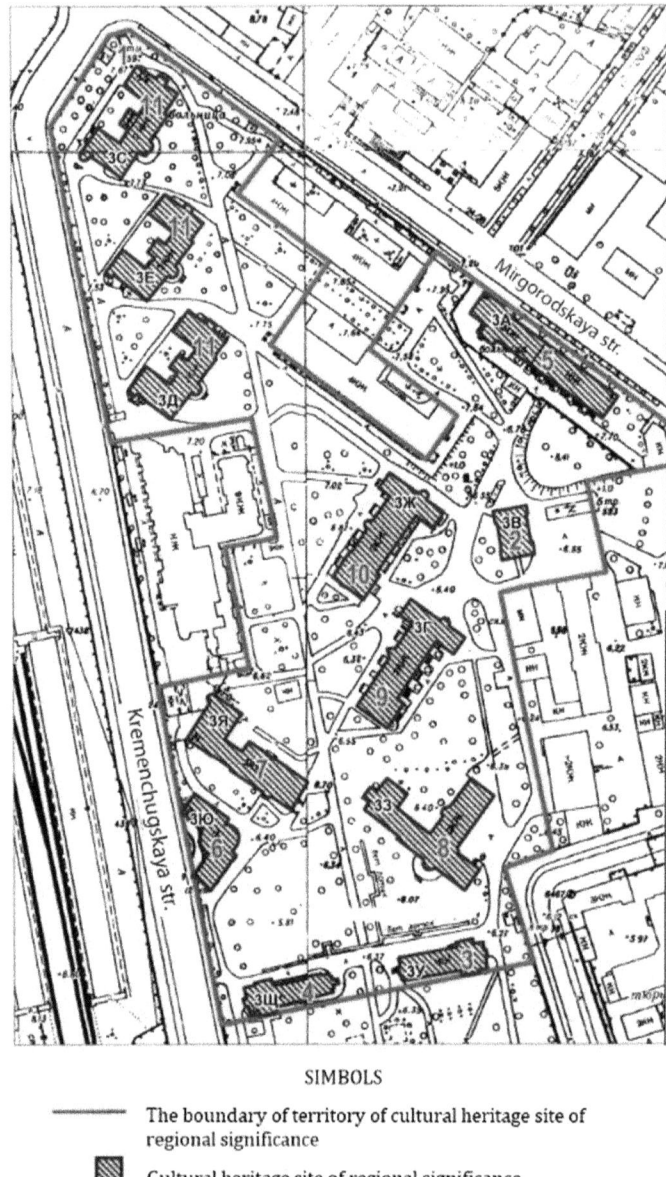

SIMBOLS

─────── The boundary of territory of cultural heritage site of
         regional significance

▨   Cultural heritage site of regional significance

Fig. 2 The boundary map of cultural heritage site of regional significance "Alexander (Botkin) City Barrack Hospital (S. P. Botkin Hospital)". 3 Mirgorodskaya Str., St. Petersburg, Russia. Composition of the object: 1—Gate; 2—Staff dwelling; 3—One-story medical building; 4—Administrative building with a reception rest; 6—Autopsy room; 7—Pharmacy and laboratory; 8—Surgical body; 9—Isolator; 10—Insulator; 11—Three pavilions of volatile infections

design but have lost a part of them after having been adapted for use as hospitals; and (3) facilities that have lost their architectural value entirely after having been adapted for use as hospitals. Examples of facilities in category one include F. M. Sadovnikov and S. G. Gerasimov Asylum and School in 66, Kamennoostrovsky Ave., which had for fifty years housed Tuberculosis Dispensary No. 3 (Fig. 3); one example of category two facility is Kushelev–Bezborodko summer estate (Elizabethan Community of Sisters of Mercy—Inter-District Tuberculosis Dispensary No. 5), whose territory was divided into plots in the second half of the nineteenth century and, later in the Soviet times, had completely lost its compositional integrity (Fig. 4); the example of category three is the compound of the Holy Trinity St. Alexander of Svir Monastery and its Alexander Nevsky Church, which was completely rebuilt in the 1950s and still houses Tuberculosis Dispensary No. 17 in Frunzensky District, St. Petersburg.

International experience offers many examples of successful adaptation of healthcare facilities and complexes for new use.

One example is Finland's Paimio tuberculosis sanatorium designed by Finnish architect Alvar Aalto, a monument of functionalism (Fig. 5). Its history as a TB sanatorium did not prevent it from being converted into a general hospital in 1960. Since 2014, this building functions as a private children's rehabilitation center. Paimio sanatorium is one of Finland's seven UNESCO World Heritage Sites [9, 10].

Around the same time as Paimio Sanatorium, numerous tuberculosis sanatoria (in fact, hospitals) were built in Switzerland and the Netherlands. These sanatoria

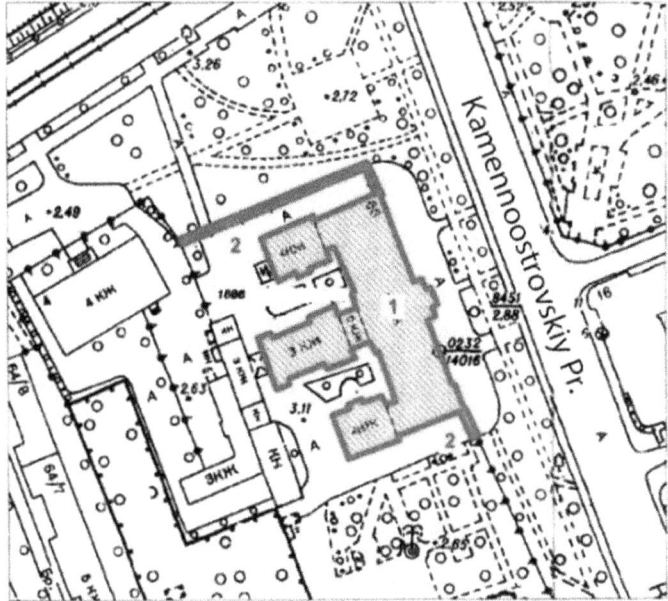

Fig. 3 The boundary map of "F. M. Sadovnikov and S. G. Gerasimov Asylum and School with wrought iron fencing" identified cultural heritage site. 66 Kamennoosrovskiy Pr., St. Petersburg, Russia

◀**Fig. 4** The boundary map of the cultural heritage site of federal significance "A. A. Bezborodko (Kushelevs-Bezborodko) Summer Estate". 40 Sverdlovskaya embankment, St. Petersburg, Russia Composition of the object: 1—Cottage with two outbuildings and transitional galleries; 2—Pier with a grotto and sphinxes; 3—Park pavilion; 4—Fence with sculptures (twenty-five lions); 5—Garden of the front yard; 6—Park; 7—Pond; 8—Complex of the Elizabethan Community of Sisters of Mercy of the Russian Red Cross Society; 8.1—House of Sisters of Mercy; 8.2—Surgical building with operating room; 8.3—First Mariinsky Surgical Pavilion; 8.4—Third Mariinsky Surgical Pavilion; 8.5 and 8.6—Treatment pavilion; 8.7—Church of St. Panteleimon the Healer

Fig. 5 Paimio Sanatorium, Finland

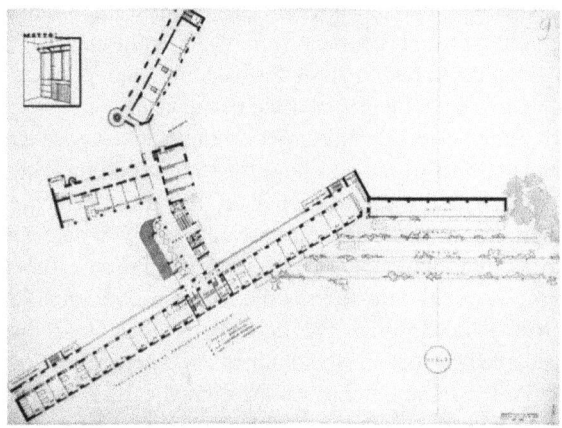

were rapidly expanding, some redesigned, demolished and rebuilt to be become hotels for athletes and tourists, since the sun and mountain air are useful not only for the sick. The Dutch tuberculosis sanatorium Zonnestraal (sunbeam), designed by architects Bernard Bijvoet and Jan Duiker and built between 1928 and 1931 near Hilversum (Fig. 6), is an outstanding specimen of architectural modernism (there is an opinion that A. Aalto's design for Paimio Sanatorium was influenced by

Fig. 6 Zonnestraal Tuberculosis Sanatorium in Hilversum, the Netherlands

Zonnestraal). Zonnestraal operated between 1928 and 1950 and became a general hospital in 1957. In 1995, it was assigned the status of a protected building and inscribed on UNESCO Tentative World Heritage List. In 1997, a master plan was developed to convert Zonnestraal into a wellness center with a mix of therapeutic and entertaining functions. Zonnestraal Estate currently houses a number of healthcare facilities. Its main building has beautifully renovated halls that are perfect venues for meetings, conferences, dinners, etc. [10, 11].

Notably, unlike infectious diseases hospitals, which had a pavilion design structure with well-thought-out ventilation systems to prevent the spread of infection, when designing tuberculosis sanatoria, the most important condition was location and the quality of its environment from the point of view of insolation. The design layout of the buildings had to allow for balconies, sun porches and terraces for patients to be able to stay outdoors any time of the year.

Tuberculosis sanatoria were being built in tsarist Russia and the Soviet Union, too, and many of which now have the status of cultural heritage sites. Those of them that changed their specialization in the 1950s (for example, to non-infectious hospitals or their departments) are still operating safely to date and their "infectious" past seems to be forgotten. N. D. Chetverikova Children's Tuberculosis Sanatorium in Sokolniki, Moscow now houses the Moscow Regional Children's Clinical Trauma and Orthopedic Hospital. The 1926–1978 tuberculosis dispensary in Ryazan is currently occupied by local medical university's department of physiology.

At the same time, as we mentioned earlier, many valuable buildings that ceased to operate as hospitals in 1990s remain abandoned. The reasons for this situation are as follows:

1. the public mind perceives such buildings as dangerous to visit and therefore unsuitable for any further use;
2. the negative media publicity over the further use of former infectious diseases hospitals leads individuals and organizations to vote for their complete or partial demolition and re-building anew;
3. the negative development of this situation is further fueled by local administrations being reluctant to adapt the former hospitals for any use other than therapeutic. The local regulations limiting the further use of former hospitals condemn them to dilapidation and oblivion;
4. the further use of medical facilities for their intended purpose causes the need for their reconstruction in order to bring them in line with today's statutory requirements; this, in turn, will either damage the monument or be contrary to monument protection legislation;
5. lack of legal frameworks that would require owners of such buildings to decide on their further use within statutorily established timelines;
6. lack of outreach and awareness campaigns to improve perception of the former infectious diseases hospitals as safe after disinfection according to current standards;
7. the current monument protection legislation is literally putting shackles on their targets by not giving them opportunities to survive in today's world.

The main approaches to organizing and conducting disinfection are established in Sanitary Rules "Sanitary-Epidemiological Requirements to Organization and Implementation of Disinfection Activities" (Sanitary rules 3.5.1378-03) and, specifically with regard to tuberculosis, in Sanitary Rules 3.1.2.3114-13 "Tuberculosis Prevention" (Section IX). It is useful to study the 2013 Manual "Infection Control System in Tuberculosis Hospitals" edited by L. S. Fedorova [12]. In addition, there are Sanitary Regulations and Standards 3.3686-21 "Sanitary and Epidemiological Requirements for Prevention of Infectious Diseases", effective since 01.09.2021, which, like previous regulations, provide, among other measures, for disinfection of premises (Section III).

4   Conclusion

Many historical hospitals owe their exceptional significance to their unique infrastructure that sets them apart from other buildings in use. The abandoned hospitals and former medical institutions that form part of the historical settlements should be evaluated in terms of their infrastructural potential [8].

The solution to the problem of preserving former hospitals obviously lies in an interdisciplinary plane and should be addressed by teams with expertise in legal regulation (amendments legislation concerning former hospitals as special type of structures; medicine (involvement of sanitary and epidemiological services in raising public awareness about the efficiency of disinfection measures, possibly with issuance of safety conformance certificate); management (investment attractiveness monitoring); as well as dedicated heritage protection committees (public awareness campaigns)).

It is believed that rebuilding a hospital is much cheaper than decontaminating it, but there is no published evidence so far. Since the problem of renovating the historical medical facilities is becoming global, the government should be able to provide, even if only partially, the funding for its solution. For this purpose, it is expedient to conduct an expert review and reassessment of the buildings to identify particularly valuable cultural heritage sites eligible for funding by the federal government. Similarly, it would be expedient to assign some of the identified monuments the status of regionally significant or remove them from the protected sites list to improve the situation. It sounds sensible to cover the buildings that have partly lost their original with selective protection measures.

References

1. Medicine of the future through the eyes of architects. Bulletin 17.04.2013 [Electronic resource]. http://vestnik.icdc.ru/world/1517-1

2. Shchukina TV (2016) The historical background of the Russian healthcare facilities: from Kievan Rus to the early 20th century. Young Scientist 7.4(111.4):44–46
3. Kisacky J (2017) Rise of the modern hospital: an architectural history of health and healing, 1870–1940. Centaurus 59(1–2):153–155
4. Nesterenko UV, Kaloshina LL (2021) The challenges of modern use of historical hospital complexes. Architectural Seasons in St. Petersburg State University of Architecture and Civil Engineering, pp 134–136
5. Karnaukhov VS, Leontiev AG (2021) The challenges of restoration and adaptation of Alexander City Barrack Hospital (S.P. Botkin Hospital). Architectural Seasons in St. Petersburg State University of Architecture and Civil Engineering, pp 117–119
6. Rybakova EY (2020) The process of repurposing historical objects in Berlin: case study of Bethanien. Architecture and architectural environment: historical and modern development. In: Proceedings of international research-to-practice conference, Tyumen, pp 389–392
7. Mordovan NA, Tsarev VI (2016) The principles of renovating the historical public facilities in Krasnoyarsk (case study of the former city hospital). Bull Tomsk State Univ Archit Civ Eng 2(55):44–58
8. Cherchi PF (2015) Adaptive reuse of abandoned monumental buildings as a strategy for urban liveability. Athens J Archit 4(1):253–270
9. Heikinheimo M (2013) Functionalism and technology. Case Paimio Sanatorium 16(4):73–79
10. Campbell M (2005) What tuberculosis did for modernism: the influence of a curative environment on modernist design and architecture. Med Hist 49(4):463–488
11. Ishida A (2017) Sanatorium Zonnestraal and a case for a critical approach to light exposure. In: Building for health and well-being: structures cities systems ACSA/ASPPH fall conference, pp 141–146
12. Fedorova LS, Yuzbashev VG, Popov SA et al (2013) The system of infection control in tuberculosis institutions: manual. Triada Publishing, LLC, Moscow

Safe Operation of Recreational Swimming Pools with Silver-Copper-Ionized Water

Andrey N. Belyaev, Vladimir O. Krasovsky, Margarita R. Yakhina, and Elena V. Kuts

Abstract Technically and hygienically safe and trouble-free operation of equipment will always be of decisive importance when it comes to socially significant facilities including public swimming pools. The paper proposes the swimming pool water disinfection facility that uses electrolytically generated oligodynamic action of silver and copper ions. The facility performance evaluation made use of the approved methodology for assessing chemical exposure risk for public health. The initial parameters were obtained experimentally. The statistical analysis of the laboratory data for six swimming pools, which have been using the electrolytically generated oligodynamic solutions for a period of five years, has shown these solutions' high level of hygienic reliability and safety. The core parameter used in the evaluation is the intake by human body of silver and copper ions from the pool water. The study has shown that this intake can lead to concentrations many times lower than the established maximum permissible concentrations. The proposed technology fully meets all sanitary and water quality regulations.

Keywords Disinfection of water · Swimming pool · Oligodynamic solutions · Water treatment · Safety · Human exposure · Dose of the substance

1 Introduction

More than 95% of Russian swimming pools use chlorine as their main disinfectant. In recent years, the issue of the appropriateness of using chlorine is gaining more and more relevance. The reason is the frequent cases of organochlorine poisoning and the increasingly frequent coverage of poisoning cases in the federal media. The study

A. N. Belyaev (✉)
Vyatka State University, Kirov, Russia
e-mail: belyaev71@list.ru

V. O. Krasovsky · M. R. Yakhina
Ufa Research Institute of Occupational Medicine and Human Ecology, Ufa, Russia

E. V. Kuts
Saint Petersburg State University of Architecture and Civil Engineering, Saint Petersburg, Russia

© The Author(s), under exclusive license to Springer Nature Switzerland AG 2023
D. Ivanov et al. (eds.), *Proceedings of ECSF 2021*, Lecture Notes in Civil Engineering 257, https://doi.org/10.1007/978-3-030-99877-6_5

conducted by Belyaev and Falaleev in 2020 [1] cites examples the public swimming pools in a number of the Russian regions with chlorine poisoning risk classified as "significant" and even "high". The main reasons for the chlorine poisoning exposure risk are malfunctioning equipment and human factor. The situation is further aggravated by the difficulty to ensure that residual chlorine content stays within the normal range of 0.3–0.5 mg/l (Sanitary Rules 2.1.3678-20 and Sanitary Rules and Regulations 1.2.3685-21) and does not exceed the MPC upper limit of 0.5 mg/l.

The use of chlorine-containing agents is associated with excessive concentrations of a whole range of other substances [10]. There is data [15] that 1.81% of swimming pool water samples contain excessive concentrations of halogen and 9.48% excessive concentrations of chloroform, the percentage of chemically unsatisfactory samples being 13.5%.

The experts at the Russian Academy of Medical Sciences' A. N. Sysin Research Institute of Human Ecology and Environmental Hygiene—the leading institution of the Russian Consumer Protection Agency (Rospotrebnadzor) in the field of human health protection—consider chlorination to be the most dangerous water disinfection method [14, 15]. Chlorine and its derivatives can lead to toxic, irritant, allergenic and even carcinogenic effects [14]. In this connection, there has been an increase in studies into chlorine-free water disinfection technologies [11].

Our analysis of the existing disinfection methods for hygienic reliability [2] has identified oligodynamic solutions as one of the most promising swimming pool water disinfection solutions from the perspective of human health, that scores highest also in "good value for money" criterion [7]. Oligodynamic solutions are aqueous solutions that have antimicrobial properties due to concentrations of metal ions. Their high antimicrobial performance is achieved by metal cations with high electrode potentials (Ag^+, Hg^{2+}, Cu^{2+}), as well as non-metallic ions capable of inactivating active enzymes that are found on the surfaces of biological substrates (bacteria, cells) [13].

Over the past decade, systems utilizing electrolytically generated copper and silver ions have become a widespread swimming pool water disinfection technology. Such systems have significantly optimized the water treatment process by all types of pools and offer a number of hygienic advantages. Silver-copper solutions form insoluble compounds with substances released by humans into water—amines and carbamides. The resultant sediments are easily removed by filtering through the sand filters of pool's water treatment system after introduction of coagulant [9]. The technological possibility to ensure constancy of subthreshold concentrations of silver and copper ions in the pool basin can ensure long-term maintenance of the bactericidal effect [8]—without the risk of forming chloramines, chlorophores, peroxides and other derivative compounds prone to chemical transformations and negative effects for human, as is the case with chlorine- and oxides-based technologies.

Despite the fact that the performance of the technology in question has been found fully meeting all current statutory requirements established for disinfection of public swimming pools and water parks, there haven't been any major publications in scientific and R&D literature on this technology over the past five years. Therefore, the systems utilizing electrolytically generated copper and silver ions remain unknown to a wide range of researchers, including scientists and hygienists, despite these systems' good potential for wider implementation. The bodies and organizations of the Russian Consumer Protection Agency (Rospotrebnadzor) do not have the standards to apply to the facilities using ion-based water disinfection treatment, for which reason this technology does not classify as a prescribed pool water treatment method. The current situation speaks to the urgency for in-depth studies into the reliability and safety of oligodynamic treatment of swimming pool water from the perspective of swimming pool users' safety.

The study object is oligodynamic solution synthesis facility AE-1 (Technical Specifications 361469-001-3091722173-2015), developed by AQUAEFFECT LLC [3] and intended for generating ionic aqueous solutions of metals used in preparation of cosmetic and disinfectant formulas, as well as swimming pool water disinfection. The choice of this facility as an object of scientific research is due to its ± 5% metal rate accuracy for both silver and copper. AE-1 generates silver at a rate of up to 3 g/h and copper up to 2 g/h.

Structurally, AE-1 facility consists of two units—ionization unit (IU) and electronic control unit (CU)—as shown in Fig. 1.

The ionization unit (IU) is a welded plastic housing with two electrolysis groups of silver and copper electrodes. The electrodes enable ionization of the water passing through the IU. The process is regulated automatically by the control unit (CU) connected to the electrolysis groups in IU. The ionization process follows the preset water treatment algorithm. The AE-1 process chart for swimming pool water treatment with silver and copper ions is shown in Fig. 2.

The oligodynamic solution in the pool bath is potable quality water enriched with silver and copper ions. The question of the effect the sub-threshold concentrations of these elements have on the pool users' health of pool users is fundamental in the

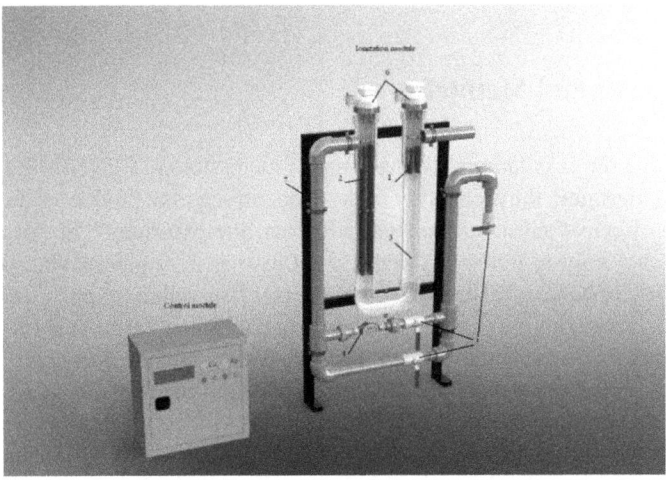

Fig. 1 Oligodynamic solution synthesis facility AE-1: 1—silver electrodes; 2—copper electrodes; 3—ionization unit housing; 4—flow meter; 5—shutoff valves; 6—electrode groups; 7—frame

Overflow grate

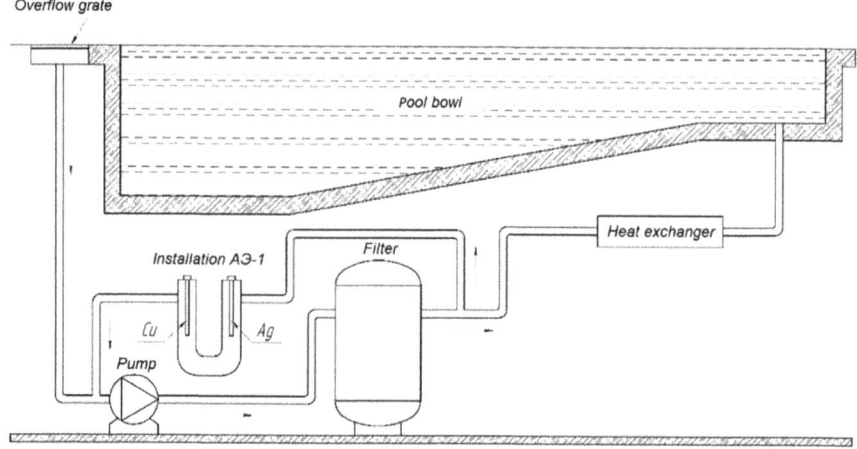

Fig. 2 Swimming pool water treatment process chart

assessment of technology [5]. Out study is novel in that it offers the laboratory data for six swimming pools that have been operating AE-1 technology for five years, analyzed statistically for reliability and safety.

Modern human consumes more than 45 tons of water over the course of 50 years [4]. During this period of time, his body consumes 16 kg of chlorine, 2 kg of nitrates, 14 g of iron and 23 g of aluminum that it receives with tap water. The task of evaluating the reliability and safety of oligodynamic solution synthesis facilities involves calculating the possible amounts of silver and copper that can enter the pool user over the given period of time.

2 Materials and Methods

In performing the calculations, we made use of methodology P 2.1.10.1920-04. This document determines the probability (risk) of occurrence as a product of the specified (standard) effective substance concentration and the exposure. The "exposure" is defined as not only in terms of duration, but also as a "multi-faceted, purposeful characteristic of contact." The input parameters, or so-called "exposure factors", to be used in calculations may contain two types of data about the process (object, etc.). The first type is instrumental (research) data which is obtained through a targeted experiment. The second type is standardized data and involves averaged reference consumption volumes, intake, doses, etc. [12].

The pathways for pool water to enter pool user's body are two—oral and dermal. The consumption can be determined based on the absorbed dose and/or the intake rate over the given time. For calculation purposes, we made use of the following data: orally consumed reference doses (RFD) of silver and copper ions, maximum permissible concentrations, as well as actual average concentrations obtained from the laboratory analysis of five-year data on six swimming pools. The calculations used 24-h period (for the convenience of calculations for year, month, hour, etc.) and 1 h as a duration of daily swimming (the average duration of a swimming class). It is logical to assume three options: the treated water is ingested intentionally, the treated water is absorbed by skin while swimming, the treated water is ingested accidentally while swimming.

The first option, when the water is ingested intentionally, assumes that the oligodynamic solution (copper-silver water) is consumed for drinking purposes and should be considered as the most hazardous situation. It is known that an adult (70 kg) consumes up to 2 l per day on average. Manual P 2.1.10.1920-04 provides the following algorithm:

$$I; J = \frac{Cw * V * EF * ED}{Bw * AT * 365}, \tag{1}$$

where I—intake rate, mg/(l day); J—dose, mg/(kg day); Cw—concentration of substance in water; V—volume of daily intake: (children—1 l/day, adults—2 l/day); EF—exposure frequency: (350 $days$/year); ED—exposure duration, years (children—6 years, adults—30 years); Bw—body weight (mg/kg)—(children—15 kg, adults—70 kg); AT—average exposure time (children—6 years, adults—30 years old).

Using the above formula (1) and based on the initial data, it is possible to calculate:

- intake rate of (I) mg/(l day) based on the concentration mg/l;
- dose (J) mg/(kg day) based on reference oral dose (RFD) mg/kg of human weight.

The volume of ions absorbed by skin was determined using the standard formula based on the daily dose and standard values for skin exposure to (potable) tap water (absorbed dose):

$$DADi; j = \frac{DAe * EV * ED * EF * SA}{Bw * AT * 365 * 1000}, \tag{2}$$

where $DADi$—intake of ions by absorption by skin, mg/(l day); $DADj$—the absorbed dose; DAe—intake of ions or the dose of ions per "event" (i.e. while in water; exposure unit: mg/cm^2); EV—frequency of events per year: 45; SA—skin surface area, cm^2: for children—6000 cm^2, for adults—^{18,000} cm^2; EF—exposure frequency: 45 $days$/year; ED—exposure duration, years: children—6 years, adults—30 years; AT—average exposure time, years: children—6 years, adults—30 years; W—body weight (mg/kg)—(children—5 kg, adults—70 kg).

Dae, mg/cm^2 is calculated according to:

$$DAe = Kp * Cw * te, \tag{3}$$

where *te*—duration of event, hour/event: 0.58 h/day (children), 1 h/day (adults); *Cw*—concentration of the substance in water; *Kp*—skin permeability coefficient logarithm of, cm^2/h.

The duration of te event and the content of ions in water are the known values. With regard to "skin permeability coefficient", it should be noted that permeability depends on many reasons, from physical activity to air temperature to water temperature, etc. Therefore, the method provides an algorithm for approximate calculation:

$$Kp = -2.8 + 0.67 * Kow - 0.0056 * MW, \tag{4}$$

where *MW*—molecular weight of substance (silver—107.8682 g/mol; copper— 63.546 g/mol), *Kow*—octanol/water distribution coefficient (silver—1.0, copper— 1.0).

We note here that octanol/water distribution coefficient characterizes the affinity of substance with organic substances. We assume its value for silver and copper equaling 1.0 nominal unit.

The unknown logarithm of skin permeability index *Kp*, calculated according to (4), is:

- for silver: 2.336 cm^2/h;
- for copper: 1.1858 cm^2/h.

Next, skin application values are calculated according to (3) and the desired volumes of ions penetrating the swimmer's body through the skin are calculated.

To determine the intake of ions in case of accidental ingestion of water while swimming, it is necessary to use the standard formula for calculating the average daily dose and standard values of factors in case of accidental ingestion of surface water (water bodies):

$$I = \frac{Cw * IR * EF * ED * ET}{AT * Bw * 365}, \tag{5}$$

where *I*—oral intake; *Cw*—concentration of the substance in water, mg/l; *IR*—intake rate, l/h (0.05 *l*/h); *EF*—exposure frequency, days/year (45 *days*/year); *ED*—exposure duration, h/day (1 *h*/day); *ET*—exposure time, h/day (1 *h*/day); *AT*—average exposure time (children—6 years, adults—30 years; carcinogens—70 years); *Bw*— body weight (mg/kg) (children—15 kg, adults—70 kg).

The results are based on the statistical analysis of 445 laboratory test protocols for water from six pools that have been using AE-1 facility for five years [6].

3 Results

The initial data for the calculations are given in Table 1. The overall result of calculations is the probabilistic semi-quantitative model describing the intake of silver and

Table 1 Initial data for calculating the intake of silver and copper by pool user's body

Parameter	Ion	
	Silver	Copper
RFD—reference oral (standardized) dose[4], mg/kg	0.0050	0.019
Maximum permissible concentrations[2], mg/l	0.0500	1.000
Actual concentrations (sample: 143 analyses)		
Minimum values, mg/l	0.0005	0.001
Maximum values, mg/l	0.2240	0.920
Average values, mg/l	0.0240	0.221

copper ions by swimmers' body from the oligodynamic solution in the pool basin. The results of calculations are expressed in the following units:

- dose estimates for oral administration: mg/(kg day);
- the rate of ion intake into the body: mg/(l day).

Table 2 shows the ions intake by body from the pool water used for "drinking" (up to 2 l per day), which is an unrealistic option, but it illustrates the level of safety of oligodynamic solutions for humans: the calculated (probabilistic) ion concentrations in the treated water used as drinking water are at "homeopathic" levels.

Table 3 shows the calculated rates of intake (concentration) of ions that can enter the swimmer's body through skin intake and by ingesting the water accidentally while swimming. The obtained values of the intake rates in case of long-term (50 years)

Table 2 Metal ions intake from the pool water used for "drinking"

Parameters analyzed	Daily dose, mg/(kg day)/Concentration, mg/(l day)	Hourly dose, mg/(kg h)/Concentration, mg/(l h)
Silver ions		
RFD—oral reference dose	1.3E−04	5.4E−06
Maximum permissible concentration	1.3E−03	5.4E−05
Maximum concentration	6.0E−03	2.5E−04
Minimum concentration	1.3E−05	6.7E−08
Average concentration	6.5E−03	2.7E−04
Copper ions		
RFD—oral reference dose	5.1E−04	2.1E−05
Maximum permissible concentration	2.7E−02	1.1E−03
Maximum concentration	2.5E−02	1.1E−03
Minimum concentration	2.7E−05	1.1E−06
Average concentration	6.0E−03	2.5E−04

Table 3 Total intake of silver and copper ions by swimming pool user body, mg/(1 h)

Parameters analyzed	Skin application	Accidental ingestion	Total
Silver ions			
Maximum permissible concentration	1.5E−07	0.6E−09	1.7E−04
Maximum concentration	6.7E−06	2.8E−07	7.1E−05
Minimum concentration	1.5E−09	6.2E−11	8.3E−07
Average concentration	7.1E−09	3.0E−10	8.3E−05
Copper ions			
Maximum permissible concentration	2.9E−06	1.2E−07	3.1E−03
Maximum concentration	2.7E−06	1.2E−07	2.9E−03
Minimum concentration	2.9E−09	1.2E−10	4.2E−06
Average concentration	6.2E−07	2.6E−08	3.1E−03

exposure do not pose any real threat to swimmers' health, since they are below the maximum permissible concentrations.

It should be noted that the rates of intake of silver and copper ions through skin and accidental ingestion are different. The copper intake rate exceeds that of silver by 8–9 times. When oligodynamic water is consumed for drinking purposes (through the stomach), the rates are identical.

4 Discussion

Based on the statistical analysis of the laboratory data on the facility intended for disinfection of swimming pool water by enriching it with silver and copper ions (six swimming pools monitored over five years), the calculation of the intake of these two metals by swimmer's body was performed. The calculations have shown that the ions enter the swimmer's body in concentrations many times lower than the established maximum permissible concentrations. The rate of ingestion of copper is 8–9 times higher that of silver. When ions of these metals consumed for drinking purposes (oral intake), the intake rates are nearly identical. This fact should be interpreted as a characteristic feature of the interaction patterns formed by silver and copper ions when entering the body through different pathways.

The novelty of the performed statistical analysis and calculations lies in the fact that oligodynamic solutions as a swimming pool water disinfection method have not yet been evaluated in terms of technical and hygienic performance, nor have they been studied sufficiently in previous research.

5   Conclusion

The swimming pool water disinfection facility is proposed that uses electrolytically generated oligodynamic action of silver and copper ions and provides long-lasting disinfection effect. To evaluate reliability and safety of the facility, the study made use of the approved methodology for assessing chemical exposure risk for public health. The initial parameters were obtained experimentally. The statistical analysis of the laboratory data for six swimming pools, which have been using the electrolytically generated oligodynamic solutions for a period of five years, has shown these solutions' high level of hygienic reliability and safety. The core parameter used in the evaluation is the intake by human body of silver and copper ions from the pool water. The study has shown that this intake can lead to concentrations many times lower than the established maximum permissible concentrations. The proposed technology fully meets all sanitary and water quality regulations, but still remains unknown to hygiene expert community due to its being omitted from the current regulatory documents and covered in technical and hygienic periodicals only poorly.

References

1. Belyaev AN, Falaleev AV (2020) Technological risks associated with the operation of swimming pools. In: Society, science, innovations (NPK-2020), XX All-Russian scientific and practical conference, vol 2, Feb 17–Apr 26, Vyatka State University, Kirov, pp 213–219
2. Belyaev AN, Fedonenko MV, Yakhina MR et al (2020) Evaluating the hygienic performance of pool water biocidal treatment methods. Occup Med Human Ecol 1:82–89
3. Belyaev AN, Flegentov IV, Peretyagin AN (2018) Device for synthesizing oligodynamic solutions. RF Patent No. 191301
4. Belyaev EN (ed) (2002) Drinking water and public health: information guide. Issue 1: Health effects of chemical composition of drinking water. Russian Ministry of Health Federal Center of State Sanitary and Epidemiological Supervision, Moscow, 63 p
5. Krasovsky VO, Yakhina MR, Belyaev AN (2020) On the advantages of swimming pool water disinfection by oligodynamic method. Natl Assoc Sci 62:16–20
6. Krasovsky VO, Yakhina MR, Belyaev AN (2021) Purification of swimming pool water with copper and silver cations and their health-improving effect. Sci Heritage 58:19–28
7. Mosin O, Ignatov I (2014) Methods for preparation of microdispersed colloid silver nanoparticles. Nanotechnol Res Pract 4(4):201–212
8. Petrov BA, Pogorelsky IP (2018) Sanitary and hygienic rationale of antibacterial activity (Ag^+) and (Cu^{2+}) achieved by AE-1 oligodynamic solutions synthesis facility (TS 361469-001-3091722173-2015): research report. State Registration No. AAAA18-118121913-9. Kirov State Medical University, Kirov, 18 p
9. Petukhova EO (2017) Swimming pool water disinfection methods. Bull PNRPU Constr Architect 8(2):36–51
10. Sokolova NF (2013) Means and methods of water disinfection (analytical review). Med Alphabet 1(5):44–54
11. Tulskaya EA (2013) Comparative evaluation of the effectiveness of water disinfection means. Public Health Environ 1(238):11–13
12. Tokarev VI (1997) Silver-based water disinfection technology: abstract of PhD thesis (engineering). Novocherkassk State Land Reclamation Academy, Novocherkassk, 27 p

13. Voznaya NF (1967) Water chemistry and microbiology: studies. In: Cherepennikov AA (ed) Handbook for universities. Vysshaya Shkola, Moscow, 324 p
14. Zholdakova ZI, Sinitsyna OO, Tulskaya EA et al (2007) On hygienic standardization of water disinfection chemicals. Hyg Sanitation 5:76–80
15. Zholdakova ZI (2010) Issues of harmonization of hygienic requirements for swimming pools with international recommendations. Hyg Sanitation 2:93–96

Application of Smart Contracts as Project-Based Approach to Innovative Healthcare Construction

Olga Yu. Bochkareva

Abstract The construction of innovative healthcare facilities has its own specific features. There exist a great number of multifaceted requirements to be met by healthcare facility projects and their construction technologies, established in statutory documents and individually developed construction designs. Highly relevant in this context is innovative approach to managing and monitoring the processes arising out of the interaction between construction participants. This paper discusses project finance and smart contracts as mechanisms for enhanced interaction between participants of innovative health facility construction projects.

Keywords Construction · Project financing · Smart contract · Innovative medicine

1 Introduction

As a priority sector of the Russian Federation, healthcare services require a significant amount of investment (RUB 100 billion rubles, the actual budgetary commitment amounting to RUB 44.7 billion). The draft National Spatial Strategy of the Russian Federation provides for several lines of development in healthcare sector that span training and promoting multidisciplinary medical centers, including national medical research centers engaging in research and training; innovative medical technologies; export of medical services; and technology-intensive medical care. Another policy paper—"Modernizing the Primary Health Care of the Russian Federation"—provides for "new construction of health facilities." As construction projects and social facilities, health care institutions involve a number of specific standards to be observed and high expenditure items. This requires, firstly, improved construction planning; secondly, ions, enhanced interaction between construction participants; and, thirdly, strengthened monitoring of the spending of the earmarked funds.

O. Yu. Bochkareva (✉)
Saint Petersburg State University of Architecture and Civil Engineering, Saint Petersburg, Russia
e-mail: olga937-308-19@mail.ru

© The Author(s), under exclusive license to Springer Nature Switzerland AG 2023
D. Ivanov et al. (eds.), *Proceedings of ECSF 2021*, Lecture Notes in Civil
Engineering 257, https://doi.org/10.1007/978-3-030-99877-6_6

Russia is actively adopting the project-based approach to construction finance, which allows to attract additional financial resources and, consequently, involved better performing construction contractors [3, 4, 13].

2 Methodology

The conceptual framework for this study relies on the papers by experts and research teams specializing on project financing [6, 7, 9, 10, 16] and digital technologies for construction sector. The data has been analyzed using general scientific methods.

The study also made use of the analytical papers of The Central Bank of the Russian Federation; policy documents of the Government of the Russian Federation—"Modernizing the Primary Health Care of the Russian Federation" and "The Digital Economy of the Russian Federation"; author's research; etc. The data has been analyzed using general scientific methods.

3 Results

In foreign countries, the project-based financing traces its history to the early twentieth century, triggered by growth in the investment activity and the scale of projects. Project-based financing was first applied in the 1930s in Texas in petroleum industry to become a common practice, since 1978, in the USA's electricity supply sector. In the early 1990s, project-based financing was adopted by the UK for its public infrastructure projects (public facilities, roads, transport, etc.) and real estate [16].

Literature offers a large number of definitions of "project finance". The most exhaustive definition is given, in our opinion, by E. R. Yescombe: "Project finance is a method of raising long-term debt financing for major projects through "financial engineering," based on lending against the cash flow generated by the project alone; it depends on a detailed evaluation of a project's construction, operating and revenue risks, and their allocation between investors, lenders, and other parties through contractual and other arrangements." [16].

It would be only logical to turn to current international practices and adapt them to the Russian situation. Thus, the project finance for healthcare construction projects can be presented as follows [16]:

Stage one is design development. This stage involves draft design solution, negotiations with potential contractors, and pooling together of finance capitals (public and private). It further involves selection of contractor general to exercise the management and monitoring of healthcare construction project throughout its life cycle. The company acting in this capacity must have the competence and expertise sufficient for integrating the technological, engineering, logistics and expert advice solutions into every stage of works, starting from the design stage (with due account of the regulatory and legislative frameworks) and all the way to furnishing the completed project with equipment and its daily operation [2, 8, 15].

Upon completion of stage one, a project company will be set up to function as the project manager.

Stage two is construction. This stage involves spending of the project finance on the construction project delivery. At this stage, the contractor may involve subcontractors for performing certain types of work. This stage usually ends on the facility commissioning date.

Stage three is operation. The delivered project is used for commercial purposes to generate the cash flow needed for loan repayment and investor income generation. In case of public healthcare facilities, this cash flow is generated by the fee-based services for population.

All things considered, the project-based approach to healthcare construction represents an appropriate method to ensure more effective planning and use of resources, as well as reduced construction and commissioning timelines.

At the same time, the efficiency of healthcare construction projects depends directly on how well the efforts of construction participants are coordinated. The lack of proper coordination can affect the cost, timing and quality of the construction in general. In this regard, it is general contractor's core task to hire those subcontractors that would see to it that the performance risk is minimal and that the facility is commissioned according to schedule. Even though participants' efforts target one common goal of delivering the project at minimum cost and in a timely manner, there may arise inconsistencies and collision of interests as the project unfolds, caused by:

- subcontractor's failure to observe the prescribed construction technology;
- subcontractor's failure to eliminate defects, deficiencies, etc. as may be identified by the contractor general;
- subcontractor's failure to observe the work/service performance deadline as specified in contract;
- subcontractor's failure to report on the expenditure of materials;
- subcontractor takes the liberty of performing extra works without the consent from the contractor general;

etc.

The risk of occurrence of these challenges enables a logical conclusion that there should be a mechanism capable of regulating comprehensively the developer-contractor interaction.

One such mechanism is offered by smart contract. The idea of a smart contract was first proposed in 1994 by Nick Szabo (USA), a computer scientist, legal scholar, and cryptographer. He described a smart contract as "a computerized transaction protocol that executes the terms of a contract." [14].

It follows from this definition that smart contract represents a contract between two or more parties, that provides for automatic execution, in the digital environment, of the provisions earlier agreed on. In simple terms, smart contract is a set of terms and sequences of actions to be executed. These terms are embedded in the hardware and software we deal with, in such a way as to make verification of their execution to be an automatic process and compliance a digital protocol.

Smart contracts enable the entire document flow and payment transactions to operate in digital environment, involving fewer people, ensuring full transparency of the responsibilities and budget, and contributing to more effective interaction between all parties to contract [12].

The main advantages of using smart contracts include [1, 5, 11]:

1. visibility of contract, i.e. easy verification of fulfillment by parties of their obligations;
2. autonomy of contract. Since smart contracts are a computer code that automatically executes all or parts of an agreement, no transaction can be "lost" as each participant has its own copy of smart contract and this copy is not susceptible to attack;
3. consistency of contract. All new or additional term of a contract will be executed only after agreement is obtained from all the parties interested in effecting it;
4. high level of protection of transaction terms from third parties.

The life cycle of a smart contract involves the following stages:

1. entering the terms and conditions, as may be agreed between the parties, into smart contract;
2. connecting the smart contract to internal and external automated systems;
3. tracking contract execution progress and confirmation of its current status;
4. self-execution upon fulfillment of all pre-set conditions (Table 1).

4   Conclusion

Our study enables a conclusion that continuous financing and mere placement of the obsolete infrastructures in healthcare sector are no longer sufficient to maintain modernization efforts. A cardinally new organizational approach is needed to upgrade construction workflows existing for innovative healthcare facility projects.

A rapidly developing mechanism, smart contracts offer some promising benefits and are suitable for use in any industry as a technical basis for changes.

Introduction of smart contracts into technology-intensive healthcare construction will facilitate better coordination of the parties' efforts and allow to avoid problems associated with materiel supply, project documentation errors, non-compliance with deadlines, etc., which ultimately leads to increased efficiency of the construction process.

Table 1 Smart contracts as a solution to the issues arising out of general contractor-subcontractor interaction during healthcare facility construction

Issue solved	Benefits of smart contracts
Automatic uploading of the required records and documents into the system	Automation of payments, reduced uncertainty and credit risk
Automatic tracking of contractor's costs under contract; payment documents/certificates can be uploaded into the system	Automated document flow and, consequently, reduced costs
Independent expert review of deficiencies in works/services performance; automatic withdrawal of penalties for deviations	All necessary documentation reposited for one smart contract
Automation of all transfers under contractual terms	Smart contract terms are written as codes that cannot be changed without the parties' consent
Impossibility of unilateral contract termination without parties' consent	Automatic verification of fulfillment of contractual terms, calculations, etc.
Impossibility of works not specified in contract without the consent of ALL the parties	
Automatic transfer of funds (extra expenses) if agreed by all smart contract parties	

References

1. Ahmadisheykhsarmast S, Sonmez R (2020) A smart contract system for security of payment of construction contracts. Autom Constr 120. https://doi.org/10.1016/j.autcon.2020.103401
2. Aleksandrova E, Vinogradova V, Tokunova G (2019) Integration of digital technologies in the field of construction in the Russian Federation. Eng Manag Prod Serv 11(3). https://doi.org/10.2478/emj-2019-0019
3. Bochkareva O (2020) Project financing in housing construction: first results. FES Financ Econ Strategy 17(1):43–48
4. Bochkareva O (2020) Project financing of housing construction: domestic and foreign experience. Bull Civ Eng 4(81):219–230
5. Carvalho A (2021) Bringing transparency and trustworthiness to loot boxes with blockchain and smart contracts. Decis Support Syst 144. https://doi.org/10.1016/j.dss.2021.113508
6. Collier NS, Collier CA, Halperin DA (2007) Construction funding: the process of real estate development, appraisal, and finance. Wiley
7. Esty BC (2003) The economic motivations for using project finance. Harvard Bus School 28:1–42
8. Faltinsky R, Tokunova G (2018) Information technologies and construction sector: why construction loses competition for innovations to other industries? In: SHS web of conferences 44. EDP Sciences, p 00033
9. Hoffman SL (2007) The law and business of international project finance. Cambridge Books
10. Levy SM (1996) Build, operate, transfer: paving the way for tomorrow's infrastructure. Wiley
11. Narayana KL, Sathiyamurthy K (2021) Automation and smart materials in detecting smart contracts vulnerabilities in blockchain using deep learning. Mater Today Proc. https://doi.org/10.1016/j.matpr.2021.04.125
12. Rosic A (2020) Smart contracts: the blockchain technology that will replace lawyers. https://blockgeeks.com/guides/smart-contracts/. Accessed 15 Sept 2021

13. Rybnov E, Akimov P, Khalvashi M et al (eds) (2021) Contemporary problems of architecture and construction. In: Proceedings of the 12th International conference on contemporary problems of architecture and construction (ICCPAC 2020), 25–26 Nov 2020, Saint Petersburg, Russia. CRC Press
14. Szabo N (2002) A formal language for analyzing contracts
15. Tokunova G, Rajczyk M (2020) Smart technologies in development of urban agglomerations (case study of St. Petersburg transport infrastructure). Transp Res Procedia 50:681–688
16. Yescombe ER (2002) Principles of project finance. Elsevier

Genetic Examination for Emergency Situations: Time, Distance, and Logistics

A. P. Gerasimov, W. A. Khachatryan, N. E. Ivanova, S. A. Kondratev, and Yu. M. Zabrodskaya

Abstract Genetic examination for emergency situations isn't standard task now. Centralization of laboratory service with the system of bio couriers changed infrastructure and process of diagnostic. But time of execution of many molecular diagnostic methods is in conflict with the temporal limits of surgical treatment and other emergency situations. "Gene search" is often the wrong question. Results of genetic tests are important for evaluation of premorbid background, dynamics rate and rehabilitation potential. Possible solutions of time conflict are depending from time limit and have some risks and organizational limitations.

1 Introduction

At this time genetic examination is considered as a planned, not urgent method [4]. But there are many emergency situations like urgent/near-urgent operation, emergencies in psychiatry and neurology (from cerebral insult up to school shooting), emergency medicine (including medicine of technogenic and nature catastrophes).

Situation in diagnostic of genetic and multifactorial diseases changed notably in recent years. Cytogenetic, biochemical and molecular genetic diagnostics are available in principle [12]. Organization of laboratory services is centralized, but the outposts (offices) are placed at the local districts near the patients. The existence of the biocourier system solved the problem of biomaterial transportation between regions [2, 10]. Using standard protocols increased importance of biobanks [6, 14]. Now in Russia we see the trend to unite "adult" and pediatric ambulatory networks into a single system. Electronic communications network from local hospital net up to internet and OMIM database help the diagnostics significantly [5].

Genetic examination results don't describe urgent situation itself (it is exogenous often). Genetic factors are forming context of situation including premorbid background, dynamics rate and rehabilitation potential [9, 13].

A. P. Gerasimov (✉) · W. A. Khachatryan · N. E. Ivanova · S. A. Kondratev · Yu. M. Zabrodskaya
Almazov National Medical Research Centre, St. Petersburg, Russia
e-mail: APGerasimow@rambler.ru

D. Ivanov et al. (eds.), *Proceedings of ECSF 2021*, Lecture Notes in Civil Engineering 257, https://doi.org/10.1007/978-3-030-99877-6_7

But in practical medicine, from our experience, both patients and doctors have incorrect expectations from genetic methods. A time limit misunderstanding is often, especially at surgery units.

Another problem exists. "Gene search" may be incorrect task. Disease may be result of copy number variation (CNV) [7] and mutations in noncoding sequences [11] Disease may be result of combination of effects of several genes, and there may be helpful gene maps and genomic nets [7]. Disease may be result from the summation in a dangerous direction of the effects of formally neutral polymorphisms. Finally, the clinical associations are described not for all genes, the databases like OMIM and ClinVar are constantly under updating.

2   Materials and Methods

Time limits of different methods of genetic diagnostics (from clinical examination up to different variants of sequencing) were analyzed and compared with the time limits of the surgical treatment.

Time limits for different molecular genetic methods including next-generation sequencing (NGS) or, more correctly, massive parallel sequencing were compared at 3 Russian molecular diagnostic companies (2 from Moscow, 1 from Saint-Petersburg), existing for several years with practical results of diagnostics. Information about time, price and technical details of diagnostics was taken from official sites.

Different logistic schemes of laboratory diagnostics were compared in the aspects of transportation object, point of getting the biomaterial and intensity of the laboratory equipment use.

3   Results

Time limits of different genetic methods are different. Duration of clinical examination is variable—from 15 min (emergency) up to 2 h (ideal situation). Biochemistry methods need several hours and potentially may be used in the day of operation. Classical cytogenetic needs 1 week, and this technical limitation is inevitable. Comparative genomic hybridization (CGH) in the modern form of array-CGH [1] needs near 1 month according official offer. The time necessary for short molecular panels is several weeks and depends from number of genes in panel. Full exome sequencing and large panels need 3–4 months, but our experience demonstrated longer time in problematic situations like combination of mutations. The described terms are in conflict with the interests of the surgical clinic and emergency medicine.

We compared terms, prices and methods of molecular genetic diagnostic in 3 laboratories (Table 1). There is small difference between big panels, clinical exome sequencing and full exome sequencing in price and time (3–4 month). Full exome

Table 1 Comparison of terms and prices in different molecular diagnostic laboratories of Moscow and Saint-Petersburg

	Company 1 (Moscow)	Company 2 (SPb)	Company 3 (Moscow)
Panel MR and ASD	35000 R 3-month MR and ASD	38900 R 3–4 month MR and ASD	27990 R 3 month NDD (622)
Clinical exome sequencing	40000 R 3 month		29900 R 2–3, 5 month (6110)
Full exome sequencing	43000 R 3 month	60000–90000 R (Consult) 2–3 month	
Full genome sequencing	99000 R 3-month	50000–60000 R (error?) 3 month	

MR Mental retardation, *ASD* Autism spectrum disorders, *NDD* Neurodegenerative diseases. For Company 3 it is indicated in brackets number of genes in panel.

sequencing presents information about all coding sequences (near 22 000 genes), it is preferable method now.

Solving the problem of distance between patient and laboratory has been changed.

"Classical scheme" was patient transportation to high-tech medical center (university clinic) and delivery of analyzes in the central laboratory. But in this situation intensity of use of laboratory equipment is low. It is impossible to equip all the hospitals by all the necessary rare diagnostic methods.

At this time logistic scheme changed. Biomaterial transportation is cheaper, faster and less problematic for patient. Point of delivering the biomaterial is placed nearer to patient. It may be local district hospital or office/outpost of laboratory service. Then biomaterial is transported into central laboratory in accordance with the requirements. This logistic scheme is sensitive to quality of nurse and bio couriers work. But intensity of use of laboratory equipment in this situation is high and this scheme is economically viable.

4 Discussion

Described terms of genetic diagnostic (cytogenetic, array-CGH, NGS/MPS methods) are in conflict with the interests of the surgical clinic and emergency medicine. In the difficult clinical situations choice the method of diagnostic may be problematic itself.

Thus, the question exists: when should be a surgical patient examined? Results of diagnostic are important (and sometimes critical) for treatment tactics. For example, for the last year we discovered one possible coagulation anomalies and one syndrome with possible intubation complication only at one unit.

Examination after the operation has no sense usually. Examination before the operation at surgery unit has high risk as result of terms conflict. Examination before the hospitalization is problematic in management aspect, but allows avoiding conflict of terms. According our opinion solution is to start an examination at the place of residence at the time of making a decision on hospitalization for the purpose of surgical treatment and get the result in the preoperative period. This method may be combined with neuroprotective preparation.

Modern logistic scheme with outpost, bio couriers and centralized laboratory service is suit for patients and doctors [15]. It is economically viable. It allows avoiding risks associated with patient transportation. But the work of bio couriers and nurses is critical in this system [10].

But these schemes don't work in random emergency situations like emergency operations, acute psychotic conditions and emergency medicine. The only solution is a mass planned screening, the first steps towards which have been taken [3]. The limiting factor is the shortage of bioinformatics. In the same time results of screening must be technically accessible for practical doctors. Now only small part of clinicians is able to use genetic dates. Educational programs for surgeons, resuscitators, emergency doctors are necessary.

5 Conclusions

The organization of genetic diagnostics in urgent situations is poorly developed. But different aspects of genetic information are important for preoperational preparation, intraoperational tactics and postoperational rehabilitation.

If it is possible to plan a surgical intervention for 1 month or more, the examination should begin at the place of residence. Combination with other forms of preoperational preparation is possible.

In case of urgent situations, the only solution is screening, combined with the accessibility of its results. Educational programs for practical doctors are necessary.

References

1. Ahn JW, Bint S, Bergbaum A et al (2013) Array CGH as a first line diagnostic test in place of karyotyping for postnatal referrals-results from four years' clinical application for over 8,700 patients. Mol Cytogenet 6(1):1–6. https://doi.org/10.1186/1755-8166-6-16
2. Beskow A (2019) Uppsala Biobank—the development of a biobank organization in a local, regional, and national setting. Upsala J Med Sci 124(1):6–8. https://doi.org/10.1080/03009734.2018.1547992
3. Chen Z, Chen J, Collins R, Guo Y, Peto R, Wu F, Li L (2011) China Kadoorie Biobank of 0.5 million people: survey methods, baseline characteristics and long-term follow-up. Inter J Epidemiol 40(6), 1652–1666. https://doi.org/10.1093/ije/dyr120

4. Church TD, Richmond FJ (2019) Biobank continuity management: a survey of biobank professionals. Biopreservation and Biobanking 17(5):410–417. https://doi.org/10.1089/bio.2018.0142
5. Dive L, Critchley C, Otlowski M et al (2020) Public trust and global biobank networks. BMC Med Ethics 21(73). https://doi.org/10.1186/s12910-020-00515-0
6. Hewitt R, Watson P (2013) Defining biobank. Biopreserv Biobank 11:09–315. https://doi.org/10.1089/bio.2013.0042
7. Iourov IY, Vorsanova SG, Yurov YB et al (2019) Ontogenetic and pathogenetic views on somatic chromosomal mosaicism. Genes 10(5):379. https://doi.org/10.3390/genes10050379
8. Iourov IY, Yurov YB, Vorsanova SG et al (2021) Chromosome Instability Aging Brain Diseases. Cells 10(5):1256. https://doi.org/10.3390/cells10051256
9. Li S, Wang M, Zhou J (2020) Brain organoids: a promising living biobank resource for neuroscience research. Biopreserv Biobank 136–143. https://doi.org/10.1089/bio.2019.0111
10. Linsen L, T'Joen V, Van Der Straeten C et al (2019) Biobank quality management in the BBMRI. be network. Front Med 6(141). https://doi.org/10.3389/fmed.2019.00141
11. Luchini C, Bibeau F, Ligtenberg MJL et al (2019) ESMO recommendations on microsatellite instability testing for immunotherapy in cancer, and its relationship with PD-1/PD-L1 expression and tumour mutational burden: a systematic review-based approach. Ann Oncol 30(8):1232–1243. https://doi.org/10.1093/annonc/mdz116
12. Mosele F, Remon J, Mateo J, Westphalen CB et al (2020) Recommendations for the use of next-generation sequencing (NGS) for patients with metastatic cancers: a report from the ESMO precision medicine working group. Ann Oncol 31(11):1491–1505. https://doi.org/10.1016/j.annonc.2020.07.014
13. Paskal W. Paskal, AM, Dębski T et al (2018) Aspects of modern biobank activity–comprehensive review. Pathol Oncol Res 24(4):771–785. https://doi.org/10.1007/s12253-018-0418-4
14. Rush A, Catchpoole DR, Ling R et al (2020) Improving academic biobank value and sustainability through an outputs focus. Value in Health 23(8):1072–1078. https://doi.org/10.1016/j.jval.2020.05.010
15. Salman A, Baber R, Hannigan L et al (2019) Qatar Biobank milestones in building a successful biobank. Biopreservation and biobanking 17(6):485–486. https://doi.org/10.1089/bio.2019.0083

Scenario Approach to Hybrid Public Spaces and Healthcare Facility Landscaping with High Level of Digital Maturity

Svetlana Danilova and Maxim Yefimov

Abstract The study analyzes the use and prospects of further development of land-scape architecture and digital technology-intensive public spaces. The main research methods include logical method, system analysis, and simulation. The study has shown that further development of interactive public spaces with high level of digital maturity can open up new pathways for more effective use of the accumulated knowledge and the technological resources towards urban healthcare facility landscaping.

Keywords Pattern · Urban morphology · Integrity · Development · BIM for landscape · Unique design · Simulation algorithm · Identity · Utilities · Technologies · Innovations · Design of alternatives · Construction · Competitive ability · Potential · Innovative development · Municipal economy · Community economy · Urban semiotics · Residential yards and gardens · County level · Digital public spaces · Neurorehabilitation · Formation of man of today · Scenario approach

1 Introduction

Modern public spaces represent a tool for transforming identities and principles of social interaction [1]. The advanced modular simulation technologies allow us to shape design concepts for our surroundings and source solutions for all kinds of landscapes [2]. The experience of implementing architectural and design solutions for public spaces landscaping (Villosi, Leningrad Region, Russia), created in 2018–2020 under the guidance of architect Svetlana Danilova, describes the modular simulation algorithm and theoretical research as a highly promising tool, used also by St. Petersburg State University of Architecture and Civil Engineering Department

S. Danilova (✉)
Saint Petersburg State University of Architecture and Civil Engineering, Saint Petersburg, Russia
e-mail: dasdanilova@gmail.com

M. Yefimov
National Institute for Integrated Support of Physical Culture and Sports, Saint Petersburg, Russia

of Architectural Landscape Design, and has many examples of integrating different kinds of fill based on function, scale, position in urban design and surrounding landscape (Fig. 1).

Modular solutions allow the use of vertical and container gardening (Fig. 2).

Importantly, modular simulation allows to create authentic, identics-based solutions with due consideration of the budget and with strong focus on detail enhancement (Fig. 3).

Using sets of standard modules and shared design code, it is possible to model all kinds of facilities—early development and rehabilitation centers, information

Fig. 1 Simulation algorithm by S. Danilova

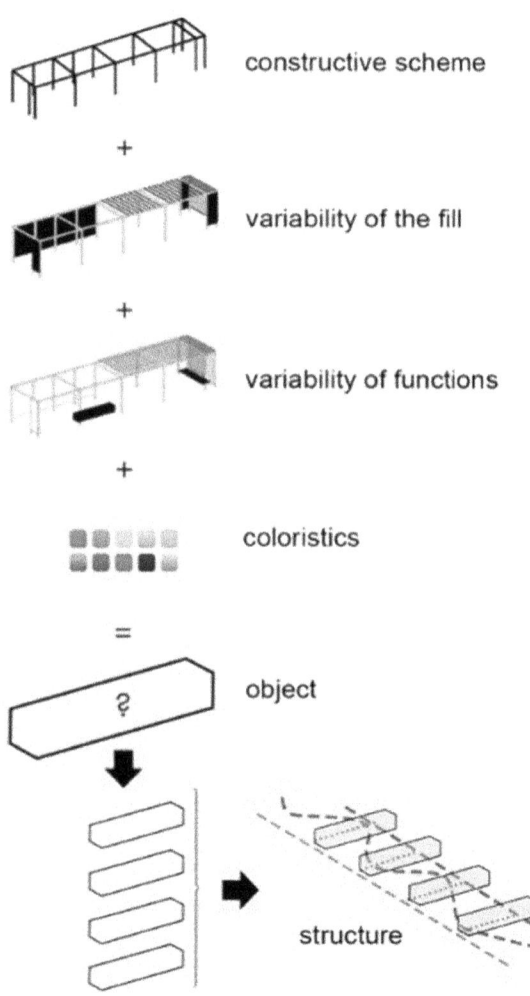

Simulation Algorithm

constructive scheme

+

variability of the fill

+

variability of functions

+

coloristics

=

object

structure

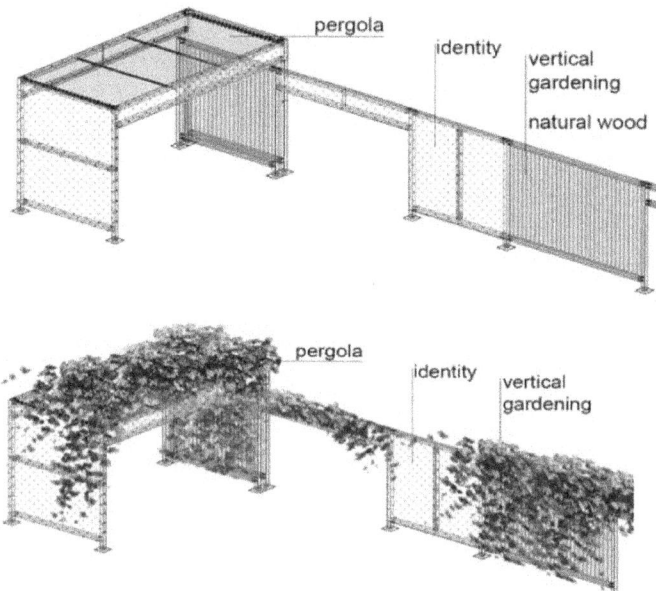

Fig. 2 Structural frame (top) and vertical gardening as a possible fill solution (bottom). Villosi Village, Leningrad Region, Russia. Author: S. Danilova

Fig. 3 Implementing the elements of integrated masterplan and design code for Villosi Village landscaping (Leningrad Region, Russia). Design developer: St. Petersburg State University of Architecture and Civil Engineering. Chief architect: S. Danilova. Design team: E. Kuznechikova, A. Yegorova, A. Soshnikova, A. Popova, E. Shcherbakova

stands, co-livings, medical offices, playgrounds in kindergartens and schools, recreation centers, sports and wellness facilities. With digital analysis systems continuing to develop rapidly, we have more possibilities for planning the processes to form individual development paths (Fig. 4).

In the context of the pandemic, integrating digital technologies into landscaping offers solutions to some of the challenges of our time. The public space of the future

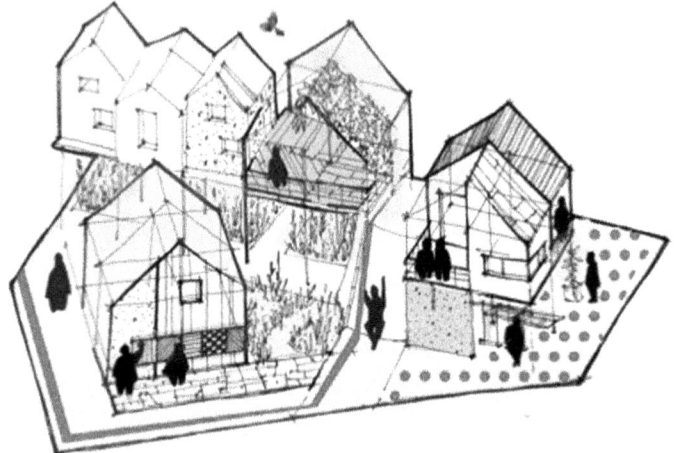

Fig. 4 Modular arrangement of split level complexes featuring the landscaping elements and combined solutions for outdoor ground cover, intuitive navigation and fill. Design supervisor: S. Danilova S. Student: K. Herman

is seen as a tool for shaping the man of today and accompanying him throughout his life while ensuring maximum comfort through, among other things, contact with nature and involvement in the interactive processes of public spaces [12].

At the level of state, our solutions offer a space that brings together the policy efforts being pursued by a whole range of governmental bodies and among them Ministry of Sports, Ministry of Education, Youth Policy Committee, Ministry of Health, Ministry of Culture, and Ministry of Defense. The process of implementing the 2024–2030 national strategies should be embedded with new patterns of state-population interaction and higher degree of harmonization of interests.

2   Materials and Methods

The study is interdisciplinary, lying on the confluence of landscape architecture, cognitive urbanism, medicine, and digital information technologies [7].

The scientific and informational kernel of the study relies on the following system modules:

- 3D learning;
- augmented reality;
- sensory-motor integration and neurodynamic gymnastics;
- neurocognitive capacity building;
- enhanced physical conditions;
- interactive communications.

At the level of more specialized work with children [11], patients and elderly citizens, a research-based framework has been designed that uses the modules targeting better neurorehabilitation and academic performance:

Sensory (tactile, proprioceptive, acoustic) development. Sensory playgrounds engage a child's touch and muscular sensations and allow to experiment with sound recognition. They offer activities that appeal to vestibular senses to develop behaviors based on how children move and position their body in space.

Coordination. Coordination playgrounds offer activities designed to develop balance and coordination.

Cognitive development. Playgrounds can benefit children by developing behaviors based on core notions of shape, color, number, volume, more-less, background-foreground, through fascinating journeys into recognition and interpretation.

Relay race. Relay races appeal to complex forms of attention and help to stimulate planning skills and self-control. They motivate families to be more active.

Civilization. Civilization module introduces children to the main achievements of civilizations in the field of culture, arts, architecture, technology and much more.

Provided that the above five modules use the prescribed protocols, they can provide the highest possible level of child development. The same is true about working with the elderly as these facilities can slow down the aging process in the brain, and patients of specialized healthcare facilities in need of rehabilitation (Fig. 5).

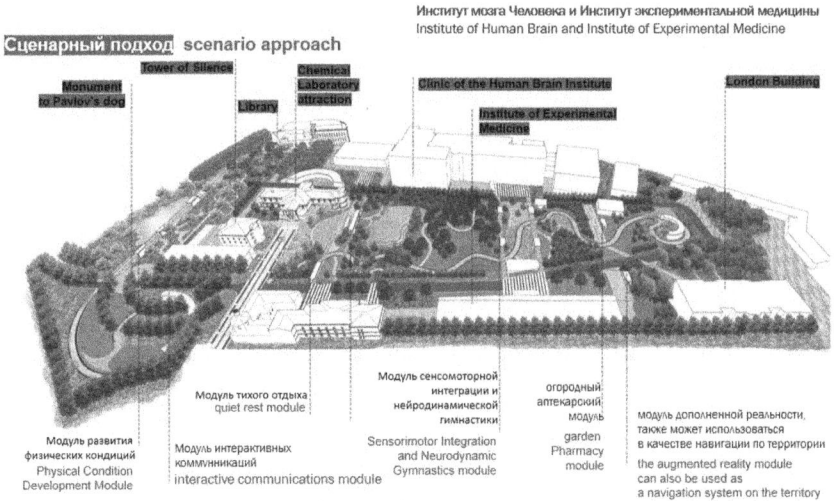

Fig. 5 Scenario approach in action: modular elements layout and landscaping solution for the Institute of Human Brain and the Institute of Experimental Medicine (St. Petersburg). Scheme: S. Danilova. Graduation thesis by A. Weitzer (research supervisors: S. Danilova, N. Kerimova, V. Gomozov V., 2021)

3 Results

Country-wide landscaping strategies are lacking in the Russian Federation. There is no comprehensive policy that would cover the surroundings of healthcare and rehabilitation facilities to embrace child development, rehabilitation of patients of different ages, monitored recreation zones, physical and neuro-cognitive correction in children, adolescents and families, and as prevention of aging [6]. Our study has enabled a number of solutions towards enhanced effectiveness of public spaces, higher public engagement, methodological guidance and digital transparency—combined in a model that can serve as a coordinating body to link together the tasks and efforts of the responsible ministries and committees.

Thanks to their design, modular systems can be integrated into public spaces of any configuration.

One more important line of research is offered by scenario approach to the architecture of hybrid spaces. This study defines *hybrid spaces* as fusions of physical territories, landscape solutions and augmented reality-assisted 3D planning. Such spaces should be solved using the *scenario approach*. On the one hand, the scenario approach to landscaping relies on a combination of techniques, elements and natural components of a given territory with account of the specifics of functional purpose, urban planning, compositional aspects, thematic content, existing and future biocenoses, ecosystem forecasting and sustainable development principles designed to achieve harmonious coexistence of man and environment. On the other hand, the efforts to build an environment conducive to comprehensive child development and rehabilitation of patients of different age groups, should seek to create the spatial–temporal patterns that would be interconnected by paths and information content consistent with their functional purpose and seasonal or all-year-round activity/event agendas. That said, in conditions of the pandemic, which means these spatial–temporal patterns are subject to rapidly changing requirements to their use and operation, it is important to be able to operate flexible solutions for adjusting to real-life situations. The scenario approach and modular simulation algorithm stands as a modern response to the challenges of the time. It allows for a variety of behaviors in accordance with individual learning, rehabilitation, recreation, sporting and leisure paths. In a world with rapidly changing social and economic conditions, to the foreground comes personal development and social acumen [4].

Thus, the scenario approach to hybrid spaces uses the following core principles:

- Forecasting
- Social interaction and identity transformation
- Architectural design planning at micro-, meso- and macro-levels of environment
- Multifunctionality with focus on facility's specific features
- Harmonization
- Routing
- Unlocking the landscape potential and integrating natural components
- Identity and authenticity enhancement

- Technology integration. Digital analysis systems offer more possibilities for planning the processes to form individual development paths
- Modularity and optimized problem-solving
- Flexibility of use
- Institutional consolidation and cross-disciplinary communication.

4 Discussion

When developing solutions for hybrid spaces of healthcare and other facilities, it is very important that sufficient attention is paid to the basic age-specific motor skills, sensory abilities, and sensorimotor integration [3]. A child unequipped with basic skills according to age may have difficulties acquiring new skills to adapt to adult life. Without a solid foundation, attempts to make progress in sports, science, military training or profession will be like colossus on the feet of clay [10].

Achieving progress in conditions of swiftly advancing technologies is possible only by integrating the flexible digital and electronic solutions into landscape environment, so as to enable more transparent child-parent-mentor or teacher-research center-state interactions [8]. Digital tools can help enable the "one child, one schedule" strategy, while also harmonizing the efforts of the ministries and departments involved in developmental strategies, especially child development strategies [9].

Thanks to the advanced technologies, the active and passive interaction patterns can be combined with augmented reality technologies as a way to support the individual paths to recreation, physical health improvement, physical and neuro-cognitive development in the landscape [5].

Notably, the issues of development and developmental problems in citizens of different ages have been seriously addressed for more than 80 years, but it is only now, with the advent of newest generation digital and electronic solutions, that the achievements of these years can be used to the maximum benefit, increasing the availability of tools created by large research centers and clinics available to every yard, social space and landscape for higher interactivity and monitoring of each process.

5 Conclusions

The study has shown that further promotion of cutting-edge technologies and interactive public spaces with high degree of digital maturity could lead to more effective use of the accumulated knowledge and resources towards better urban landscape designs forhealthcare facilities.

References

1. Androulaki M, Frangedaki E, Antoniadis P (2020) Optimization of public spaces through network potentials of communities. Proced Manufact 44:294–301
2. Ferdani D, Fanini B, Piccioli MC et al (2020) 3D reconstruction and validation of historical background for immersive VR applications and games: the case study of the forum of augustus in Rome. J Cult Herit 43:129–143
3. Jens K, Gregg JS (2021) How design shapes space choice behaviors in public urban and shared indoor spaces-a review. Sustain Cities Soc 65. https://doi.org/10.1016/j.scs.2020.102592
4. Lai LW, Ho DC, Chau KW et al (2021) Property rights and the perceived health contribution of public open space in Hong Kong. Land Use Policy 107. https://doi.org/10.1016/j.landusepol.2021.105496
5. Lak A, Aghamolaei R, Baradaran HR et al (2020) A framework for elder-friendly public open spaces from the Iranian older adults' perspectives: a mixed-method study. Urban Forest Urban Greening 56. https://doi.org/10.1016/j.ufug.2020.126857
6. Lambe S, Knight I, Kabir T et al (2020) Developing an automated VR cognitive treatment for psychosis: gameChange VR therapy. J Behav Cogn Therapy 30(1):33–40
7. Lu X, Wang X, Wu R (2020) Urban garden landscape design based on VR technology and internet of things system. Microprocess Microsyst. https://doi.org/10.1016/j.micpro.2020.103432
8. Pawlowski CS, Schmidt T, Nielsen JV et al (2019) Will the children use it?—A RE-AIM evaluation of a local public open space intervention involving children from a deprived neighbourhood. Eval Prog Plann 77. https://doi.org/10.1016/j.evalprogplan.2019.101706
9. Richardson EA, Pearce J, Shortt NK et al (2017) The role of public and private natural space in children's social, emotional and behavioural development in Scotland: a longitudinal study. Environ Res 158:729–736
10. Senda M (2015) Safety in public spaces for children's play and learning. IATSS Res 38(2):103–115
11. Timperio A, Giles-Corti B, Crawford D et al (2008) Features of public open spaces and physical activity among children: findings from the CLAN study. Prev Med 47(5):514–518
12. Wei ZHU, Jiejing WANG, Bo QIN (2021) Quantity or quality? Exploring the association between public open space and mental health in urban China. Landscape Urban Plann 213. https://doi.org/10.1016/j.landurbplan.2021.104128

Evaluating and Ensuring the Environmental Safety of Buildings

Tamara Datsyuk, Yulia Leontieva, Alexander Sokolov, and Timur Mellekh

Abstract The paper discusses the integrated methodology for ensuring the environmental safety of buildings (ESB). The building and its surrounding environment are considered as one dynamic system. ESB evaluation uses numerical modeling and experimental determination of structural performance. The study provides cases that can be recommended for ESB purposes.

Keywords Methodology · Environmental safety · Numerical modeling · Experimental research

1 Introduction

The adoption by Russia of the concept of sustainable development has led to more intense development of environmental engineering as an emerging field. Environmental engineering has among its tasks not only to reduce the negative environmental impact of construction sector, but also to ensure the environmental safety of buildings and ecologically safe environment. The efforts to evaluate and assess the environmental safety should use a system of criteria analogous to LEED and BREEAM. Domestic and international literature offers some prominent papers that explore the principles of environmental evaluation of buildings and their ecological impact [1, 5, 8, 10, 13]. By observing all commonly recognized requirements for buildings, it is possible to achieve a safe and comfortable living environment throughout buildings' entire life cycle with minimized impact on the ecosystem.

While there exist regulatory documents (GOST R ISO 14001–2007, GOST R 54,964–2012) in the Russian Federation, the system of generally accepted criteria for assessing the environmental safety of buildings, structures and construction materials is unfortunately still lacking.

T. Datsyuk (✉) · Y. Leontieva · A. Sokolov · T. Mellekh
Saint Petersburg State University of Architecture and Civil Engineering, Saint Petersburg, Russia
e-mail: tdatsuk@mail.ru

© The Author(s), under exclusive license to Springer Nature Switzerland AG 2023
D. Ivanov et al. (eds.), *Proceedings of ECSF 2021*, Lecture Notes in Civil Engineering 257, https://doi.org/10.1007/978-3-030-99877-6_9

2 Materials

There has been developed a methodology for evaluating the environmental safety of new construction projects and existing buildings, contributed by the team of Saint Petersburg State University of Architecture and Civil Engineering [4, 11]. This methodology operates a sufficient number of environmental safety indicators (ESI) that characterize the environmental performance of buildings—a total of 54 ESIs grouped into 6 clusters. The proposed methodology is supported by evaluation criteria and mathematical techniques. The ESIs are quantified on a 10-point scale, score 1 corresponding to maximum permissible value, score 5 to consistency with current regulatory standards, and score 10 to green construction. For the purpose of ranking buildings according to environmental safety class, a comprehensive environmental safety indicator (CESI) is proposed. Buildings with CESI score higher than 6 can be considered as environmentally safe.

ESB is evaluated both at the design stage and throughout the life cycle. Unfortunately, the design-stage ESB may not guarantee that the building will stay consistent with environmental safety standards during its operation, since the design stage normally uses the current regulatory documents (5-point scale) that do not incorporate green standards. In this regard, new construction projects, and especially healthcare facilities, should be made subject to stricter standards so as to rule out the occurrence of negative situations during construction and operation stages. The construction process must use the materials with confirmed environmental compatibility. To increase the efficiency of design developments, a comprehensive methodology is proposed for ensuring the ESB. It is a design validation tool and can help to obtain the missing information and quantify the main ESIs using numerical methods or experimentally [2].

The core set of ESIs adopted by the proposed ESB assessment methodology can be presented as three interconnected complexes (Fig. 1). The building and its surrounding environment are considered as one dynamic system, while the indoor

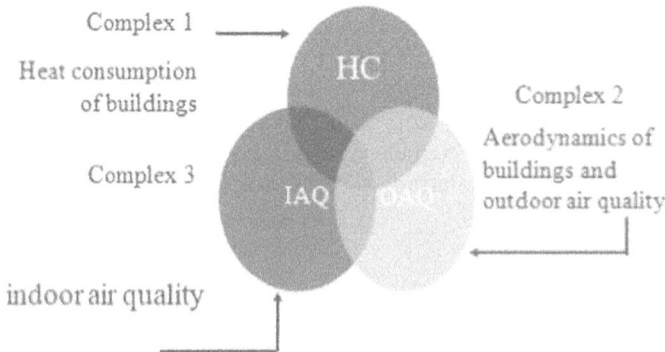

Fig. 1 Components of ecologically safe and comfortable environment

environment is seen as the main subsystem responsible for safe and comfortable environment and operating according to the pre-set parameters (Complex 3).

Building's surrounding environment is defined as zone within which external factors (wind speed and pressure, concentrations of harmful substances) are affecting the heat and mass transfer through the building envelope and the quality of indoor air. The heat and mass transfer within one single dynamic system (SDS) is interpreted in terms of dynamic connections [3].

Complex 1. The actual energy consumption by buildings will depend on the accuracy of thermal calculations and the reliability of thermophysical performance of building materials and structures, especially translucent ones.

For Complex 1, the ESB calculations should experimentally determine:

- thermophysical properties of insulation;
- heat transfer resistance of translucent structures.

Using the experimentally determined thermal characteristics of building materials, it is now possible to create numerical models of the temperature patterns in stand-alone structures and façade components, paying special attention to interfaces.

Complex 2. The parameters of indoor climate and their spatial distribution depend on the aerodynamic conditions of the given neighborhood and the levels of outdoor air pollution. These parameters often deviate from their design values, especially in case of natural ventilation. The layout of a neighborhood—the shape, size and location of its constituent buildings—forms its aerodynamic conditions. The areas near facades develop wind shadows. They are traps for harmful substances and pathways for them to enter the buildings with incoming air.

In many neighborhoods, the main sources of pollution are near surface (transport, self-sufficient boilers, ventilation exhausts, etc.) Heavy traffic, underground garages and parking lots area sources of air and noise pollution. Contributing to near-surface background pollution levels are also the near-by industries, whose emissions can, under certain conditions, increase the air pollution levels within healthcare facilities.

For Complex 2, the ESB calculations should be preceded by neighborhood layout analysis in order to assess the local ecological situation in the area of the projected building and identify possible sources of air and noise pollution. It is advisable to use mathematical modeling to:

- evaluate the aerodynamic conditions in the neighborhood – velocity and pressure fields at facades and air induction shafts for the most unfavorable and prevalent wind directions;
- identify location and size of wind shadow near the facades of the building under analysis;
- check the location of air intake points and exhaust emissions.
- Experimentally:
- measure the noise pollution level (traffic and other sources);
- measure the content of harmful substances, their quantitative and qualitative composition.

Complex 3. For the purpose of achieving high-level ESIs and comfortable indoor climate in medical premises, it is recommended that the calculations following the current standards and regulations are complemented by the outcomes of Complex 1 and 2 analyses.

Using the data obtained at the design stage during the analysis of the external factors and the calculated heat and mass transfer through the building envelope, it is recommended to use mathematical modeling technique to analyze the project for:

- distribution of the indoor climate parameters and concentrations of harmful substances in the volume of typical rooms;
- stagnant zones and zones of increased air mobility;
- efficiency and positioning of air supply devices;
- changes in air exchange rate due to the changes in pressure at the facades of the building;
- expected noise level during the day and at night;
- duration of insolation and daylight factor (DF).
- Experimentally, for ES conditions verification purposes:
- justify the choice of air supply device;
- determine the soundproofing of the translucent structures that provide acceptable noise levels;
- determine the soundproofing of the floor framings and room partitions.

3 Methods

Our ESB evaluation technique made use of software-assisted (STAR-CCM+) mathematical modeling and specially developed codes [6, 7, 9, 14, 15].

The mathematical models of Complexes 1 to 3 used the Reynolds-averaged Navier–Stokes equations:

Equation for conservation of momentum:

$$\frac{\partial \rho u_i}{\partial t} + \frac{\partial}{\partial X_j}\left(\rho u_j u_i - \tau_{ij}\right) = -\frac{\partial P}{\partial X_j} + s_i \tag{1}$$

where t is time, X_i are spatial coordinates, u_i are components of the velocity vector, P is piezometric pressure, ρ is density, τ_{ij} are stress tensor components, s_m is mass source, s_i are pulse source components.

The stress tensor is expressed as:

$$\tau_{ij} = 2\mu s_{ij} - \frac{2}{3}\mu \frac{\partial u_k}{\partial X_k}\delta_{ij} - \overline{\rho u_i' u_j'} \tag{2}$$

where s_{ij} is strain rate tensor.

Equation for conservation of mass:

$$\frac{\partial \rho}{\partial t} + \frac{\partial}{\partial X_j}\left(\rho u_j\right) = s_m \tag{3}$$

To obtain the data on temperature fields and concentration of harmful substances, the system will be supplemented with the following equations:

Equation for conservation of heat quantity:

$$\frac{\partial \rho C_p T}{\partial t} + \frac{\partial}{\partial X_j}\left(\rho C_p T u_j\right) = \frac{\partial}{\partial X_j}\left[\lambda \frac{\partial T}{\partial X_j} - \rho C_p \overline{u'_j T'}\right] \tag{4}$$

Equation for conservation of passive impurities:

$$\frac{\partial \rho c}{\partial t} + \frac{\partial}{\partial X_j}\left(\rho c u_j\right) = -\frac{\partial}{\partial X_j}\left[\rho D \frac{\partial c}{\partial X_j} - \rho \overline{u'_j c'}\right] \tag{5}$$

The oscillatory member $\overline{u'_i u'_j}$, following the ideas of Boussinesq, is represented as:

$$-\rho \overline{u'_i u'_j} = 2\mu_t s_{ij} - \frac{2}{3}\mu_t\left(\frac{\partial u_k}{\partial X_k} + \rho k\right)\delta_{ij} \tag{6}$$

where μ_t is turbulent viscosity. The additional members in the energy transfer equations and the impurity components can be modeled as follows:

$$-\rho \overline{u'_i c'} = \frac{\mu_t}{Sc_t}\frac{\partial c}{\partial X_i} \tag{7}$$

$$-\rho C_p \overline{u'_i T'} = C_p \frac{\mu_t}{Pr_t}\frac{\partial T}{\partial X_i} \tag{8}$$

Different turbulence models are used to close the hydrodynamics equations. Among the commonly used ones is the standard k-ε.

Experimental verification of thermal, soundproofing and environmental performance of building materials and structures is carried out by accredited laboratories in accordance with the current GOST.

4 Discussion

Below are the examples of applying the proposed comprehensive methodology for ensuring ESB across Complexes 1 and 3.

The commonly recognized systems (LEED, BREEAM) use as key criterion energy consumption. At design stage, when modeling the thermal performance of external enclosing structures, the heat and mass transfer processes are often neglected, unfortunately. Nor do the humidity conditions receive due attention. This leads to underestimated energy consumption at the operation stage. To give one example, the heat-insulating inserts, that are widely used in the construction industry to increase heat transfer resistance of floors, appear to have their estimated performance deviating from the experimental findings.

Heat-insulating inserts lead to redistributing of heat flow due to the significant difference in the thermal conductivity of concrete and polystyrene. The issues relating to the influence of heat-insulating inserts often arise during the construction phase, requiring adjustments to design solutions and calculation of the temperature fields, especially for structural connections. Figure 2 shows 2D and 3D models of the temperature field of the reinforced concrete floor with heat-insulating inserts.

The heat transfer resistance values of obtained by modeling the heat and mass transfer processes, differ from the calculated values by more than 50%.

As an example of the experimental assessment of thermal properties, Fig. 3 shows a translucent structure component being tested in the climate chamber.

The testing is carried out according to GOST 26,602.1–99. The structure is a fragment of the facade with a two-chamber energy-saving double-glazed window. The error in determining the heat transfer resistance does not exceed 10%. It should be noted that the standard heat transfer resistance calculations allow for the inaccuracy of 20% to 50%.

In healthcare facilities, the quality of indoor climate is a special focus. For example, for the optimal choice of ventilation and air conditioning design solutions, it is necessary to analyze the situation as a whole, i.e. to identify the spatial relationship of the dynamic processes occurring indoors and in the atmosphere surrounding the building.

The use of modern insulated translucent structures requires a more in-depth assessment of the air exchange rate in buildings with natural ventilation, and well-reasoned choice of air supply devices. With mechanical ventilation, the processes to be analyzed are air distribution and temperature conditions maintenance.

Given below are the results of implementing the comprehensive ESB assurance methodology for better efficiency of natural ventilation and the results of testing of window units and fresh air valves.

The results of air permeability testing of window units, performed at Saint Petersburg State University of Architecture and Civil Engineering Test Center over the period of 10 years, show that at a pressure of 50 Pa, which is the pressure used in air exchange rate calculations, the premise receives as a result of infiltration not more than 6 $m^3/(m^2 h)$ of air. For example, a premise with an area of 30 m^3 and 2.4 m^2 window units will have the air exchange rate of 0.38 $^{h-1}$.

The construction materials market offers a wide range of window vents and fresh air valves. Their performance characteristics are determined experimentally. The air supply devices must meet the following requirements:

- the amount of incoming air should be sufficient enough to provide the required air exchange;

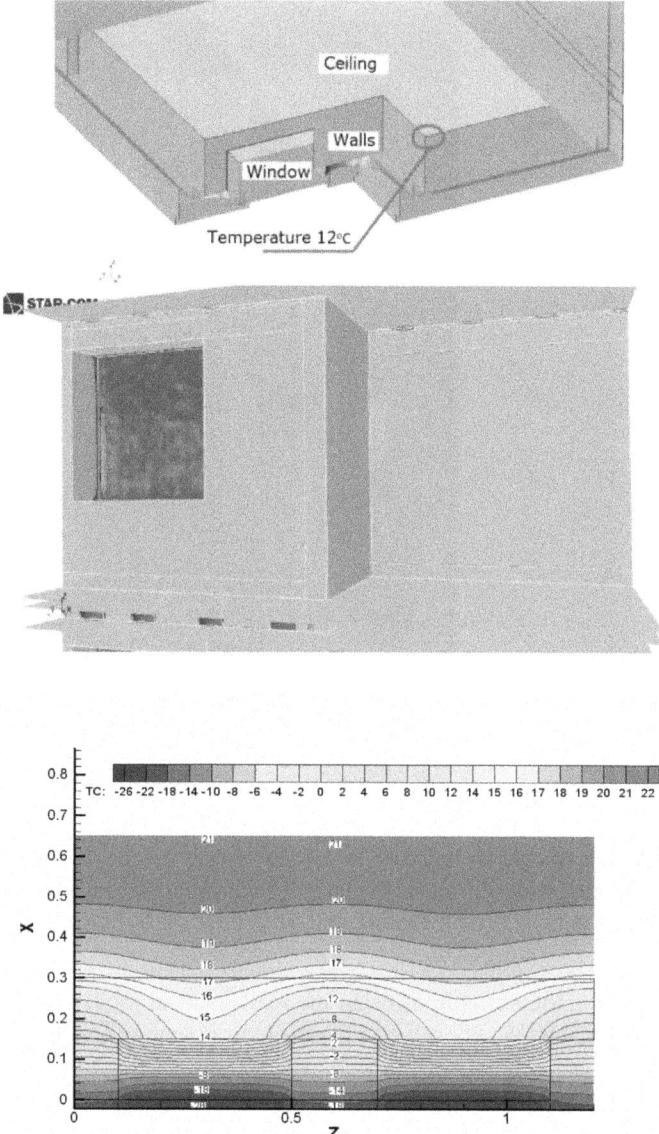

Fig. 2 2D and 3D temperature field calculation example

- the soundproofing of valved structures should ensure the acceptable sound pressure levels in the premises, especially at night;
- the temperature on the valve surfaces should be above the dew point to avoid condensation;

Fig. 3 Determining the heat transfer resistance of a translucent structure component in climatic chamber at Saint Petersburg State University of Architecture and Civil Engineering

- the distribution of cold air in the room with the supply device open should ensure sufficient mixing and rule out sharp changes in the temperature gradient. Conformance with the above requirements will be confirmed by relevant certification tests.

Figure 4 shows the example of using the mathematical modeling technique to identify the optimal air supply valve and evaluate its operating conditions.

The temperature field analysis is used for developing measures to prevent moisture condensation and low temperature zones.

5 Conclusions

The reviewed methods are suitable for applying at design and operation stages to predict and evaluate the environmental safety of public and residential buildings.

For the purpose of verifying the conformance with particular environmental safety indicators, it is advisable to use the comprehensive ESB assurance methodology at the design stage and during construction.

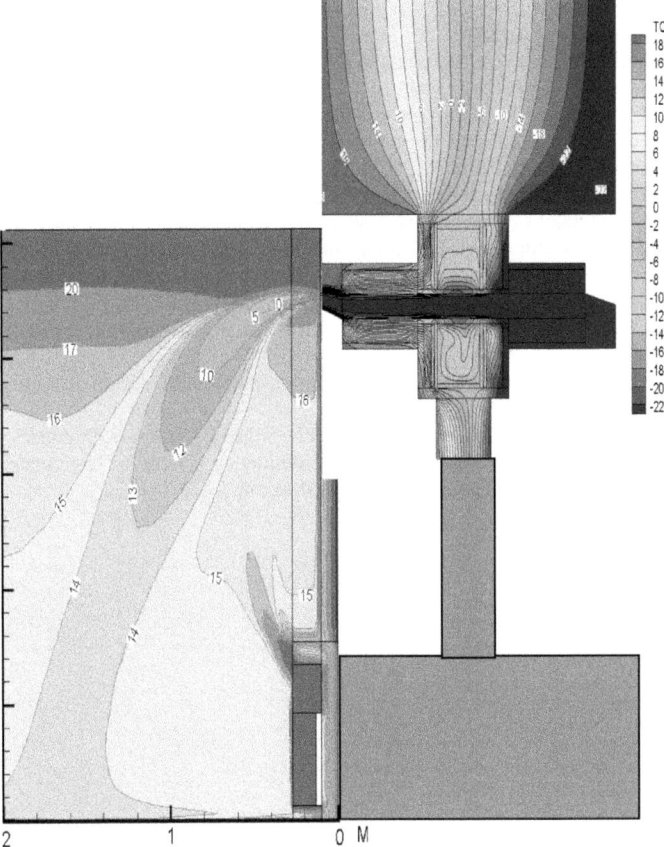

Fig. 4 The temperature field formed by a 0.1 m diameter fresh air valve installed at the top of the window unit

References

1. Bakaeva NV, Natarova AY, Igin AY (2017) Green construction criteria for assessing the environmental perfomance of residential and public buildings. Proc Southwestern State Univ 21(1):57–68. https://doi.org/10.21869/2223-1560-2017-21-1-57-68
2. Datciuk TA, Denisikhina DM, Anshukova EA (2019) Predicting air quality in underground structures. In: Geotechnics fundamentals and applications in construction: new materials, structures, technologies and calculations. CRC Press, pp 54–58
3. Datsuk T, Pukhkal V, Ivlev U (2014) Forecasting of microclimate in the course of buildings design and reconstruction. In: Advanced materials research, vol 1020. Trans Tech Publications Ltd. pp 643–648
4. Datsyuk TA, Smirnov EB, Pinkevich IK (2013) The methodology of comprehensive evaluation of the environmental safety of projected residential buildings. Bullet Civ Eng 5(40):219–226
5. Dias WPS, Chandratilake SR, Ofori G (2017) Dependencies among environmental performance indicators for buildings and their implications. Build Environ 123:101–108

6. Ferziger JH, Perić M, Street RL (2002) Computational methods for fluid dynamics, vol 3. Springer, Berlin, pp 196–200
7. Grimitlin AM, Datsyuk TA, Denisikhina DM (2013) Mathematical modeling of ventilation processes. AVOK SEVERO-ZAPAD, St. Petersburg
8. Haapio A, Viitaniemi P (2008) Environmental effect of structural solutions and building materials to a building. Environ Impact Assess Rev 28(8):587–600
9. Lu S, Wang YH, Zhang RF et al (2011) Numerical study on impulse ventilation for smoke control in an underground car park. Procedia Eng 11:369–378
10. Sidorenko VF (2006) Comprehensive evaluation of residential neighborhoods' environmental performance as a factor of living environment optimization. Ecology of Urban Areas 1:42–49
11. Smirnov EB, Datsyuk TA, Taurit VR (2017) Evaluating the environmental safety of new construction projects. Water Ecology: Challenges Solut 3(71):83–99
12. Datciuk T (2016) Forecasting of ecological situation in course of buildings design. Architectural Eng 1(2):19–22
13. Vakili-Ardebili A, Boussabaine AH (2010) Ecological building design determinants. Architectural Eng Des Manage 6(2):111–131
14. Wilcox DC (1994) Turbulence modeling for CFD, 3rd edn. California. Industries, USA
15. Zhao R, Zhou L, Ma J (2018) CFD design of ventilation system for large underground bus terminal in Macau Barrier Gate. J Wind Eng Ind Aerodyn 179:1–13

Modern Challenges of Healthcare Construction

Marina Egorova

Abstract Healthcare construction has an important role in shaping the image of a country. The Constitution of the Russian Federation defines the Russian Federation as a welfare state with a system of "occupational and human health protection measures." In 2019, the human health protection was threatened by the COVID-19 pandemic. According to Timur Andrbaev, Construction Director at Moscow International Medical Cluster Foundation, "the pandemic will lead to a greater demand in healthcare construction expertise consistent with the international standards and allowing for rapid re-designing of facilities depending on the tasks and scope of work to be achieved." The innovative approach to healthcare construction can lead a quality improvement of the life of population, as well as make life easier for patients in palliative care. Innovative healthcare development is led by Japan, America, Great Britain and Germany, followed by China and India with. Russia ranks among the least progressive countries as new health technologies appear in its facilities much less often.

1 Introduction

The domestic healthcare construction currently faces a great number of challenges. Innovations in healthcare play a huge role in improving people's quality of life and can significantly increase life expectancy. Healthcare innovations unfold through IT systems, the latter being actively used in the fight against genetic and oncological diseases. In pursuit of higher quality of life, healthcare efforts target also the expanded construction of new facilities.

M. Egorova (✉)
Saint Petersburg State University of Architecture and Civil Engineering, Saint Petersburg, Russia
e-mail: marina-332@mail.ru

© The Author(s), under exclusive license to Springer Nature Switzerland AG 2023
D. Ivanov et al. (eds.), *Proceedings of ECSF 2021*, Lecture Notes in Civil
Engineering 257, https://doi.org/10.1007/978-3-030-99877-6_10

2 Materials and Methods

The issues of innovative medicine and healthcare construction have been explored by researchers in many countries with focus on their legal frameworks. The domestic research relies on prominent papers by I. P. Pavlov, N. F. Gamaleya, D. K. Zabolotny, V. M. Bekhterev, N. N. Burdenko, M. P. Konchalovsky, E. N. Pavlovsky, A. N. Sysin, L. A. Tarasevich, among others. The regulatory framework for the domestic health-care construction is comprised by Sanitary Regulations and Standards 2.1.3.2630-10 and Construction Standards and Regulations II-L.9–70.

3 Results

There exist three types of hospital construction designs—decentralized (pavilion hospitals), centralized, and mixed. Decentralized type offers better prevention of nosocomial infections, since it allows for separating different patient categories, and has park areas on its territory. The downsides of decentralized hospitals lie in their departments being disconnected from one another, duplication of medical devices and medical and diagnostic rooms, increased size of service personnel, longer distances between units, longer delivery of food from the nutrition unit, and high land consumption [9].

The centralized type makes up for all disadvantages of the pavilion hospital by occupying a smaller area, with more space for parks and landscaping, faster travel between units, faster nutrition delivery, and more convenient operation of sanitary equipment. But, like any centralized system, the centralized hospital has a number of disadvantages. It has a poorer protection system against nosocomial transmission and limited access to walking areas for patients in upper-floor departments.

The mixed type is comprised of the main building with admissions unit, main somatic departments, clinical and diagnostic departments (X-ray, physiotherapy, clinical laboratory), and a pharmacy; its outpatient clinic, maternity unit, children's diseases, infectious diseases, pathology and anatomical departments, administrative and economic services are housed in separate buildings. Construction in single units has modernized the healthcare construction by connecting all stand-alone hospital buildings, with exception of infectious diseases and radiological department, into one with the help of passages [11].

Structurally, the innovative hospital consists on:

- admissions unit;
- treatment sections;
- surgical unit;
- treatment and diagnostics department;
- pharmacy;
- nutrition unit;
- training, administrative, auxiliary premises, utility rooms;

- hotel, day habitation.

There is a number of hygienic considerations to be accounted for when designing healthcare facilities, such as optimal conditions for treatment of patients and work of the staff; optimal air exchange and ventilation; accommodation of wards and diagnostic departments; convenient connection between treatment sections and diagnostic unit, auxiliary services, lobby, park; nonspecific prevention of nosocomial infections [2–3].

Today's trends in healthcare design and construction are towards specialization and centralization of functionally homogeneous departments and hospital services—intensive care, organ transplantation, laser treatment, hyperbaric oxygenation, radionuclide diagnostics, etc. There appear large, consolidated diagnostic centers equipped with unique, costly equipment, and facilities with focused research specialization—perinatal, cardiovascular, oncological, and psychiatric centers [7].

Another hospital design trend is a 9-to 12-storey monoblock of wards connected, via a covered way or underground passage, to a 2-to 3-storey building of treatment and diagnostic services. Many modern hospitals are multi-unit, with surgical and therapeutic departments housed in two separate buildings and the treatment and diagnostic unit close to them. The existing hospitals receive expansions—new buildings housing wards. High-rise multidisciplinary centers occupy smaller land area; their personnel don't have to cover long distances between units; the time of delivery of patients, medicines and nutrition is reduced; treatment sections are close; the upper floors have higher chemical and microbial air pollution levels; increased vertical mobility (elevator, stairs); elevator wait times are long; patients on the upper floors may experience limited access to walks; upper floors may experience high wind exposure in winter and are unsuitable for patients with fear of heights; noise [15].

The first floor of high-rise healthcare centers is occupied by trauma department and surgical purulent infections, the latter being a separate unit with induced ventilation. Aseptic operating unit and departments of anesthesiology, intensive care and therapy, chronic hemodialysis, postoperative wards should be accommodated not lower than the second floor—close to the surgical unit and isolated from each other. X-ray rooms and radiology departments with ionizing radiation sources must not be adjacent, either horizontally or vertically, to wards for pregnant women and children [1].

With multi-unit hospitals, the number of floors is limited to 5–7; surgical units can occupy a separate building; there are premises for fee-based services and less use of elevators and, hence, lesser likelihood of accidents in elevator cabins, but still longer delivery of medicines and patients to necessary sections. Hospitals with fewer floors enjoy better conditions for treatment, sanitary and hygienic measures [5].

One example of innovative healthcare facility is arthrology center in Moscow for patients with joint and musculoskeletal diseases. Its patients undergo 7-to 10-day examination, receive treatment for exacerbation and follow the prescribed treatment pattern. Pharmacological therapy takes place at home, following which the patient returns to the center for remedial treatment. Such a treatment scheme allows an increased patient capacity, is cost-effective and conducive to higher therapeutic and preventive effect.

4 Discussion

In overseas countries, healthcare construction has a number of distinctive features, one being that patients are distributed among sections according to the severity of their condition, not disease itself. There are departments of intensive care for the critically ill; intermediate care for moderately severe conditions; convalescents and patients under extensive testing; patients with chronic diseases requiring longer treatment periods; home care; outpatient treatment [7].

Unlike the Russian hospitals, the hospitals abroad are custom-built; the majority of them are private, independent, unique, self-sufficient facilities (not state-owned); they don't have admissions units; catering is organized according to restaurant system; hygienic recommendations are advisory, not compulsory, as they are in Russia [11].

The hospitals abroad are products of a more "industrialized" approach, with three core lines of operations—"Examination and diagnostics," "Treatment", "Recovery; working capacity examination; monitoring of long-term effects; treatment efficiency evaluation." At each stage, the patient is managed by designated team of specialists and doctors.

5 Results

Sanitary Regulations and Standards (SRS) 2.1.3.2630-10 for healthcare facilities.

SRS 2.1.3.2630-10 establishes sanitary and epidemiological requirements to arranging, equipping maintaining of healthcare facilities; their epidemiological conditions, preventive measures, occupational conditions, catering and nutrition for patients and personnel.

The structure, layout and equipment of hospital premises should ensure streamlined performance and exclude the possibility of crossing of flows with different degrees of epidemiological risk.

The architecture and design layout of healthcare premises should be conducive to optimal occupational conditions, therapeutic and diagnostic processes, and compliance with sanitary and epidemiological regime.

Sections (premises) subject to special aseptic regime, ward sections, radiology and therapy departments, and units of closed process cycle (laboratories, nutrition unit, decontamination unit, pharmacy, laundry) should not have through passages.

Medical organizations that serve as training and research facilities will have classrooms for students and teachers, auxiliary rooms (locker rooms, toilets, storerooms). New projects and healthcare facilities under reconstruction will have patient wards equipped with bathrooms (sink, toilet, shower). Bathroom doors should open outwards. Rooms for accommodating wheelchaired (wheeled bed) patients must have doorways with a width of 120 cm minimum (110 cm for existing medical facilities) [8].

Hospitals in densely built up areas and in-patient facilities without restorative treatment and care units may be expanded at the expense of 10–15% of their park and garden zones. Green spaces and lawns will account for at least 50% (previously 60%) of in-patient facility's total area [13].

Ceiling height mustn't be lower than 2.6 m (earlier 3 to 3.3 m). Of the three air microbial purity indicators, only one remains valid, CFU/m3: pre-use 200 and in-operation 500 for cleanliness class A premises; total microbial count is not rated for class B conditionally clean premises (previously 750 and 1000 CFU). Class B includes boxes, box-like wards, adult wards, neonatal anterooms, doctors' offices, etc.

New projects and healthcare facilities under reconstruction can offer tailored diet (tablet nutrition)—a system where hospitals' nutrition units prepare for patients (and staff) individual lid trays based on individual menus. Nutrition is delivered to wards in special thermal containers on trolleys with special compartments for use containers. All used containers are delivered back to the nutrition unit.

Healthcare facilities with tablet nutrition are allowed to do without ward section canteens; their pantries are one-room premises with hand wash sink, dish sanitizing tub (epidemiological measure), household refrigerator, microwave oven, electric kettles. Dish washing is carried out centrally in the nutrition unit's premises for washing kitchen utensils and patient and staff tableware, and treatment of tablet nutrition trolleys.

6 Conclusions

The Russian government and medical community are faced with the choice of development path for healthcare system. One option is towards reduced state-funded services and increase in privately owned facilities. The other option is towards strengthening the role of the state in healthcare management and financing to fully assure citizens' constitutional rights to free medical care. This choice will be decisive for the fate of the Russian Federation as a welfare state and its capability of creating the conditions conducive to higher living standards and human development.

References

1. Alexandrova OA, Yarasheva AV, Nenakhova YS (2020) Preparing nursing departments for metropolitan healthcare facilities: challenges and solutions. problems of social hygiene and healthcare sector. History of Med 28(S1):680–6
2. Bonkalo TI, Veprentsova SY, Nikitina N et al (2020) Factors of the development of fear of disease progression in patients with breast cancer. Open Access Macedonian J Med Sci 8(E):74–80
3. Bonkalo TI, Shmeleva SV, Zavarzina OO et al (2020) Integrated identity as a psychological determinant of a person'S harmonious eating behavior. Revista Inclusiones 7(S2–2):409–422

4. Czech M, Baran-Kooiker A, Atikeler K et al (2020) A review of rare disease policies and orphan drug reimbursement systems in 12 Eurasian countries. Front Public Health 7:416
5. Czech M, Hołownia-Voloskova M, Baran-Kooiker A (2020) Policy developments of health technology assessment in the European Union. Postępy Biochemii 1:318–321
6. Gribkova IV, Galstyan GM, Polyanskaya TY et al (2020) Kaolin, used to trigger coagulation in thrombin generation test, increases sensitivity of the method in hemophilia patients. Blood Coag Fibrinol 31(3):193–197
7. Hołownia-Voloskova M, Tarbastaev A, Golicki D (2021) Population norms of health-related quality of life in Moscow, Russia: the EQ-5D-5L-based survey. Qual Life Res 30(3):831–840
8. Kholovnya-Voloskova ME, Tolkushin AG, Kornilova EB, Zavyalov AA (2020). Comparative analysis of medical evaluation technologies. problems of social hygiene and healthcare sector. History of Med 28. Special Issue
9. Koltsova E, Lukina G, Shmidt E et al (2020). AB0303 predictors of serious infections in patients with Rheumatoid arthritis receiving target therapy. 79(1S):1451
10. Poliakova K, Kornilova E, Holownia-Voloskova M et al (2020) PSS5 cost of illness analysis: drug therapy of patients with glaucoma in Moscow. Value in Health 23:S362–S363
11. Semenova VG, Ivanova AE, Sabgaida TP et al (2020) Evolution of Moscow's loss of able-bodied population from injuries of vague intention in the 2000s. problems of social hygiene and healthcare sector. History of Med 28 (S1):1075–1080
12. Shkrumyak AR, Kamynina NN, Aksenova EI (2020) Basic issues of organizing doctors' training during the coronavirus pandemic. problems of social hygiene and healthcare sector. History of Med 28 (S1):851–856
13. Vosheva NA, Kamynina NN, Voshev DV et al (2020) Modern domestic and international methods for protecting medical personnel from new coronavirus infection (COVID-19): analytical review. Bullet Russian Acad Med 75(5S):386–394
14. Yudina NN, Chernyshev EV, Melgunova MS et al (2020) Evaluating the export potential of the healthcare system of the Russian Federation: case study of one Russian region. Problems of social hygiene and healthcare sector. History of Med 28(S2):1186–1189
15. Yarasheva AV, Aleksandrova OA, Medvedeva EI, Alikperova NV, Kroshilin SV (2020) Problems and prospects of personnel support of the Moscow healthcare system. Econom Soc Changes: Facts Trends Forecast 13(1):174–190

Legal Framework for Shaping and Promoting the Biomedical Research Infrastructure

Dmitry Ivanov

Abstract The paper discusses the current issues of the legal frameworks for creating and promoting biomedical research infrastructure. Particular attention is paid to the experience of the European Union in building long-term development strategies for biomedical research infrastructure. Using the research-based criteria to be met by modern research facilities, a comparison is made between the European and the Russian concepts underlying legal regulation of the biomedical research infrastructure.

1 Introduction

Our research capacity and availability of technology-intensive medical care are largely contingent on the state of our infrastructure for biomedical research. This infrastructure is, in turn, dependent on the complex linkages involving other material and social elements that shape the setting for medical research and development, diagnosis and patient care. This setting is evolving swiftly, influencing the ongoing research [8] and inducing changes in the research infrastructure.

The rapid development of biomedicine, biopharmaceutics and biotechnologies forces us to revise, from time to time, the seemingly entrenched approaches to biomedical research and its supporting infrastructures. There has been much discussion recently about the role of biomedical research infrastructure in the fight against COVID-19 [16], way to achieve laboratory biosafety and biosecurity [11], and how the state of research infrastructure can affect research performance [6], among other issues.

In this context, the relevance of efforts to develop the existing research infrastructures, build new biomedical facilities, and finance the infrastructure projects are beyond doubt. The careful monitoring of the developments being achieved in Russia and abroad indicates a significant increase in the lawmaking efforts in the field of biomedical research. Activities related to the reconstruction of existing biomedical

D. Ivanov (✉)
Saint Petersburg State University of Architecture and Civil Engineering, Saint Petersburg, Russia
e-mail: dmivanov@lan.spbgasu.ru

© The Author(s), under exclusive license to Springer Nature Switzerland AG 2023
D. Ivanov et al. (eds.), *Proceedings of ECSF 2021*, Lecture Notes in Civil Engineering 257, https://doi.org/10.1007/978-3-030-99877-6_11

facilities and the engineering of new ones are becoming increasingly subject to regulation. This regulation unfolds along two main paths, one involving the public relations arising in connection with the conduct of research, and the other targeting to set standards and requirements for physical conditions and individual research technologies.

The time has come for an in-depth study of the Russian regulatory framework existing for the biomedical research infrastructure, and benchmarking it against the progress achieved internationally. Of particular note in this regard are the strategy building efforts of the European Union (EU), designed to lead to major breakthroughs in biomedical research.

2 Materials and Methods

The study examines the regulatory frameworks existing in Russia and the EU for research and development sector, civil engineering and healthcare services.

Its empirical basis comprises the data obtained from previous sociological research, published studies and other sources.

The methodological framework of the study encompasses general dialectical method of inquiry, analysis and synthesis, abstraction and concretization, induction, deduction and analogy; legalistic approach, comparative legal analysis; and structural analysis and system design methods. The study relies on the cross-disciplinary approach to prospects of the biomedical research infrastructure.

3 Results

In the EU, the systematic work towards improved biomedical infrastructure has been in progress for many years. The resultant infrastructure is marked by highly integrated nature and increased adaptability and flexibility.

The EU owes its impressive results not only to sufficient funding. No less important are its research efforts towards shaping the standards and requirements for biomedical facilities. Besides, the exercise of legal regulation in the field of biomedical research allows the European research infrastructure to arrive at more effective performance patterns.

The issues of biomedical infrastructure development receive considerable attention also in Russia. However, the Russian approach differs greatly from the European concept, which manifests itself prominently in the regulation of public relations. At the same time, the potential of the legal regulation remains under-utilized.

4 Discussion

Many studies explore biomedical infrastructure as an independent object of scientific research. This is not surprising, since the success, and sometimes the very possibility of scientific and applied research, depends on its condition. The definition of what seamless research infrastructure should be like evolves in response to the increase in the researcher toolkit, and so do the complexity and ambition of the research tasks. Advanced biomedical research clearly requires an effective integration of physical and biological sciences, as well as a developed infrastructure to facilitate communication between disciplines that speak their long-established languages [14].

The two essential requirements to be met by modern research infrastructure are self-sufficiency and interaction. Self-sufficiency is key to safety. However, the major bulk of the advanced research is interdisciplinary in nature, calling for well-coordinated interaction among research teams. Unlike those early laboratories with experimenting tables, high windows for ventilation and lighting, chemical agent bottle stands and wash sinks, the exemplary biological laboratory of today is expected to meet much higher safety and technology standards. It should offer a comfortable working environment. The more comfortable the lab is and the fewer distractions it has, the more likely scientists are go deeper into research for higher performance [10].

The development of biomedical research infrastructure is driven not only by our material capabilities, but also our ideas about what research facilities should be like. The modern concepts value quality design as what can lead to safer, healthier environments for conducting biomedical research [4]. Healthcare facilities should be adaptable and flexible enough to be repurposed, if needed, and to meet the expectations of their users and managers. This requirement is due to rapid improvements being experienced by medical technologies [5].

It is not surprising in this context that each new medical facility design is seen as a complex task associated with high risks [13]. In fact, the range of standards to be met by modern biomedical research infrastructure facilities can be quite specific in terms of transport accessibility, zoning, uninterrupted power supply, lighting, ventilation, water supply, sanitation, indoor climate parameters maintenance, etc.

One trend is towards decentralized research systems [2], which have their research facilities distributed among focal points for easy integration of resources. Another trend is towards quality transformation of biomedical research facilities. One commonly cited example of distributed research infrastructure is BBMRI-ERIC (Biobanking and BioMolecular Resources Research Infrastructure—European Research Infrastructure Consortium). But now BBMRI-ERIC is increasingly linked with serving as a network tool for the existing biobanks, not as a creator of new biobanks and bioresources for Europe, whilst biobanks themselves now serve as national projects and venues for better interaction between the state, the public, the research community and the economic agents [1].

Contrary to publications arguing that biological research, unlike research in domains other than live sciences, does not require large-scale and expensive infrastructure [7], practice speaks to a different reality. A good example is the progress achieved in biobanking infrastructure over the past decades. Large-scale biobanking has become a noticeable trend in genomics, with new social dimensions emerging with regard to communication, scientific debate, regulation and public monitoring of the biobanks as valuable assets [3].

The above examples clearly demonstrate how the research communities can shape demand in research infrastructure. Where there's demand, there's supply. For example, the EU—planet's leader in research and innovation—has for many years been working towards coordinated policy aimed at implementing the European Research Area concept [15]. This concept has as its integral component the strengthening of Europe's scientific and technical base. To this end, the long-term framework programs have been consistently implemented in the EU, providing, in particular, for the researchers based in different countries to have access to the advanced research infrastructures outside of their home countries, as well as for better coordination and integration of these infrastructures across Europe [12]. This practice is designed to ensure high quality research and to lead to favorable environments conducive to scientific progress and new technologies.

The legal framework for above mentioned multi-year framework programs is Article 182 (formerly Article 166) of the Consolidated Versions of the Treaty on European Union and the Treaty on the Functioning of the European Union (2016/C 202/01)—the document envisaging, among other things, the long-term research and development planning.

One prominent example of the European framework programs is Horizon 2020, implemented in accordance with Regulation (EU) No 1291/2013 of the European Parliament and of the Council of 11 December 2013. This Regulation, in particular, defines "research infrastructure" as objects, resources and services to be applied by the research communities for their research purposes and stimulating innovations in their chosen fields. The elements comprising research infrastructures include extensive scientific facilities and tool kits; knowledge-based resources such as collections, archives and scientific data; electronic assets such as information and computing systems and communication networks; any other infrastructure of unique nature for excellence in research and innovation. The document further classifies the assets of the research infrastructures into "self-sufficient", "virtual" and "distributed".

As follows from Annex I to the above Regulation, the research infrastructures are the yardsticks for evaluating the EU's competitiveness across the entire spectrum of scientific fields and are necessary for scientific innovation. In this regard, there is a concrete goal: create, using the generally accepted criteria, a solid foundation for generating, servicing and operating the research infrastructures that would be able to maintain research endeavors at the globally leading level. To achieve this goal, joint research infrastructures are being created so that scientists in different parts of Europe could have an integrated, transnational access to regional and national research infrastructures and be able to conduct research at the highest level, supported by newly deployed electronic infrastructures to unlock the full potential of networks, computing and scientific data, and by active introduction of information and communication technologies.

Like any other legislation, the European legislation can't be all-embracing. In some cases, gaps are filled by adopting constituent documents and local acts. Thus, BBMRI-ERIC operates in accordance with the Charter, duly approved by the Commission Implementing Decision of 22 November 2013 on Setting Up the Biobanks and Biomolecular Resources Research Infrastructure Consortium (BBMRI-ERIC) as a European Research Infrastructure Consortium (2013/701/EU). The Charter defines the biobank (Biomolecular Resource Center) as a collection, repository and distribution center of all types of human biological samples, such as blood, tissues, cells or DNA, and/or related data, such as clinical and research data, as well as biomolecular resources, including models and microorganisms that can contribute to the understanding of human physiology and human diseases. In addition, the BBMRI-ERIC Charter provides for the creation and development of a pan-European research infrastructure of biobanks and biomolecular resources.

Let us now consider the situation in Russia. The regulatory provision concerning the long-term research and development planning is missing at the level of the Eurasian Economic Commission. Although the Board of the Eurasian Economic Commission does have the authority to plan for the research activities for the reporting year.

At the level of the Russian national legislation, every citizen is guaranteed freedom of scientific, technical and other types of creativity (sec. 1, Article 44 of the Constitution of the Russian Federation); activities promoting human health are encouraged (sec. 2, Article 41 of the Constitution of the Russian Federation).

It is obvious that for these constitutional norms to take full effect in the field under consideration, it is necessary to provide it with adequate material resources—research laboratories, biobanks, preclinical research centers, etc. And in this respect, the Russian approach is in line with the European one: the strategic planning, particularly in the field of R&D and public health protection, is established in Federal Law No.172-ФЗ dd. 28.06.2014 "On Strategic Planning in the Russian Federation".

At the same time, there are considerable differences in the infrastructure development approaches practiced in the Russian Federation and in the EU. Thus, in contrast to the principle of decentralized research infrastructure, which is pursued in the EU, the Russian Federation Strategy for Research and Technological Development, endorsed by RF President Decree N 642 dated 01.12.2016, proceeds from the principle of resource concentration (concentration of intellectual, financial, organizational and infrastructural resources). This implies creation of centers for shared use of equipment; preferential focus on large research infrastructures on the territory of Russia; preferential channeling of the support to areas (regions) with a high concentration of research, development, and innovation infrastructure. This, however, does not rule out opportunities for network-based development of R&D and innovation-driven cooperation and research consortia.

Resource concentration has its pros and cons. The logic behind this approach is that government-funded biomedical research and healthcare studies are expected to yield maximum profit for taxpayers and society as a whole. Most often, this results in positive economies of scale or permanent return on the scale of biomedical research. At the same time, individual research units, laboratories and projects usually account for fewer studies; the lack of authenticity of their research raises questions. Besides, reduced economies of scale and/or volume may be products of increasing coordination difficulties when research teams hire more people or researchers with different backgrounds [9].

The RF Healthcare Development Strategy 2025 (endorsed by RF President Decree N 254 dated 06.06.2019) has among its high-priority tasks construction and reconstruction of healthcare facilities; development of healthcare infrastructure; optimal access to primary health care; and better transport accessibility of healthcare services through expanding the existing public transport networks and laying paved roads. But, given its purpose of a strategic planning document, the Strategy does not specify the standards to be met by infrastructure projects.

In accordance with para. c.1, Article 114 of the Constitution of the Russian Federation, the Government of the Russian Federation is committed to promoting Russia's potential for research and technological development. This provision is explained in more detail in the legislation on research and state scientific and technical policy, which grants the Government of the Russian Federation the authority to identify the lines of innovation-driven activities to be supported first.

Russia's sector-specific legislation (particularly, on healthcare and drug circulation) contains provisions regulating individual, and in many cases interdisciplinary, issues of conducting biomedical research and using appropriate infrastructure, leaving many legal issues unregulated or requiring clarification. It is not always clear what requirements the infrastructure projects are expected to meet. In case of biobanks, it is not clear in principle, since the current legal framework does not provide for their definition.

5 Conclusions

Despite the considerable attention being given in Russia to the issues of biomedical infrastructure, relevant regulatory mechanisms remain underutilized.

In our opinion, the current regulatory instruments should regulate not only the expediency of biomedical research and its infrastructure, but also the qualitative characteristics for new projects to meet. Their scope should go beyond mere technical regulation.

At the level of federal law, there is a number of unregulated issues that relate to conduct of the biomedical research and the use of relevant infrastructure (for example, to promote network-based research and the distributed infrastructure principle, the issues to be solved first relate to processing of donors' personal data of donors, informed consents, biological material turnover, etc.). Thus, the efforts should be targeted at shaping the dedicated legal framework for biomedical research and its infrastructure.

References

1. Argudo-Portal V, Domènech M (2020) The reconfiguration of biobanks in Europe under the BBMRI-ERIC framework: towards global sharing nodes?. Life Sci Soc Policy 16(9). https://doi.org/10.1186/s40504-020-00105-3
2. Bot BM, Wilbanks JT, Mangravite LM (2019) Assessing the consequences of decentralizing biomedical research. SAGE Publications Ltd., Big Data and Society. https://doi.org/10.1177/2053951719853858
3. Cambon-Thomsen A, Ducournau P et al (2003) Biobanks for genomics and genomics for biobanks. Comp Funct Genomics 4:628–634. https://doi.org/10.1002/cfg.333
4. DiBerardinis LJ, Baum JS, First MW (2013) Guidelines for laboratory design: health, safety, and environmental considerations: fourth edition. In: Guidelines for laboratory design: health, safety, and environmental considerations. 4th edn. Wiley. https://doi.org/10.1002/9781118633816
5. Domingo N (2015) Assessment of the impact of complex healthcare features on construction waste generation. Buildings 5(3):860–879. https://doi.org/10.3390/buildings5030860
6. Gallo F, Seniori Costantini A, Puglisi MT et al (2020) Biomedical and health research: an analysis of country participation and research fields in the EU's Horizon 2020. European J Epidemiol. Springer Science and Business Media B.V. https://doi.org/10.1007/s10654-020-00690-9
7. Gannon F (2006) Life science infrastructures are different. EMBO Rep 7:347. https://doi.org/10.1038/sj.embor.7400664
8. Goldsworthy C (2019) The effect of dynamic social material conditions on cognition in the biomedical research laboratory. Phenomenol Cogn Sci 18:241–257. https://doi.org/10.1007/s11097-018-9600-0
9. Hernandez-Villafuerte K, Sussex J, Robin E et al (2017) Economies of scale and scope in publicly funded biomedical and health research: evidence from the literature. Health Res Policy Syst 15:3. https://doi.org/10.1186/s12961-016-0167-3
10. Kornberg K (2016) You, Me, and We: Biolabs for the 21st Century. Cell 164:1097–1100. https://doi.org/10.1016/j.cell.2016.02.017
11. Mueller S (2021) Facing the 2020 pandemic: what does cyberbiosecurity want us to know to safeguard the future? Biosafety Health 3:11–21. https://doi.org/10.1016/j.bsheal.2020.09.007
12. Péro H, Faure JE (2007) European research infrastructures for the development of nanobiotechnologies. Trends Biotechnol 25:191–194. https://doi.org/10.1016/j.tibtech.2007.03.001
13. Petrosoniak A, Hicks C, Barratt L et al (2020) Design thinking-informed simulation: an innovative framework to test, evaluate, and modify new clinical infrastructure. Simulation in Healthcare 15(3):205–213. https://doi.org/10.1097/SIH.0000000000000408
14. Ravid K, Faux R, Corkey B et al (2013) Building interdisciplinary biomedical research using novel collaboratives. Acad Med 88(2):179–184. https://doi.org/10.1097/ACM.0b013e31827c0f79
15. Slepak VY, Amiyants AA (2019) Formation of the European research area in the context of the evolution of the legal framework for European research. Current Challenges Russian Law (9):142–152. https://doi.org/10.17803/1994-1471.2019.106.9.142-152
16. Tarasova EV, Makarevich PI, Efimenko AY et al (2021) European biomedical research infrastructures and the fight against COVID-19 pandemic. Sovremennye Tehnologii v medicine 13(1):6–16. https://doi.org/10.17691/stm2021.13.1.01

Analysis of Indoor Air Quality in the Swimming Pool in Ulan-Ude

Alexander A. Dmitriev, Yulia V. Ivanova, and Voldemar R. Tayrit

Abstract Comfortable indoor climate benefits our physical condition and mental health, inducing the desire to visit again, which is especially important for shopping and entertainment centers. In swimming pools, the structural engineering solutions often fail to provide even the lowest minimum air quality. Their poor quality leads to condensation exposure, mold and structural decay. In order to analyze the ventilation performance in indoor swimming pools, a full-scale study was performed of the air quality in one of the pools. The paper presents its methodology and results, as well as recommendations for assuring indoor climate standards for swimming pools.

1 Introduction

1.1 Design Solutions

The subject under analysis is the ventilation system of the swimming pool located in a shopping and entertainment center in Ulan-Ude.

The pool has two basins, one for entertainment, measuring 25×5.6 m, and the other for swimming, measuring 25×4.9 m. The larger basin is installed with waterfall, hydro massage and a 5.1×4.9 m Jacuzzi.

The fragment of the basement floor plan showing the swimming pool is given in Fig. 1.

The assimilation of heat and moisture in provided by supply and exhaust ventilation system (SE1) with air supply rate $L_r = 11{,}500$ m^3/h. The ventilation system is shown in Fig. 2.

For air exchange calculation, the number of visitors was assumed to equal 36. There are no seats in the pool for spectators.

The ventilation unit represents a special air conditioner for swimming pools with two-stage heat recovery for dehumidification and general ventilation. The unit is

A. A. Dmitriev · Y. V. Ivanova (✉) · V. R. Tayrit
Saint Petersburg State University of Architecture and Civil Engineering, Saint Petersburg, Russia
e-mail: juliavit@mail.ru

D. Ivanov et al. (eds.), *Proceedings of ECSF 2021*, Lecture Notes in Civil Engineering 257, https://doi.org/10.1007/978-3-030-99877-6_12

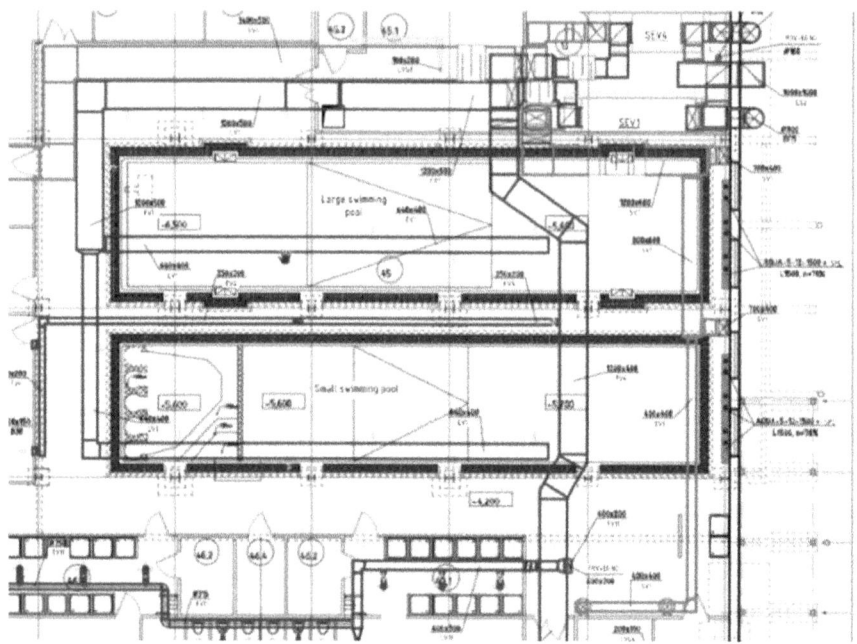

Fig. 1 The fragment of the basement floor plan showing the swimming pool

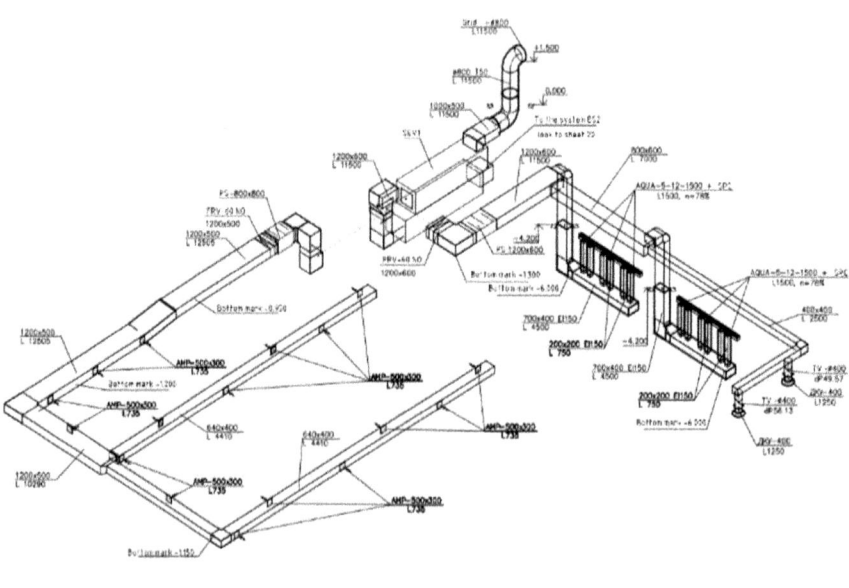

Fig. 2 Ventilation system diagram

comprised of air damper, heater, recuperative heat exchanger, mixing chamber, filters, supply and exhaust fans, and heat pump.

The air exchange uses "upward-downward" supply scheme. The air is supplied through the floor-mounted slit diffusers below the windows and through the conical-jet ceiling air diffusers above the windows and escapes through the adjustable grilles in the zone above the water surface.

1.2 Actual Solutions

Figure 3 shows the location of water entertainment facilities in the larger basin.

The initial inspection has revealed the following design deviations:

- absence of heated poolside paths;
- the air ducts are galvanized steel, not stainless steel;
- the actual number of supply air distribution devices is less than the design one;
- ceiling air diffusers shut off;
- the partition between the pool room (Room 45 in Fig. 1) and the sauna (Room 46.2, Fig. 1) has two black windows (900 × 1400 mm, 500 mm floor height) that get heated due to the high sauna temperature (Fig. 4);
- absence of devices to protect visitors from possible burns from contact with the hot surfaces of the windows and heat sources, which are prescribed by Construction Regulation 31-112-2004 "Swimming pools" (Fig. 5);
- the water temperature in the smaller basin is 26 °C and in the larger one 29 °C instead of the design temperature of 30 °C;
- the indoor air temperature measured 28–29 °C at the time of the inspection;

Fig. 3 The larger basin

Fig. 4 General view of the larger basin

– the pool area houses a hammam which is not mentioned in the pool design
  documentation (Fig. 4).

1.3   Findings

The initial inspection of the pool has revealed a whole number of discrepancies and
inconsistencies with the its original design.

   The indoor air and water temperature differs from the design temperature. The air
temperature should be 1 °C or 2 °C higher than the water temperature, i.e. not lower
than 30 °C according to the sanitary and epidemiological norms of the Russian Federation Construction Regulation 2.1.3678-20 "Sanitary and epidemiological requirements to operation of premises, buildings, structures, equipment and transport; operating conditions of economic entities engaged in the sale of goods, performance of
works or provision of services." Air temperatures lower than the standard temperature can lead to body cooling when leaving the larger basin, while the water in the
smaller basin will feel cold.

   Due to the discrepancy between the actual solutions and the design ones, there
are heat and moisture releases in the pool area:

– the hammam generates hot water vapors that inevitably flow into the pool area
  with every opening of its door;
– the difference between the actual water and air temperatures and the design
  temperatures adds to the moisture content according to the nonlinear dependence formula given in the Association of German Engineers manual according
  to VDI-Richtlinien. VDI 2089. Blatt 1.03.2005.

Fig. 5 Heating devices and air supply ventilation grilles

- the wall separating the pool area (45) and the sauna rooms (46.2) has two internal windows (900 × 1400 mm, floor height 500 mm) that release the heat into the pool area.

The actual heating and ventilation systems are inconsistent with their original design.

The actual number of the supply air distribution devices is less than specified in the design documentation. There are no devices to protect visitors from possible burns from contact with the hot surfaces of the sauna windows and heat sources.

Thus, the actual measurements of the pool area climate have shown its inconsistency with the sanitary and epidemiological requirements applicable to air environments of swimming pools. Ventilation system recommendations require a more in-depth analysis of the indoor climate in this swimming pool.

2   Methodology

2.1   Actual Measurements Rationale

The expediency of the original design solutions has been confirmed by mathematical modeling [5].

The mathematical modeling technique involves the use of complex identical equations and computer-assisted calculations [6, 12, 13].

The modelling was performed in STAR-CCM+ and involved:

- creating the geometry of the pool area;
- building the computational grid (1 800 000 elements);
- model setup (boundary and initial conditions);
- calculation and result visualization;
- results analysis.

The efforts to develop the ventilation recommendations for the swimming pool in question involved the analysis of actual conditions [1, 3, 10, 14]. For the purposes of mathematical modeling, the boundary conditions should be set with account of the amount of heat and moisture from conditional hazard sources. Since the heat and moisture from the hammam-generated air masses could not be determined by calculation, actual measurements had to be taken.

2.2   Description of the Methodology

The actual measurements involved the use of special engineering devices to measure the pool's indoor climate parameters and obtaining the array of data sufficient enough for processing and averaging the results obtained.

The performance of actual measurements followed GOST 30494-2011 "Interstate standard. Residential and public buildings. Indoor climate parameters" and GOST

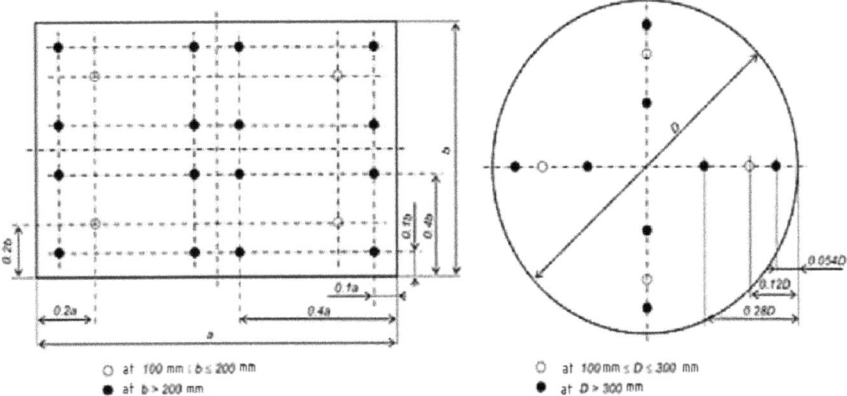

○ at 100 mm ≤ b ≤ 200 mm
● at b > 200 mm

○ at 100 mm ≤ D ≤ 300 mm
● at D > 300 mm

Fig. 6 The fragment of the basement floor plan showing the measurement points

12.3.018-79 "Ventilation systems. Aerodynamic testing methods". The measurements were made at the time of season change in cloudy weather during daylight hours (10:00–15:00). Outdoor air conditions: temperature +3.5 to +6.4 °C, relative humidity 37.6 to 46.0%.

The measurements were made under normal operating conditions, with visitors (up to 10) actively using the hydro massage, Jacuzzi and waterfall.

Changes were being recorded manually three times every 5 min using Testo 435 measuring device.

Testo 435 has been duly certified and tested, and has its measuring range and permissible inaccuracy meeting the requirements.

The measurement of temperature, humidity and air velocity covered equal-sized areas of not more than 100 m² and were taken at points indicated in Fig. 6 at a height of 0.1, 0.4 and 1.7 m.

The coordinates of the velocity measurement points on ventilation grilles, as well as the number of measurement points are determined by the shape and size of the dimensional section. The maximum deviation of the measurement points coordinates from those indicated in Fig. 7 should not exceed ±10%. The number of measurements to be taken at each point must be at least three and the duration of measurement should be at least 10 s.

3 Results

The actually measured averaged indoor climate parameters and the regulatory performance values are given in Table 1.

Similarly, the remaining indoor climate parameters were calculated as averages of the measured values.

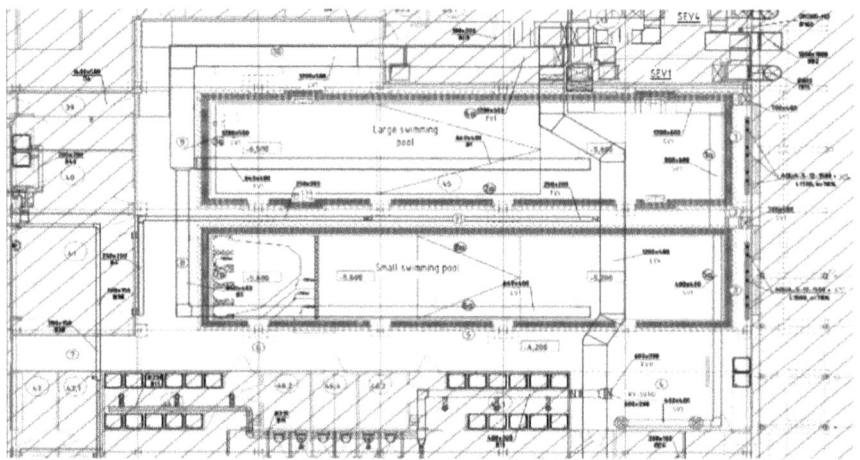

Fig. 7 The coordinates ventilation grilles measurement points

The actual indoor climate parameters are as follows:

- Volumetric average air temperature in the pool area: $t_a = 27.9$ °C;
- Volumetric average humidity in the pool area: $\varphi_a = 51.4\%$;
- Volumetric average indoor air velocity: $v_a = 0.12$ m/s;
- Average temperature of inner glazing surface: $t_g = 27.5$ °C;
- Average temperature of the inner surface of the outer wall: $t_{ow} = 27.5$ °C;
- Average temperature on inner walls: $t_{iw} = 28.1$ °C;
- Water temperature in smaller basin (swimming basin): $t_{sb} = 25.8$ °C;
- Water temperature in larger basin (Jacuzzi basin): $t_{lb} = 29.2$ °C;
- Actual air exchange: $L_\phi = 5690$ m³/h;
- Supply air temperature: $t_s = 29.0$ °C;
- Relative humidity of supply air: $\varphi_s = 22.0\%$.

The board ceiling space, transit ducts, pipes and structural elements have been found to have no condensation moisture.

4 Conclusion

The study has found the actual indoor climate parameter values to be deviating from the regulatory values.

The indoor air temperature does not correspond to the required one, which leads to visitors complaining they feel cold when leaving the larger basin, whose temperature is higher than the indoor air temperature.

Although the volumetric average humidity generally does meet the standard, its readings at points 6, 7 and 8 are higher than normal. It should also be borne in

Table 1 Indoor climate parameter values measured at control points versus regulatory values

Measurement point	Actual readings	Regulatory performance values					
	Elevation (m)	t (°C)	φ (%)	v (m/s)	t (°C)	φ (%)	v (m/s)
1	0.1	27.4	45.8	0.18	30.2	50–65	≤0.2
	0.4	27.9	40.5	0.13			
	1.7	27.8	39.5	0.14			
2	0.1	28.4	35.5	0.20	30.2	50–65	≤0.2
	0.4	28.3	43.7	0.11			
	1.7	28.3	38.9	0.12			
3	0.1	28.4	49.9	0.11	30.2	50–65	≤0.2
	0.4	28.4	44.9	0.14			
	1.7	28.5	45.2	0.19			
4	0.1	28.0	46.2	0.09	30.2	50–65	≤0.2
	0.4	28.2	48.6	0.06			
	1.7	28.2	49.0	0.12			
5	0.1	28.1	53.1	0.07	30.2	50–65	≤0.2
	0.4	28.2	52.3	0.11			
	1.7	28.2	53.3	0.08			
6	0.1	28.1	58.0	0.12	30.2	50–65	≤0.2
	0.4	28.2	60.4	0.18			
	1.7	28.2	60.5	0.12			
7	0.1	27.4	59.6	0.21	30.2	50–65	≤0.2
	0.4	27.2	60.1	0.11			
	1.7	27.3	60.7	0.06			
8	0.1	27.8	64.8	0.09	30.2	50–65	≤0.2
	0.4	27.7	60.7	0.19			
	1.7	27.8	59.4	0.11			
9	0.1	27.7	55.6	0.15	30.2	50–65	≤0.2
	0.4	27.7	55.7	0.17			
	1.7	27.7	56.1	0.11			
10	0.1	27.6	52.0	0.08	30.2	50–65	≤0.2
	0.4	27.7	53.2	0.07			
	1.7	27.7	54.2	0.07			
11	0.1	27.9	46.2	0.09	30.2	50–65	≤0.2
	0.4	27.9	42.7	0.12			
	1.7	27.9	49.1	0.08			

mind that the measurements were made on a weekday during the daytime, when the occupancy is low. The highest moisture gain occurs on weekend evenings and holidays as the busiest time.

5 Recommendations

To maintain the indoor temperature at the required level, the supply air should be higher than the required indoor air temperature, since high humidity induces in the pool area the polytropic process of simultaneous cooling and moisturization.

The supply air being able to pass through the dehumidifier and into the pool area with low moisture content means the automation is working properly and, consequently, there is a lack actual air exchange to achieve complete assimilation of moisture. To maintain the relative humidity within 50–60% range at full occupancy, it is necessary to increase the air exchange or use stationary (indoor) dehumidifiers [2, 4, 9].

With an increase in air exchange, there is a risk of increased air velocity. To avoid it, it is recommended the supply ventilation grilles are increased to the design size [7, 8, 11].

The visual inspection of the air ducts serving the pool and the transit air ducts passing through the pool has revealed paint swelling, since high humidity exposure makes structural elements susceptible to corrosion. The most optimal solution is stainless steel ducts, as originally designed, but in order to avoid replacement costs, it is recommended to treat the existing air ducts with anti-corrosion AZ 185 aluminum zinc. Anti-corrosion coating should be applied to all the metal structures in the pool area.

Also, as a corrosion prevention measure, it is recommended to install air extraction system in the area with highest humidity (point 7 in Fig. 6).

Heat sources (Fig. 5) and sauna windows should be shielded otherwise enclosed to avoid possible burns and injuries.

References

1. Antonov PP (2005) Methods of calculation and design of microclimate systems in swimming pools. SI-TES-Air conditioner LLC, Moscow
2. Asdrubali F (2009) A scale model to evaluate water evaporation from indoor swimming pools. Energy Build 41(3):311–319
3. Bogoslovsky VN, Pirumov AI, Posokhin VN (1992) Internal sanitary devices in 3 parts. Part 3. Book 1. Ventilation and air conditioning. Designer's Handbook
4. Comino F, de Adana MR, Peci F (2018) Energy saving potential of a hybrid HVAC system with a desiccant wheel activated at low temperatures and an indirect evaporative cooler in handling air in buildings with high latent loads. Appl Therm Eng 131:412–427

5. Dmitriev AA (2020) Analysis of fun pool ventilation system. Engineering systems and urban economy. In: II regional scientific and practical conference of Saint Petersburg State University of Architecture and Civil Engineering
6. Felgueiras F, Mourão Z, Morais C et al (2020) Comprehensive assessment of the indoor air quality in a chlorinated Olympic-size swimming pool. Environ Int 136.https://doi.org/10.1016/j.envint.2019.105401
7. Gabriel MF, Felgueiras F, Mourão Z, Fernandes EO (2019) Assessment of the air quality in 20 public indoor swimming pools located in the Northern Region of Portugal. Environ Int 133.https://doi.org/10.1016/j.envint.2019.105274
8. Kharitonov VP (2007) Design of ventilation systems for indoor pools in cottages. AVOC 6:52
9. Kokorin OY (2003) Modern air conditioning systems. Publishing House of Physical and Mathematical Literature, Moscow
10. Krasnov YS, Borisoglebskaya AP, Antipov AV (2004) Ventilation and air conditioning systems. Design recommendations for industrial and public buildings. Thermocool, Moscow
11. Lahdensivu J, Aromaa J (2015) Renovation of an alkali–aggregate reaction damaged swimming pool. Case Stud Constr Mater 3:1–8
12. Lochner G, Wasner L (2017) Ventilation requirements for indoor pools. ASHRAE J 59(7):16–24
13. Shah MM (2012) Improved method for calculating evaporation from indoor water pools. Energy Build 49:306–309
14. Tertichnik EI, Kamenev PN (2015) Ventilation. ACB, Moscow

Clean-Room Class D Air Distributor Performance Evaluation: Case Study of RTC Polisan

E. A. Ilin and A. M. Grimitlin

Abstract This study investigates the actual performance of Trox VDW-Q-Z/600× 24 air distributors in the warmer period of the year using the example of clean-room Class D premises of RTC Polisan. The obtained performance values are compared against the standard values for indoor climate parameters to be achieved by swirl diffusers. Further comparison is made with regulatory documentation, based on the specific design features of the ventilation systems. Main factors are identified that determine the clean room class of pharmaceutical production facilities.

1 Introduction

The current regulatory framework for pharmaceutical production cleanliness grades comprises EU GGMP, RF Ministry of Industry and Commerce Regulation N 916 dd. 14.06.2013, and GOST ISO 14644-1-2002, the latter having been abolished as a national standard according to the records in the electronic register of legal standards, regulations and specifications. The EU GGMP (Guide to Good Manufacturing Practice) was developed by the European Union and the USA to regulate the organization and quality control of medicinal products manufacture. RF Ministry of Industry and Commerce Regulation N 916 dd. 14.06.2013 repeats the EU GGMP standard: its Russian version contains a number of additions adapted for use in the Russian Federation. The comparison with EU GGMP clean-room classes, provided in RF Ministry of Industry and Commerce Regulation N 916 dd. 14.06.2013, refers to the classification in GOST ISO 14644-1-2002. The maximum allowable particle counts are shown in Table 1. The clean-room class depends on concentration of suspended particles of certain size, usually in the range from 0.5 microns to 5 microns per 1 m^3 [7].

The need for suspended particles count and classification is associated with the concept of sterility and asepticism [6]. Depending on the purpose of the room and

E. A. Ilin (✉) · A. M. Grimitlin
Saint Petersburg State University of Architecture and Civil Engineering, Saint-Petersburg, Russia
e-mail: egor.ilin.98@bk.ru

© The Author(s), under exclusive license to Springer Nature Switzerland AG 2023
D. Ivanov et al. (eds.), *Proceedings of ECSF 2021*, Lecture Notes in Civil Engineering 257, https://doi.org/10.1007/978-3-030-99877-6_13

Table 1 Permissible concentrations of aerosol particles

Zone	Maximum allowable particle count per 1 cubic meter of air with particle size equal to or greater than			
	Background concentration		Concentration while in operation	
	0.5 microns	5.0 microns	0.5 microns	5 microns
A	3520	20	3520	20
B	3520	29	352,000	2900
C	352,000	2900	3,520,000	29,000
D	3,520,000	29,000	–	–
Comparison with GOST ISO 14644-1-2020				
ISO4	3520	29	–	–
ISO7	352,000	2930	–	–
ISO8	3,520,000	29,300	–	–

the operation performed, it is recommended to use the clean-room classes presented in Table 2.

When it comes to operation of pharmaceutical production facilities, one factor to be borne in mind is microbiological monitoring of clean zones, which may affect the air environment parameters and human well-being, as well as the quality of the products being manufactured [10].

To achieve the desired clean-room class, sterile and aseptic manufacturing conditions, as well as recommended CFU, it is important to:

(1) isolate the work surface from all external influences of aggressive environment (White 2002);
(2) use displacing ventilation for air exchange [1],
(3) obtain the output air masses with microclimate parameters consistent with Sanitary Rules and Regulations 2.2.4.548-96.2.2.4;
(4) achieve the right geometry of the supply air jet for removal of suspended particles from walls, ceilings, and work surfaces [4], and
(5) organize all necessary stages of air purification [2].

Table 2 Operation to be performed

Class	Sterile operations
A	Filling of products that mustn't be exposed to contamination risk
C	Preparation of solutions that mustn't be exposed to contamination risk. Product filling
D	Preparation of products
Class	Aseptic conditions
A	Aseptic preparation and filling
C	Preparation of solutions to be filtered
D	Material handling after washing

2   Materials and Methods

The study relies on the scientific papers by E. V. Chernyakov on enhancing the energy efficiency of clean room air preparation and distribution systems [4]. A methodology is proposed for evaluating the air distribution performance using CFD analysis (Jianping et al. 2021).

This paper discusses the pharmaceutical production facility housed in cleanroom Class D premises of Polisan Research and Technology Center (hereinafter, "RTC Polisan"). The object of the study is room 104 (intended for preparing dosage forms under aseptic conditions) and its Trox WDV-Q-2/600×24 air distributors. The premise is a manufacturing facility with dry and normal modes of operation as defined in Construction Rules 50.13330.2012. The computational models of the air distributor, the experimental room, and the manufacturing room with equipment and personnel were created in Solid Works and imported into Star CCM+. The actual conditions were modeled for warm season.

Object 1 (Fig. 1).

The lamellae of the air distributor are directed at an angle of 30° with respect to the lid they are attached to. The lamellae are 24 and directed clockwise. This 3D model is exclusive of the perforated flap for air flow stabilization and equal distribution;

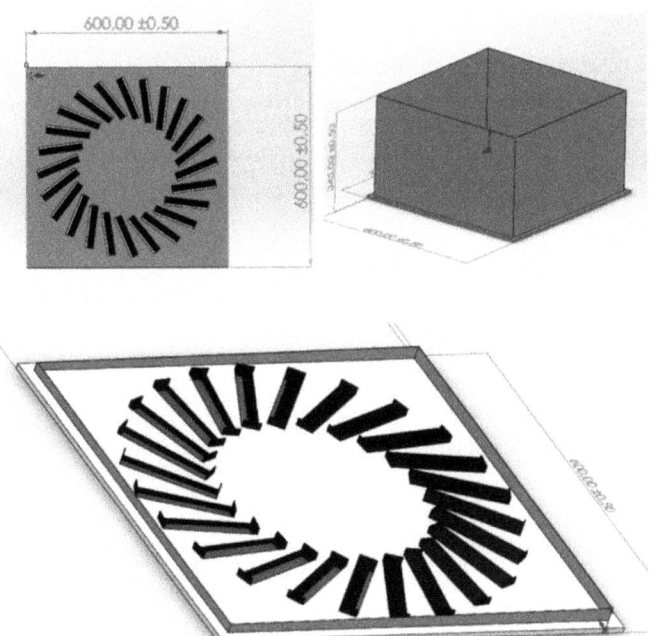

Fig. 1 Solid works 3D model of Trox WDV-Q-2/600×24 air distributor

this parameter is considered in the air supply distribution over the entire top cover of the static chamber [9].

Object 2.

The model of the $10 \times 10 \times 10$ experimental room was created in Solid Works. Its 200×100 mm exhaust openings are 100 mm below the ceiling.

Object 3.

The 3D model of the manufacturing room was created also in SolidWorks with observance of all design dimensions and location of its equipment, swirl diffusers and exhaust system. Notably, the lamellae are flush with the ceiling. The five swirl diffusers are located on the ceiling right above the personnel area. The exhaust hood is at the bottom of the side walls. This is because the laminar flow creates unidirectional air flow to displace the harmful substances to be removed from the working area (Launder 1974). The air exchange is designed to assimilate heat gain from personnel, equipment and lighting, moisture emissions and suspended particles.

The air exchange values are according to the design inflow and exhaust rates. The heat emissions from equipment and lighting are according to the technical design assignment. The human heat gain and moisture release are according to Table 1 for IIb energy consumption category at an ambient temperature of 25 °C. Table 3 shows the physical parameters affecting indoor climate.

Table 4 shows the supply and exhaust air parameters for boundary conditions setting.

Table 3 The boundary conditions

Item	Heat emission (W)	Moisture emission (kg/h)	CO_2 concentration (kg/s)
Facility 1	1748	–	–
Facility 2	1000	–	–
Facility 3	1000	–	–
Lighting	500	–	–
Personnel	252	0.185	0.0489
Total	4500	0.185	0.0489

Table 4 Supply and exhaust air parameters

Object	WDVs	Consumption m³/h (kg/s)	t_a (°C)	d (kg/m²*C)	CO_2 (kg/m²*C)
Supply air parameters					
1	1	621 (0.207)	20	–	–
2	5	3100 (1.036)	20	0.075	0.00061
Exhaust air parameters					
1	1	621 (0.207)	25	–	–

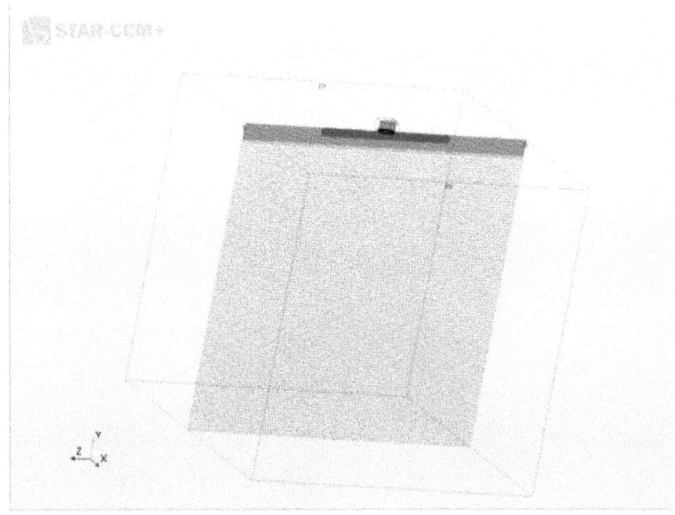

Fig. 2 Study Object 1 computational mesh

For visual verification of the ceiling effect under the given physical parameters of the air and for higher accuracy of the physical boundary conditions and mesh refinement, Study Object 2 was created in Star-CCM+.

Calculations in Star-CCM+ should be preceded by building the mesh for each object of study and subsequent setting of the boundary conditions. For Study Object 2, the basis mesh size is set at 0.1 m. The ceiling effect is presented in volume mesh with a fineness of 25% and 25/2% (Fig. 2) [5].

Object 3 has a mesh size is 0.1 m. Similarly, the ceiling effect is presented in volume mesh with a fineness of 25% and 25/2% (Fig. 3).

To determine the parameter values at which the supply air ceiling effect will actually occur and convergence achieved, we have imported the 3D model of the experimental room into Star CCM+. The boundary conditions are the parameters in Tables 3 and 4; the volumetric heat emissions are set at 1500 W. Under non-isothermal conditions, the supply air will mix with the internal air and, due to density difference, the ceiling effect will occur.

3 Results

3.1 Study Object 2

See Figs. 4 and 5.

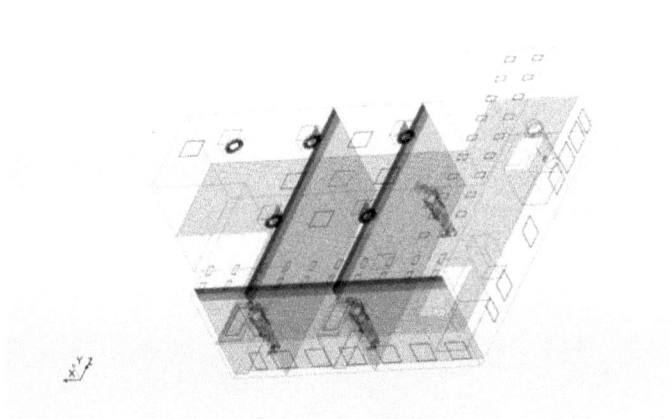

Fig. 3 Study Object 2 computational mesh

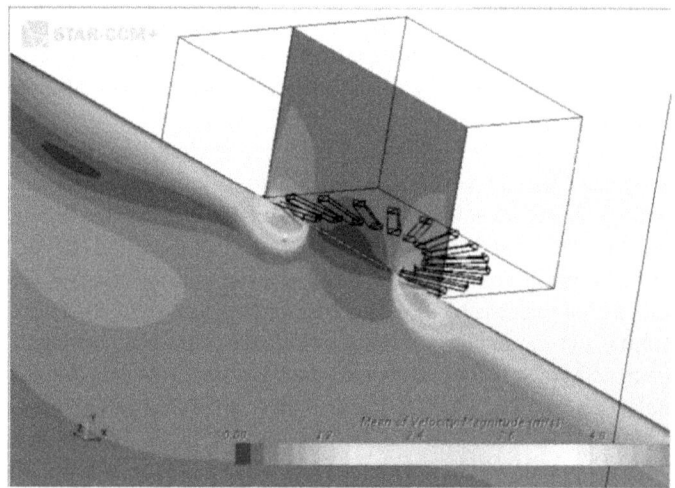

Fig. 4 The air distributor velocity field

3.2 Study Object 3

See Figs. 6, 7, 8, 9, 10 and 11.

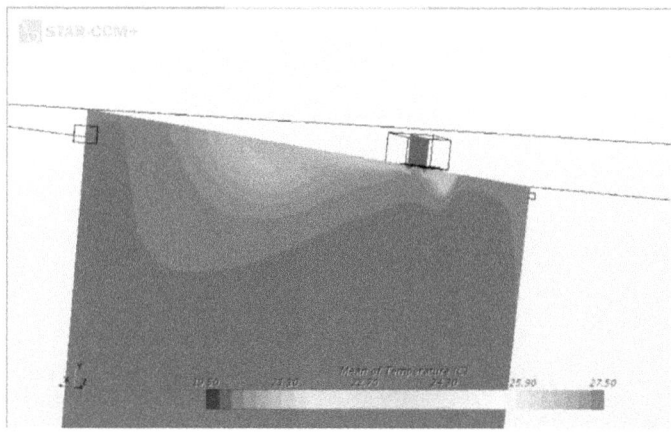

Fig. 5 The air distributor temperature field

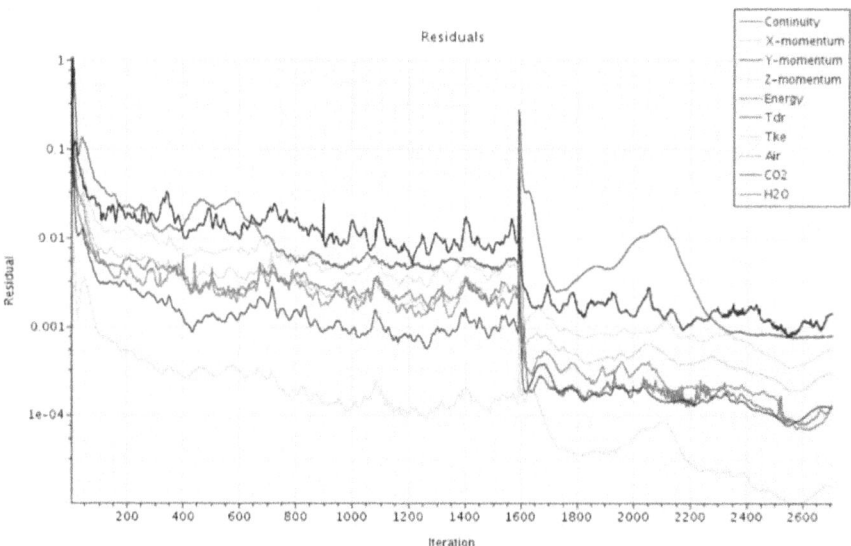

Fig. 6 Convergence diagram

4 Discussion

After 2700 iterations in Study Object 2 Residuals graph, the point velocities are equal and the temperature at the exhaust hood does not exceed the difference between the supply and exhaust air for mixing ventilation of 5–10 °C. The air velocity values at computational points do not exceed those of the permissible velocities at the lamella

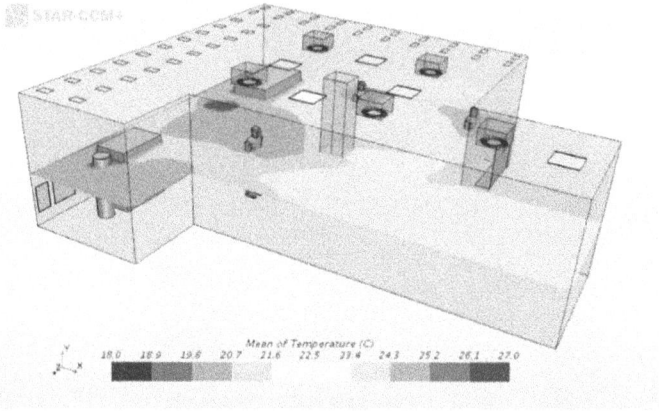

Fig. 7 The working area temperature field

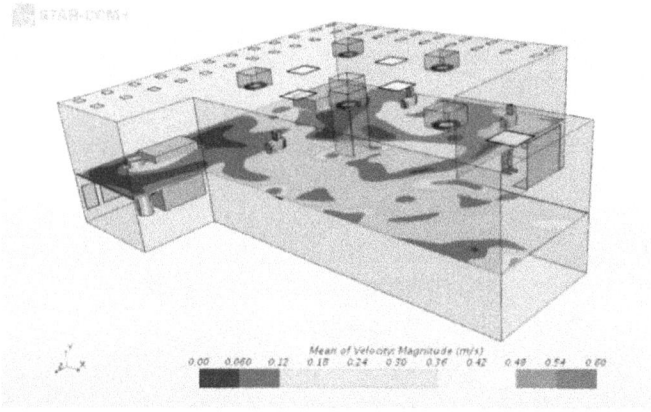

Fig. 8 The working area velocity field

outlet and in the working area, which disrupts the even distribution of the supply air across the ceiling volume by creating a turbulent flow (Figs. 4 and 5).

Table 5 shows the air velocities at computational points.

Visual representation of the velocity field being uniformly distributed at the outlet of air distributor (Fig. 4). Visual representation of the distribution of the temperature field at the outlet of the air distributor and over the entire area of the ceiling (Fig. 5).

It can be concluded that at this given air flow rate, the swirl diffuser performs in accordance with the specifications set for one air distributor. It performs within permissible ranges of velocity and noise level, and its air jets are distributed evenly over the ceiling surface. This means that the mesh values and the boundary conditions have been identified correctly and modeling can be started of the actual design conditions at these given values.

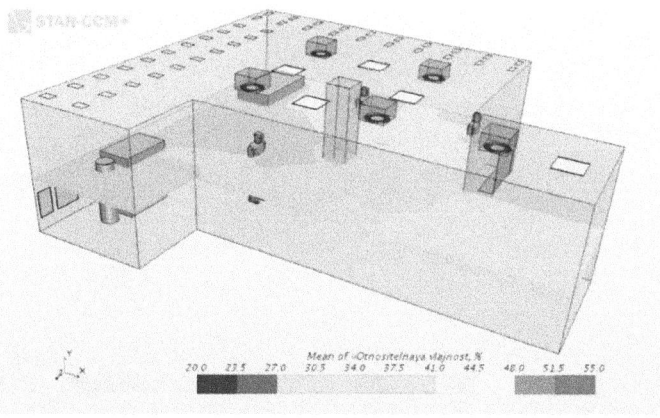

Fig. 9 Relative humidity

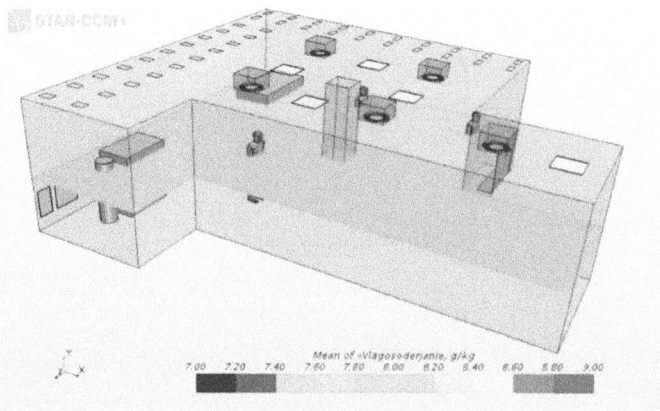

Fig. 10 Moisture content

After 2700 iterations of the Residuals graph (Fig. 6), convergence is achieved in the exhaust device for Study Object 3's temperature and CO_2 (Pozin and Ulyasheva [13]. Convergence results are presented in Table 6.

The air temperature is found to vary in the working area between 20 and 27 °C, its distribution being uneven relative to the volume of the room (Fig. 7). The volumetric air velocity is under 0.6 m/s. In the working area, the air velocity does not exceed 0.3 m/s. Due to the ceiling effect, the air velocity in the working area near the walls is 0.6 m/s (Fig. 8). The volumetric relative humidity varies in the range of 27–48% (Fig. 9) and the moisture content 7.4–7.8 g/kg (Fig. 10). CO_2 concentration is in the range of 400–550 ppm (Fig. 11).

Based on the comparative analysis, the results obtained for the working area do fall within the acceptable range for temperature, relative humidity and velocity prescribed

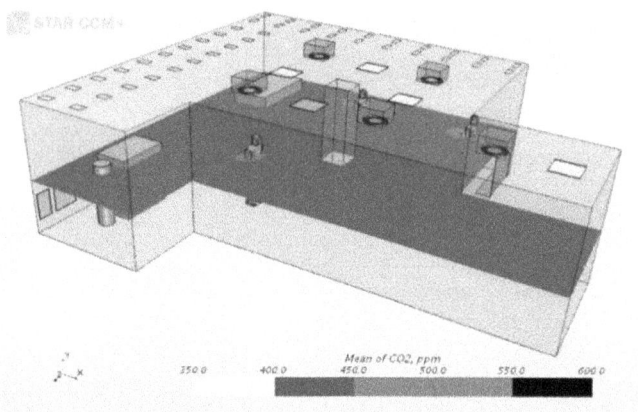

Fig. 11 CO$_2$ concentration

Table 5 The air velocities at computational points

Computational point	Velocity (v) (m/s)
Point 1	3.46
Point 2	0.45
Point 3	0.28

Table 6 Determining of convergence

Convergence determining		
	Star CCM+	Analytical calculation
t_B (°C)	24.16	14.38
CO$_2$ (ppm)	425	426
d (g/kg)	7.66	7.65

by Sanitary Rules and Regulations 2.2.4.548-96.2.2.4, but do not correspond to the optimal values. The actual parameter values in the working area correspond to their design values with exception of velocity. For economic reasons, however, the actual indoor climate parameters can be considered acceptable. The comparative analysis results are presented in Table 7.

For clean rooms, the permissible indoor climate conditions are established according to the criteria of permissible thermal and functional state of a person for 8-h shift. They are unlikely to cause any health problems but can lead to general or local sensation of thermal discomfort, tension of thermoregulation mechanism, feeling unwell or decreased performance [14].

To achieve most optimal indoor climate parameters, it is necessary to increase the number of air distributors. It is advisable to change direction of the lamellas to increase the volume of swirl flow, or to install an air distributor with more guide lamellas.

Table 7 Comparative analysis

Indoor climate parameter	Design values	Optimal values as per Sanitary Rules	Allowed range as per Sanitary Rules	Star CCM+
t_a (°C)	19–25	19–21	21.1–27.0	20–27
φ_a (%)	20–60	40–60	15–75	27–48
v_a (m/s)	0.07–0.2	up to 0.2	0.2–0.5	up to 0.6
d (g/kg)	–	–	–	7.4–7.8
CO_2 (ppm)	–	–	–	400–550

5 Conclusion

The performed comparative analysis has provided data on whether:

1. the actual indoor climate parameter values are consistent with the standards;
2. the supply air jet spreads throughout the room and whether the ceiling effect actually occurs;
3. there is temperature difference between the supply air and the indoor air;
4. suspended particles are removed from surfaces [12].

It can thus be concluded that the air exchange system under analysis does satisfy requirements established for clean-room Class D, and that the CFD-based method is suitable for verifying if air distributors performance is conducive to prescribed indoor climate parameters and air exchange.

References

1. Borisoglebskaya AP (2009) The operating rooms of hospitals. Air flow control. AVOK, Moscow
2. Chen JJ, Lan CH, Jeng MS et al (2007) The development of fan filter unit with flow rate feedback control in a cleanroom. Build Environ 42(10):3556–3561
3. Chernyakov EV (2013) Supply air velocity effects on transfer and removal of aerosol contaminants from working area. Cleanliness Technol:23–27
4. Chernyakov EV (2014) Improving the energy efficiency of clean room air preparation and distribution systems. North Caucasus Federal University Stavropol
5. Denisikhina DM (2010) Methodology for mathematical modeling of air conditioning and ventilation problems. In: VI international conference "Air 2010", pp 54–56
6. Fedotov AE (2016) Control of microbial contamination in air. Technol Cleanliness 3:5–11
7. Ilyin EA (2020) Design features of air conditioning and ventilation systems of clean rooms in aseptic pharmaceutical production. In: Innovative methods of designing building structures of buildings and structures, pp 123–129
8. Long J, Zhang B, Yang B et al (2021) Review of researches on coupled system and CFD codes. Nuclear Engineering and Technology
9. Karki KC, Patankar SV (2006) Airflow distribution through perforated tiles in raised-floor data centers. Build Environ 41(6):734–744
10. Krasnov YS, Borisoglebskaya AP, Antipov AV (2004) Ventilation and air conditioning systems. Design recommendations for industrial and public buildings. Thermocool, Moscow

11. Launder BE, Spalding DB (1983) The numerical computation of turbulent flows. In: Numerical prediction of flow, heat transfer, turbulence and combustion. Pergamon, pp 96–116
12. Saidi MH, Sajadi B, Molaeimanesh GR (2011) The effect of source motion on contaminant distribution in the cleanrooms. Energy Build 43(4):966–970
13. Pozin G, Uljaševa V (2013) Convergence of numerical modeling of heat-air-exchange processes in a ventilated room. World Appl Sci J 23(13):117–121
14. Tabunshchikov YA (2008) Microclimate and energy saving: it's time to understand priorities. AVOC: ventilation, heating, air conditioning, heat supply and construction thermophysics 5:4–11
15. Whyte W (2010) Cleanroom technology: fundamentals of design, testing and operation. John Wiley & Sons

Estimation Residual Resource of Reinforced Stone Structures by Changing the Parameters of the Section Masonry

D. I. Korolkov, M. V. Gravit, G. I. Bolod'yan, and E. A. Meshalkin

Abstract In this article, the authors describe a method for calculating the limiting service life of reinforced stone structures by changing the geometric parameters of the section. Formulas for calculating the limiting service life of reinforced stone structures for a number of design cases are derived. The advantages and disadvantages of this method are indicated. A number of assumptions are described for calculating the ultimate service life. It describes how to calculate the residual life after determining the service life limit. The authors proposed to calculate the rate of change of geometric parameters using regression equations.

1 Introduction

Currently, there are no methods for assessing the residual resource of reinforced-stone structures based on the data of instrumental examination. As a rule, the residual resource is determined by indirect methods based on an assessment of the amount of physical wear and tear [3, 5–9, 14, 19, 21]. Since most of the stone buildings were built quite a long time ago, there is an urgent need to assess the remaining service life. This article will consider the cases of calculating the residual resource or the limiting service life of reinforced stone structures for the first group of limit states according to SP 15.13330.2012 Stone and reinforced stone structures.

D. I. Korolkov (✉)
Saint Petersburg State University of Architecture and Civil Engineering, Saint Petersburg, Russia
e-mail: korol9520@yandex.ru

M. V. Gravit
Peter the Great St.Petersburg Polytechnic University, Saint Petersburg, Russia

G. I. Bolod'yan
All-Russian Research Institute for Fire Protection of Ministry of Russian Federation for Civil Defense, Emergencies and Elimination of Consequences of Natural Disasters, Balashikha, Russia

E. A. Meshalkin
LLC "Gefest Group", Moscow, Russia

© The Author(s), under exclusive license to Springer Nature Switzerland AG 2023
D. Ivanov et al. (eds.), *Proceedings of ECSF 2021*, Lecture Notes in Civil Engineering 257, https://doi.org/10.1007/978-3-030-99877-6_14

Evaluation of the performance and service life of building structures, adapted to the objects of innovative medicine, is a very important task. The calculation method presented in this article and the derived formulas will solve this problem for buildings with reinforced masonry walls.

2 Methods

Among the many approaches to assessing the value of the residual resource, the simplest, from a mathematical point of view, can be considered the approach that allows you to determine the value of the residual resource by the service life. The general form of the equation with the introduction of a number of assumptions will be:

$$T_{res} = A\left(-\frac{1}{B}t_{expl} + T_{ult}\right) \tag{1}$$

Coefficient A takes into account the state of the construction object during the commissioning of the object, i.e. the quality of the construction and installation work performed.

The coefficient B takes into account the operating conditions of the facility.

In general, these coefficients are functions of the form $f(x_1; x_2; \ldots; x_n)$, where $x_1, x_2, \ldots, x_n$—parameters on which this coefficient depends.

The assessment of the value of the coefficient B is made empirically based on the study of operational documentation and the results of a survey of the capital construction object.

In order to assign the coefficient A, it is necessary to give a comprehensive assessment of the quality of construction and installation work. This can be done in two ways.

The **first method** is based on the method of expert assessments.

$$P = \frac{\alpha_1 Q_1 + \alpha_2 Q_2 + \cdots + \alpha_n Q_n}{\alpha_1 + \alpha_2 + \cdots + \alpha_n}, \tag{2}$$

where $\alpha_1, \alpha_2, \ldots, \alpha_n$—coefficients that take into account the significance of individual parameters (significance coefficients); $Q_1, Q_2, \ldots, Q_n$—assessment of the degree of compliance of individual parameters with the requirements of the project, regulatory documents and standards; n—number of estimated parameters.

Depending on the value of the complex assessment, the coefficient A will be equal (see Table 1).

The advantages of this approach:

1. Rapid assessment of the quality of construction and installation work.

Table 1 Dependence of the coefficient A on the complex indicator P

Complex indicator P	Coefficient A	Approximation equation
<3	0	–
3.00...3.50	0.50...0.69	A = 0.38P − 0.64
3.51...4.50	0.70...0.89	A = 0.192P + 0.026
4.51...5.00	0.90...1.00	A = 0.2041P − 0.0205

2. Ease of performance assessment of the quality of construction and installation work. A minimum of instrumental control is required.

The disadvantage of this method of assessing the quality of construction and installation work can be considered the fact that it is based mostly on subjective assessments of certain parameters. Accordingly, the value of the complex indicator for different specialists assessing the quality of the completed construction and installation work can vary greatly. If we refer to the inspection of buildings and structures that have served for many years, the problem becomes even more acute, because, as statistics show, the executive documentation for such capital construction objects is either completely absent, or it is not complete.

In order to minimize the amount of error when using this estimation method, it is necessary:

1. Conduct ultrasonic flaw detection to detect internal defects, which could be unambiguously attributed to the period of erection of a building or structure; determine the strength of the structure and, if necessary, calculate the bearing capacity, etc.
2. As much as possible, involve as many experts as possible in assessing the quality of the construction and installation work performed.
3. Apply well-known methods for assessing the physical deterioration of building structures.

These measures can significantly reduce the uncertainty of the results when conducting a survey of buildings and structures, especially when the facility does not have a complete set of design and as-built documentation.

The **second method** for assessing the quality of construction and installation work was proposed by A. Kh. Bayburin. Unlike the previous method, here the indicator of the quality of construction and installation works takes values from 0 to 1. Therefore, this indicator can be initially considered a coefficient A.

The equation will look like:

$$A = \frac{0.3K_{\mathrm{CK}} + 0.15(K_{\mathrm{D}} + K_{\mathrm{T}}) + 0.05(K_X + K_S) + 0.5(K_R + K_P)}{1.7} \quad (3)$$

where K_{CK} is an indicator that evaluates the construction quality system; K_{D}—indicator of defect-freeness of technological processes of construction and installation works; K_{T} is an indicator of the accuracy of the processes; K_X and K_S are indicators of the stability of processes in relation to systematic and random errors, respectively (the ratio of the number of stable parameters to their total number); K_R—indicator of the reduction in bearing capacity; K_p is an indicator of the decrease in structural reliability as a result of defects.

The main disadvantage of using such an indicator when conducting a survey can be considered the complexity of its application. In fact, the indicators themselves included in the formula are functions of the form f (x1; x2; ...; xn), where x1, x2, ..., xn are the parameters on which this indicator depends. Therefore, the application of this approach is possible only when there is reliable information that makes it possible to unambiguously evaluate each of the parameters.

The maximum service life will be determined by the formula:

$$T_{ult} = \frac{P_{actual} - P_{min}}{v} \qquad (4)$$

P_{actual} value of the parameter currently monitoring the technical condition;
P_{min} minimum value of the parameter;
v rate of change of the parameter value over time.

3 Results and Discussion

Crumple (local compression)

When calculating the crushing of masonry with mesh reinforcement, the formulas look like:

$$T_{ult} = \begin{cases} \dfrac{\left[\psi \cdot (1.5 - 0.5 \cdot \psi) \cdot \sqrt[3]{\frac{A}{A_c}} \cdot A_c\right]_{actual} - \frac{N_c}{R}}{v} \\ \dfrac{N_c - \left[\psi \cdot (1.5 - 0.5 \cdot \psi) \cdot \left(R + \frac{p \cdot \mu \cdot R_s}{100}\right) \cdot A_c\right]_{actual}}{v} \end{cases} \qquad (5)$$

$$T_{ult} = \begin{cases} \dfrac{\left[\psi \cdot \sqrt[3]{\frac{A}{A_c}} \cdot A_c\right]_{actual} - \frac{N_c}{R}}{v} \\ \dfrac{N_c - \left[\psi \cdot \left(R + \frac{p \cdot \mu \cdot R_s}{100}\right) \cdot A_c\right]_{actual}}{v} \end{cases} \qquad (6)$$

R   calculated resistance to compression of the masonry;
A   sectional area of the element.
N_c longitudinal compressive force from local load;
Ψ   coefficient of completeness of the pressure diagram from local load.
A_c crumple area to which the load is transferred.

When the strength of the solution is less than 2.5 MPa, formulas (5) and (6) will take the form:

$$T_{ult} = \begin{cases} \dfrac{\left[\psi\cdot(1.5-0.5\cdot\psi)\cdot\sqrt[3]{\frac{A}{A_c}}\cdot A_c\right]_{actual} - \frac{N_c}{R}}{v} \\[4mm] \dfrac{N_c - \left[\psi\cdot(1.5-0.5\cdot\psi)\cdot\left(R+\frac{p\cdot\mu\cdot R_s}{100}\cdot\frac{R_1}{R_{25}}\right)\cdot A_c\right]_{actual}}{v} \end{cases} \tag{7}$$

$$T_{ult} = \begin{cases} \dfrac{\left[\psi\cdot\sqrt[3]{\frac{A}{A_c}}\cdot A_c\right]_{actual} - \frac{N_c}{R}}{v} \\[4mm] \dfrac{N_c - \left[\psi\cdot\left(R+\frac{p\cdot\mu\cdot R_s}{100}\cdot\frac{R_1}{R_{25}}\right)\cdot A_c\right]_{actual}}{v} \end{cases} \tag{8}$$

p   coefficient taken with a brick (stone) voidness up to 20% inclusively equal to 2, with a voidness from 20 to 30% inclusive—equal to 1.5, with voidness above 30%—equal to 1;

R_1 design resistance of unreinforced masonry in the considered period of solution hardening;

R_{25} design resistance of the masonry with a mortar grade of 25.

μ percentage of reinforcement by volume:

$$\mu = \frac{V_s}{V_k} \cdot 100 \tag{9}$$

For meshes with square meshes made of reinforcement with a section A_{st} with a mesh size C and a distance between the meshes S.

$$\mu = \frac{2 \cdot A_{st}}{C \cdot S} \cdot 100 \tag{10}$$

V_s and V_k—respectively the volume of fittings and masonry (Fig. 1).
With central compression

$$T_{ult} = \frac{N - \left[\varphi\cdot\left(1-\eta\cdot\frac{N_g}{N}\left(1+\frac{1.2\cdot e_{0g}}{h}\right)\right)\cdot\left(R+\frac{p\cdot\mu\cdot R_s}{100}\right)\cdot A\right]_{actual}}{v} \tag{11}$$

When the strength of the solution is less than 2.5 MPa.

$$T_{ult} = \frac{N - \left[\varphi\cdot\left(1-\eta\cdot\frac{N_g}{N}\left(1+\frac{1.2\cdot e_{0g}}{h}\right)\right)\cdot\left(R+\frac{p\cdot\mu\cdot R_s}{100}\cdot\frac{R_1}{R_{25}}\right)\cdot A\right]_{actual}}{v} \tag{12}$$

When the strength of the mortar is more than 2.5 MPa, the ratio R_1/R_{25} is taken equal to 1.

Design of eccentrically compressed elements with mesh reinforcement at small eccentricities that do not go beyond the core of the section.

Fig. 1 Cross (mesh) reinforcement of stone structures: 1—reinforcing mesh; 2—release of reinforcing mesh to control its laying

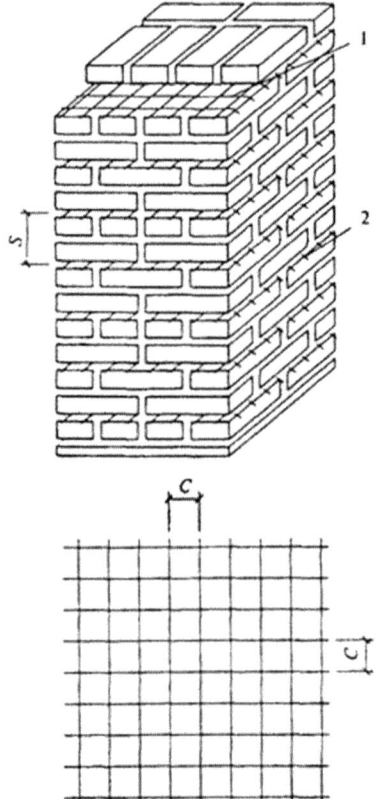

$$T_{ult} = \frac{N - \left[\varphi_1 \cdot \left(1 - \eta \cdot \frac{N_g}{N}\left(1 + \frac{1.2 \cdot e_{0g}}{h}\right)\right) \cdot \left(R_1 + \frac{p \cdot \mu \cdot R_s}{100} \cdot \left(1 - \frac{2 \cdot e_0}{y}\right)\right) \cdot A_c \cdot \omega\right]_{actual}}{v}$$

(13)

For a rectangular section:

$$T_{ult} = \frac{N - \left[\varphi_1 \cdot \left(1 - \eta \cdot \frac{N_g}{N}\left(1 + \frac{1.2 \cdot e_{0g}}{h}\right)\right) \cdot \left(R_1 + \frac{p \cdot \mu \cdot R_s}{100} \cdot \left(1 - \frac{2 \cdot e_0}{y}\right)\right) \cdot A \cdot \left(1 - \frac{2 \cdot e_0}{y}\right) \cdot \omega\right]_{actual}}{v}$$ (14)

When the strength of the solution is less than 2.5 MPa.

$$T_{ult} = \frac{N - \left[\varphi_1 \cdot \left(1 - \eta \cdot \frac{N_g}{N}\left(1 + \frac{1.2 \cdot e_{0g}}{h}\right)\right) \cdot \left(R + \frac{p \cdot \mu \cdot R_s}{100} \cdot \frac{R_1}{R_{25}} \cdot \left(1 - \frac{2 \cdot e_0}{y}\right)\right) \cdot A_c \cdot \omega\right]_{actual}}{v}$$

(15)

For a rectangular section:

$$T_{ult} = \frac{N - \left[\varphi_1 \cdot \left(1 - \eta \cdot \frac{N_g}{N}\left(1 + \frac{1.2 \cdot e_{0g}}{h}\right)\right) \cdot \left(R + \frac{p \cdot \mu \cdot R_s}{100} \cdot \frac{R_1}{R_{25}} \cdot \left(1 - \frac{2 \cdot e_0}{y}\right)\right) \cdot A \cdot \left(1 - \frac{2 \cdot e_0}{y}\right) \cdot \omega\right]_{actual}}{v}$$

$$(16)$$

N   calculated longitudinal force;

N_g calculated longitudinal force from continuous loads;

η coefficient taken according to table 21 according to state Standard of Russia SP 15.13330.2012 Stone and reinforced masonry structures;

e_{0g} eccentricity from long-term loads;

h   height of the section in the plane of the bending moment;

e_0 eccentricity of the calculated force N relative to the center of gravity of the section;

ω coefficient determined by the formulas given in table 20 according to state Standard of Russia SP 15.13330.2012 Stone and reinforced masonry structures.

Since the calculation of the maximum service life of stone structures is carried out at once according to several formulas, then the minimum of those obtained is taken as the final result.

4 Conclusion

Now let's talk about some points of calculating the maximum service life.

The first thing that should be addressed is the presence in a number of formulas of such a value as the calculated length of the element. When calculating the limiting service life, we proceed from the assumption that this value remains unchanged throughout the entire service life, i.e., firstly, the fastening scheme of the element does not change and, secondly, the change in the actual free length of the element is negligible. This means that the change in length occurs only due to the influence of the external environment.

The second thing you should pay attention to is that the calculation does not imply a change in geometric parameters as a result of artificial strengthening or weakening of the element cross-section.

Third, you should pay attention to the fact that the calculation assumes that the action of transverse and longitudinal forces, bending moments remains unchanged during the entire service life.

The advantage of this method of calculation can be considered that we obtain objective values of the value of the maximum service life and, accordingly, the residual resource, since the initial data are obtained during the instrumental examination.

The disadvantage of this approach can be considered its complexity or the impossibility of implementation in some cases in practice.

This method can be implemented directly when it is possible to directly measure the change in geometric parameters.

When considering the rate of change of geometric parameters, the question arises about the lack of data for the past period of operation. Since the rate of change in geometric parameters is not the same at different time intervals, in the absence of observation data, the estimate of the rate of change in geometric parameters will be incorrect and, accordingly, we will receive an incorrect value for the maximum service life.

To solve this problem, it is important to record the current values of geometric dimensions when organizing regular technical inspections. In this case, having a number of data, you can apply regression analysis and build equations for the dependence of changes in geometric parameters [1, 2, 4, 10–13, 15–18, 20, 22–24].

To solve this problem, the authors propose to use regression equations. The most commonly used multivariate regression equations are:

(1) Multiple Linear Regression:

$$y = \alpha_0 + \alpha_1 \cdot x_1 + \cdots + \alpha_n \cdot x_n + \varepsilon \tag{17}$$

(2) Polynomial Regression:

$$y = \alpha_0 + \alpha_{n,i} \cdot x_i^n + \alpha_{n-1,i} \cdot x_i^{n-1} + \cdots + \alpha_{1,i} \cdot x_i + \cdots + \varepsilon \tag{18}$$

(3) Power regression:

$$y = \alpha_0 \cdot x_1^{\alpha_1} \cdot \ldots \cdot x_n^{\alpha_n} + \varepsilon \tag{19}$$

(4) Significant regression:

$$y = \alpha_0 \cdot \alpha_1^{x_1} \cdot \ldots \cdot \alpha_n^{x_n} + \varepsilon \tag{20}$$

(5) Exponential regression:

$$y = e^{(\alpha_0 + \alpha_1 \cdot x_1 + \cdots + \alpha_n \cdot x_n)} + \varepsilon \tag{21}$$

(6) Logarithmic regression:

$$y = \alpha_0 + \alpha_1 \cdot \ln x_1 + \cdots + \alpha_n \cdot \ln x_n + \varepsilon \tag{22}$$

$$y = \alpha_0 + \alpha_1 \cdot \lg x_1 + \cdots + \alpha_n \cdot \lg x_n + \varepsilon \tag{23}$$

(7) Semilogarithmic regression:

$$y = \alpha_0 + \alpha_1 \cdot \log x_1 + \cdots + \alpha_n \cdot \log x_n + \varepsilon \tag{24}$$

In conclusion, I would like to note that despite the fact that the proposed method was considered for stone structures, it can be applied to other types of structures (reinforced concrete, metal, stone and composite).

References

1. Aven T (2011) Interpretations of alternative uncertainty representations in a reliability and risk analysis context. Reliab Eng Syst Saf 96(3):353–360. https://doi.org/10.1016/j.ress.2010.11.004
2. Aven T, Zio E (2011) Some considerations on the treatment of uncertainties in risk assessment for practical decision making. Reliab Eng Syst Saf 96(1):64–74. https://doi.org/10.1016/j.ress.2010.06.001
3. Braila NV, Khazieva LF, Staritcyna AA (2017) Results of technical inspection monitoring of the operation object. Mag Civil Eng 74(6):70–77. https://doi.org/10.18720/MCE.74.7
4. Choi H, Popovics JS (2015) NDE application of ultrasonic tomography to a full-scale concrete structure. IEEE Trans Ultrason Ferroelectr Freq Control 62(6):1076–1085. https://doi.org/10.1109/TUFFC.2014.006962
5. Chernykh A, Korolkov D, Nizhegorodtsev D et al (2020) Estimating the residual operating life of wooden structures in high humidity conditions. Archit Eng 5(1):10–19. https://doi.org/10.23968/2500-0055-2020-5-1-10-19
6. Erokhina OO (2019) Analysis of publications about the methods of static friction definition for bulk material. IOP Conf Ser Mater Sci Eng 560(1). https://doi.org/10.1088/1757-899x/560/1/012014
7. Gravit M, Antonov S, Nedviga E et al (2016) Strength test of high-precision tunnel lining blocks. Procedia Eng 165:1658–1666. https://doi.org/10.1016/j.proeng.2016.11.907
8. Gravit M, Dmitriev I, Ishkov A (2017). Quality control of fireproof coatings for reinforced concrete structures. IOP Conf Ser Earth Environ Sci 90(1). https://doi.org/10.1088/1755-1315/90/1/012226
9. Gravit M, Nedviga E, Vinogradova N et al (2017) Fire resistance of prefabricated mono-lithic slab. In: MATEC web of conferences, vol 106.https://doi.org/10.1051/matecconf/201710602025
10. Garnier V, Piwakowski B, Abraham O et al (2013) Acoustic techniques for concrete evaluation: improvements, comparisons and consistency. Constr Build Mater 43:598–613. https://doi.org/10.1016/j.conbuildmat.2013.01.035
11. Hassan AMT, Jones SW (2012) Non-destructive testing of ultra high performance fibre rein-forced concrete (UHPFRC): a feasibility study for using ultrasonic and resonant frequency testing techniques. Constr Build Mater 35:361–367. https://doi.org/10.1016/j.conbuildmat.2012.04.047
12. Karaiskos G, Deraemaeker A, Aggelis DG et al (2015) Monitoring of concrete structures using the ultrasonic pulse velocity method. Smart Mater Struct 24(11). https://doi.org/10.1088/0964-1726/24/11/113001
13. Karaiskos G, Tsangouri E, Aggelis DG et al (2016) Performance monitoring of large-scale autonomously healed concrete beams under four-point bending through multiple non-destructive testing methods. Smart Mater Struct 25(5). https://doi.org/10.1088/0964-1726/25/5/055003
14. Korolkov D, Chernykh A, Gravit M (2019) Method for determining the residual resource of building structures by the terms of their operation. In: International scientific conference on energy, environmental and construction engineering. Springer, Cham, pp 389–402
15. Moradi-Marani F, Rivard P, Lamarche CP et al (2014) Evaluating the damage in reinforced concrete slabs under bending test with the energy of ultrasonic waves. Constr Build Mater 73:663–673. https://doi.org/10.1016/j.conbuildmat.2014.09.050

16. Payan C, Abraham O, Garnier V (2018) Ultrasonic methods. In: Non-destructive testing and evaluation of civil engineering structures. Elsevier, pp 21–85. https://doi.org/10.1016/B978-1-78548-229-8.50002-9
17. Planès T, Larose E (2013) A review of ultrasonic coda wave interferometry in concrete. Cem Concr Res 53:248–255. https://doi.org/10.1016/j.cemconres.2013.07.009
18. Schabowicz K (2014) Ultrasonic tomography—the latest nondestructive technique for testing concrete members—description, test methodology, application example. Arch Civil Mech Eng 14(2):295–303. https://doi.org/10.1016/j.acme.2013.10.006
19. Soldatenko VS, Smagin VA, Gusenitsa YN et al (2017) The method of calculation for the period of checking utility systems. Mag Civil Eng 70(2):72–83. https://doi.org/10.18720/MCE.70.7
20. Tsioulou O, Lampropoulos A, Paschalis S (2017) Combined non-destructive testing (NDT) method for the evaluation of the mechanical characteristics of ultra high performance fibre reinforced concrete (UHPFRC). Constr Build Mater 131:66–77. https://doi.org/10.1016/j.con buildmat.2016.11.068
21. Volkov M, Kibkalo A, Vodolagina A et al (2016) Existing models residual life assessment of structures and their comparative analysis. Procedia Eng 165:1801–1805. https://doi.org/10.1016/j.proeng.2016.11.925
22. Vu QA, Garnier V, Chaix JF et al (2016) Concrete cover characterisation using dynamic acousto-elastic testing and Rayleigh waves. Constr Build Mater 114:87–97. https://doi.org/10.1016/j.conbuildmat.2016.03.116
23. Wolf J, Pirskawetz S, Zang A (2015) Detection of crack propagation in concrete with embedded ultrasonic sensors. Eng Fract Mech 146:161–171. https://doi.org/10.1016/j.engfracmech.2015.07.058
24. Xu Y, Jin R (2018) Measurement of reinforcement corrosion in concrete adopting ultrasonic tests and artificial neural network. Constr Build Mater 177:125–133. https://doi.org/10.1016/j.conbuildmat.2018.05.124

Integration of Hospital Therapeutic Gardens into the Green Frame of Sevastopol

E. E. Krasilnikova, I. V. Zhuravleva, L. I. Uleyskaya, I. A. Zaica, and A. A. Goncharik

Abstract The past decade has witnessed a series of major international events dedicated to the "return of nature to cities." The modern processes of urban development and planning are associated with creation of comfortable and safe urban environments, the quality of life being the main criterion of the effectiveness of urban planning strategies. To the foreground come the tasks of providing the population with an urban environment that would contribute to longer lifespan and greater comfort of living for all categories of citizens. Healthy ageing is increasingly seen as the main element in urban environment quality evaluation. The comfortable urban setting with increased health improvement opportunities is an important tool to mitigate the negative consequences of urbanization, burdened by globalization and COVID-19. The urban planning and development strategies are currently undergoing changes. The green frames of urban localities, destroyed in the 1990s by unregulated development of various functional purpose, are regaining their role as the key marker of the balanced urban development and spatial planning approach. The study has been conducted within the framework of Grant 29/06-31.

1 Introduction

A system allowing to integrate natural landscapes into urban settings of big and smaller cities serves the basis for creating the urban green infrastructures. The need to integrate and link together the diversity of the facilities to be provided with green spaces into a one single ecological network, forms an essential element of the sustainable, healthy cities of today and of the future. We have many good examples of successful urban green infrastructures projects [1, 2, 7].

Therapeutic landscapes are currently represented mostly by therapeutic gardens belonging to healthcare facilities (the experience of America, France, Spain, Singapore [3], among other countries). Therapeutic gardens classify as a structural element of therapeutic landscapes because of their garden-type layout. With trees and shrubs,

E. E. Krasilnikova (✉) · I. V. Zhuravleva · L. I. Uleyskaya · I. A. Zaica · A. A. Goncharik
Sevastopol State University, Sevastopol, Russia
e-mail: landurbanizm@gmail.com

the therapeutic gardens have their design layouts developed from the perspective of the physical, psychological and social needs of their visitors—primarily the patients undergoing treatment and rehabilitation [16]. Aptly selected and arranged, the assortments of trees and shrubs can turn a location into an attractive scenery for physical exercise, improving physical health and mental state, reducing chronic pain and stress, and enhancing concentration and attention.

Intended for patients (with autism, psychiatric disorders, tuberculosis, oncological diseases, etc.), healthcare facilities' therapeutic gardens are seen by doctors as important elements contributing to the recovery and convalescence [7]. A hospital's therapeutic garden is more than just a park or public space; it is that final touch that makes the image of its hospital complete [11]. The species for planting in a therapeutic garden are selected based on the locality's natural and climatic characteristics, healing properties and survivability of beautification function during the vegetation season. Modern therapeutic gardens serve as passages between buildings and structures, recreation facilities and walking routes; they do not intersect with transport routes and approach roads, and, importantly, they are a connection to adjacent urban space. One example of how therapeutic gardens can be integrated into urban texture and urban green infrastructures is Parc Sanitary Pere Virgili in Gracia district of Barcelona. Formerly an old military hospital, which was, in 1999, converted into a healthcare facility with a wide range of medical, research and training services, this park is an excellent specimen of Spain's therapeutic gardens of social importance [10].

Another equally interesting example of integrating therapeutic landscapes into healthcare setting is the Kansai Rosai, Osaka, Japan. Laid out as part of reconstruction timed to the 50th anniversary of Kansai Rosai Hospital, this hospital garden is the first of its kind in the country, designed by Yoshisuke Miyake of SEN Inc. and planted in April 2004. Kansai Rosai is an intensive care hospital and is located in an industrially developed part of Amagasaki, Hyogo Prefecture in the south of Japan, bordering on private buildings and a road with heavy traffic. The acoustic screens are hidden behind the artificial hills, positioned according to a well-thought-out pattern, to make the industrial noise almost inaudible on an area of 5000 m^2—the size of the garden. Kansai Rosai garden has nine functional zones, designated in accordance with the needs of different groups of visitors [9].

Integrating natural elements into the architecture of healthcare facilities is currently a trend in many countries. Their prime purpose is to serve as a nondrug treatment to facilitate convalescence. In the time of increased vulnerability to the spread of COVID-19, such trend is particularly beneficial.

Landscape urbanism, based on an integrated, interdisciplinary approach to the formation of a comfortable, socially oriented and environmentally sustainable urban structure, defines key ecosystem-based solutions towards healthier urban environments. The perception of a city as a landscape is one of the ideological concepts of landscape urbanism.

The principles of horticultural therapy were first proposed by English nurse F. Nightingale in 1860. She noted the positive effect the picturesque views from windows had on the process of recovery and had thus formulated the core principles of horticultural therapy—a practice that uses contact with plants to improve mental and physical health. In the late nineteenth century, the positive influence of therapeutic gardens on patients' health was attributed mostly to the emotional impact of their picturesque scenery: admiring the colors of greenery and the beauty and flowers had a beneficial effect on mental well-being.

The idea that humans possess an innate tendency to seek connections with nature has led to the emergence of biophilic design in artificial environments: the artificially created connectivity to natural environment is designed to enhance people's fitness and wellbeing [8, 10, 17].

2 Materials and Methods

The main objectives of the study are to promote creation and further integration of wellness and therapeutic landscapes into the green framework (green infrastructure) and the urban landscaping system of Sevastopol; and to promote health-improving effects on hospital patients and citizens—through more extensive use of wellness gardens and, particularly, therapeutic gardens, which have proven instrumental in advancing people's psychological and mental state.

The environmental significance of the surveyed territory relates to the presence of natural landscapes, zonal vegetation types, unique phytocenoses, rare species, as well as valuable park communities and green zones. The vegetation cover has been shaped by the area's natural conditions and is characterized by high degree of floral diversity. Descriptions of the vegetation cover growing within the boundaries of the facilities under analysis were made in the course full-scale fieldwork. Descriptions of the flora species composition used the floristic and geobotanical data and The Crimean Species Indicator. The floristic and geobotanical studies were carried out in accordance with the generally accepted methods of field research. The descriptions of the local vegetation rely on the data obtained during the field studies.

The study results define the vegetation of the surveyed health facilities' in Sevastopol as an artificially created community of introduced species (cultural phytocenoses). Cultural phytocenoses are less durable and often lack clearly manifest structural composition; they are one- or two-tiered and have mostly aesthetic and landscaping functions, requiring constant maintenance and care. In the parks around Sevastopol's healthcare facilities, the cultural phytocenoses represent natural habitats where the introduced plants are combined with the primary vegetation, representing mixed-type communities. Structurally and in terms of species composition, mixed communities are similar to background cenoses and have their introduced species occupying the corresponding tiers.

Initially, therapeutic gardens became widespread solely due to their aesthetic and decorative properties. Later, as hospital design solutions evolved towards isolated spaces, the hospital gardens started to be seen not only as decorative elements. As known, contemplative landscapes, even if contemplated over a short time, can relieve muscle tension and normalize blood pressure, mental and cardiac activity.

The therapeutic effect of a hospital garden is based on its orderliness and ability to restore in patients the sense of connectivity to natural environment. Modern landscape architecture is well aware of this fact and recommends that medical and psychological considerations be taken into account when developing landscape designs.

When designing modern therapeutic gardens, it is necessary to consider elements such as hard landscape, vegetation, water bodies and paths [11].

Therapeutic gardens should be planted with species that can provide visual interest all year round; there should be herbaceous plants, shrubby plants and tree species, the latter being that element that enlivens the landscape with the game of light and shade and the rustling sound of leaves.

It is recommended that hard elements of therapeutic gardens should occupy no more than one third of their area and use natural materials. Water bodies on flat surfaces facilitate relaxation and concentration in patients. Path design should be conducive to increased motor activity, which is achieved not only by design itself but also by the use of appropriate materials [12, 15].

The therapeutic gardens of today are outdoor spaces designed to benefit their users' physical, psychological and social needs, to which there exists proven evidence. The majority of non-specialized therapeutic gardens offer multisensory experience to revitalize patients' sensory organs. The species to be planted in therapeutic gardens should be aroma-friendly and safe to touch [6].

Multisensory experience is contributed in therapeutic gardens by the sounds of water, birds, insects and small animals, which must be safe for patients. The benefits of color therapy are achieved through coloristic health effects of natural vegetation colors—the colors of foliage, flowers, tree trunks [9].

Therapeutic gardens are designed not only for use by hospitals patients. Horticultural therapy is often provided in public gardens with activity zones, one being the gardening zone where visitors are invited to plant seedlings, create flower combinations, study ornamental plants, etc. These activities are aimed to provide experience with nature; promote social interaction and physical activity in visitors and patients; stimulate sensory activity through contact with nature; improve physical and mental well-being in visitors and patients; and stimulate interest in plants and gardening.

3   Results

The terrain under analysis is hilly. Dasha Sevastopolskaya Hospital No.3 is 40 m above sea level; its eastern side faces a steep slope leading to Ushakova Balka Nature Reserve. The hospital is located on the south side of Sevastopol Bay in number 15, Nadezhdintsev Street, Nakhimov District, Sevastopol. Its territory borders on a railway in the North; private housing in Zheleznodorozhnaya Street in the West; medium-height housing and Nadezhdintsev motor road in the South; and Ushakova Balka Nature Reserve in the East. The hospital has its entrance in Nadezhdintsev Street (Fig. 1).

Fig. 1 The hospital layout
and current borders

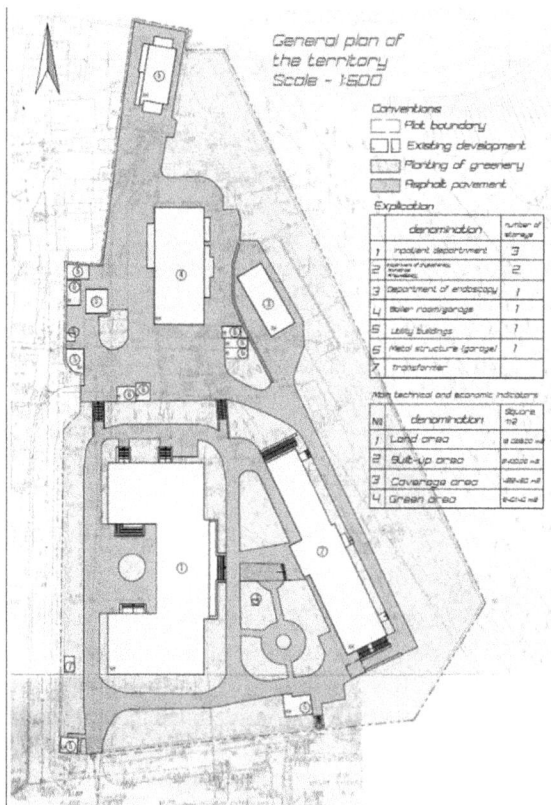

The main building is located in the central part of the hospital territory; its main entrance faces Zheleznodorozhnaya Street. The physiotherapy department is located in the eastern part and faces Ushakova Balka Nature Reserve and the sea. The northern part of the territory borders on railway and has boiler room and outbuildings.

The total area of the hospital is 1.20 ha.

The landscaped area takes up 0.5 ha, which indicates an inconsistency with landscape gardening standards. The established norm is 30 m^2 of greenspace per patient, inclusive of paths and garden facilities. There are 85 trees and 93 shrubs on the territory of the hospital. The tree-shrub ratio is found inconsistent with the standard 1:6 and requires planting of additional 417 shrubs. In the southern part of the territory, the planting rate is 330 trees per 1 ha, requiring 80 additional trees.

The territory of the hospital is home to European common spruce *Picea abies*, Crimean pine *Pinus nigrasub sp. Pallasiana*, cypress *Cupressus sempervirens Pyramidalis*, white cedar *Thuja occidentalis*, the coniferous species accounting for 21.35% of the total landscaped area.

The dominant species is locust *Robinia pseudoacacia* (about 34%), followed by horse chestnut *Aesculus hippocastanum* (14%), cherry *Prunus subg. Cerasus*

(8.3%), Japanese scholar-tree *Styphnolobium japonicum* (8.2%), common ash *Fraxinus excelsior* (7%), silk tree *Albizia julibrissin* (3.5%), common almonds *Prunus dulcis* (2.3%), Japanese medlar *Eriobotrya japonica* (1.1%) and European holly *Ilex aquifolium* (1.1%).

The overall condition of the planted area is generally satisfactory, but a lot of trees and shrubs planted near buildings and utilities and there are dry trees that need to be replaced. It is necessary in this regard to carry out landscape tending, sanitary felling and thinning.

The condition of the lawns can be described as unsatisfactory. There are sections with missing lawns (possibly due to local soil and climatic conditions). It is recommended to plant the lawns with groundcover species—common periwinkle *Vinca,* ivy *Hedera helix,* or cottoneaster *Cotoneaster perpusillus.*

Aesthetically, landscape composition is lacking in plant and flower clusters in its lower tier.

It is recommended to install this therapeutic garden with small architectural forms.

4 Discussion

For more effective integration of this therapeutic garden into Sevastopol's green infrastructure, it is proposed to expand its range of species, create tree-shrub compositions, and convert it into a modular therapeutic garden with outdoor zones for recreation, relaxation and wellness. This will help to isolate the hospital garden from the surrounding streets and create favorable conditions for convalescence.

There are flower beds on the territory of the hospital, mostly roses. The diversity of perennial and annual flowers and herbaceous plants is low (Fig. 2). It is recommended that the free plot in front of the main entrance is laid out with a therapeutic rosary.

Dendrological matrices have been developed for the therapeutic garden, based on the chosen coloristic solution (Fig. 3).

The study involved testing of 21 domestic and 71 foreign varieties of roses, sourced from Nikitsky Botanical Garden. Fragrance testing was largely subjective; the notes and accords were tested in the morning time (no more than ten varieties per day). The intensity of fragrance was determined according to a three-point scale: fragrance free varieties (with very weak aroma)—1 point; fragrant varieties (aroma is felt near the flower)—2 points; intense fragrance varieties (aroma is felt at a distance of 25 m from the flower)—3 points. A total of 45 three-point and 22 two-point varieties were selected for the therapeutic rosary. All the varieties have high decorative effect and are resistant to a whole range of fungal diseases.

The therapeutic effect of color is achieved through the use of one dominant color in the flower landscape composition.

The matrix for rose (magenta) compositions is represented by roses, loosestrife, sedum, panicum Shenandoah, common globe thistle, reed grass, common majoran, heather, eupatorium.

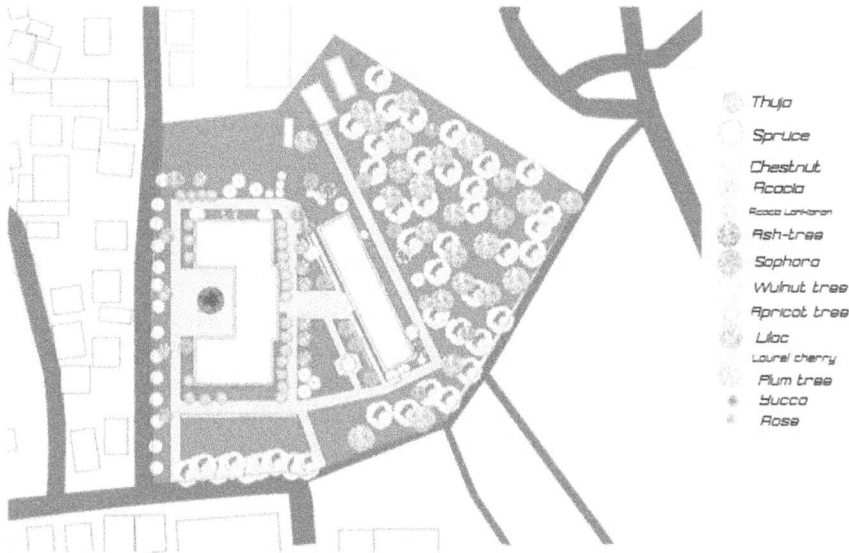

Fig. 2 The hospital vegetation layout map (developed by authors)

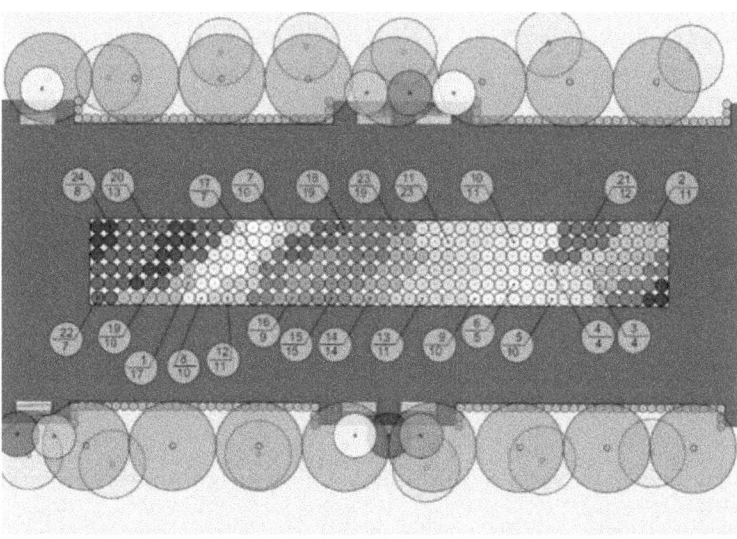

Fig. 3 Example of a shrub rose bed using coloristics (dendrological matrix). Author: E.E. Krasilnikova ©

5 Conclusions

Cities need their urban development strategies to be consistent with the need for comfortable urban environment in conditions of high-density development; adjustment of environmental and economic development goals; strengthening of social ties between community members. Planners should be able to the natural component in such a way as to support, not slow down, urbanization by forming comfortable, healthy urban environment. Integration of therapeutic landscapes into cities' green infrastructure is a natural progression of the global trend towards integrated, socially-oriented and environmentally sustainable greenspace extension.

Given the current refunctionalization of urban landscaping systems, comfortable and multifunctional green frames can be created through full or partial integration of therapeutic and wellness landscapes. The majority of city hospitals are gated premises that classify as landscaped areas of limited access. For these limited-access areas to be harmoniously integrated into the urban green infrastructure and the inner-city greening system, they should be added with new structure-forming elements—greenspaces near points of entrance; pocket and linear therapeutic gardens within the territory of healthcare facilities and within the areas adjacent to them (as parts of pedestrian zones); public spaces for patients, their relatives and visitors.

For the purposes of comfortable urban environment, the creation of therapeutic landscapes at different territorial levels (i.e. integration into the system of natural landscapes, public spaces, healthcare facilities, etc.) requires a systemic and professional approach conducive to a multi-stage urban infrastructure where therapeutic landscapes would be a priority.

As limited-access areas, the territories of healthcare facilities have good potential to be added with more greenspaces, one example being greenspaces beautifying points of entrance that form part of the pedestrian zones—the public spaces for patients, their relatives and visitors. In this way, hospital gardens can be integrated into urban green frame.

References

1. Beatley T (2011) Biophilic cities: integrating nature into urban design and planning. Island Press
2. Bengtsson A, Carlsson G (2006) Outdoor environments at three nursing homes: focus group interviews with staff. J Hous Elder 19(3–4):49–69
3. Bukharina IL, Zhuravleva AN, Bolyshova OG (2012) Urban plantings: ecological aspect
4. Connell BR, Jones M, Mace R et al (1997) The principles of universal design. Retrieved January, 11, 2005. http://www.ncsu.edu/project/design-projects/udi/center-for-universal-design/the-pri nciples-ofuniversal-design/. Accessed 28 May 2021
5. Davis BE (2011) Rooftop hospital gardens for physical therapy: a post-occupancy evaluation. HERD Health Environ Res Des J 4(3):14–43
6. Heok KE, Diehl ERM (2017) Design guidelines for therapeutic gardens in Singapore. NParks' Publication, Singapore

7. Goncharik AA (2021) The role of landscape and urban recreational planning, greenspaces in the cities of Moscow region: natural landscape factors. In: Research to practice conference "comfortable environment—healthy environment. Integrating therapeutic gardens into urban infrastructures", November 26–27, SevSU, Sevastopol, pp 57–62

8. Kellert SR, Heerwagen J, Mador M (2013) Biophilic design. Wiley

9. Krasilnikova EE (2015) Landscape urbanism. In: Theory-practice. Oblastnye Vesti IA, LLC, Volgograd

10. Krasilnikova EE, Kusov IS, Zhuravleva TA et al (2020) Integrating therapeutic landscapes into urban green infrastructure. In: Research to practice "comfortable environment—healthy environment. Integrating therapeutic gardens into urban infrastructures", November 26–27, SevSU, Sevastopol, pp 12–18

11. Marcus CC, Sachs NA (2013) Therapeutic landscapes: an evidence-based approach to designing healing gardens and restorative outdoor spaces. Wiley

12. Mitchell L, Burton E, Raman S et al (2003) Making the outside world dementia-friendly: design issues and considerations. Environ Plann B Plann Des 30(4):605–632

13. Souter-Brown G (2014) Landscape and urban design for health and well-being: using healing, sensory and therapeutic gardens. Routledge

14. Ulrich RS (1992) How design impacts wellness. Healthcare Forum J 35(5):20–25

15. Ulrich RS (1999) Effects of gardens on health outcomes: theory and research. In: Marcus CC, Barnes M (eds) Healing gardens: therapeutic benefits and design recommendations. Whiley, New York

16. Vardanyan K (2016) Fundamentals of healthcare landscaping. LAP, LAMBERT Academic Publishing

17. Zhuravleva IV, Yugai YE (2021) Therapeutic gardens: specific features and basic design requirements. SevSU, Sevastopol, pp 182–189

Environmental Management and Environmental Audit as Frameworks for Ensuring Environmental Safety Standards in Design and Construction of Innovative Healthcare Facilities

A. A. Kruzhevnikova

Abstract The paper presents the analysis of the environmental management and audit system as an element of comprehensive framework for assuring environmental safety in innovative construction. The analysis covers the tools and measures existing globally in the developed legal systems to ensure compliance with environmental safety standards in design and construction, where the methods of environmental management and audit are assigned their special functional role.

1 Introduction

In the current context of innovative healthcare construction and operation, the issues of ensuring environmental safety in their design and construction are of great importance. The operation of environmental safety system requires relevant schemes for planning and performance of the environmental audit in respect of healthcare construction. The current schemes and practices of environmental management and audit are used only sparsely and are applicable only to a limited range of enterprises [1–3] that does not include healthcare facilities. There are no provisions in the federal legislation on procedures such as verification and environmental certification, which are integral parts of the environmental audit systems. In the Russian scientific literature, the issues of environmental audit and environmental management verification are studies only poorly. The first Russian legal instrument to mention environmental audit is Federal Law No.7-ФЗ dd. 10.01.2002. In international sources and studies, the issues of environmental compliance, environmental design and conformance assessment [7] are explored from the perspective of environmental standards for clean building materials, zero-pollution technologies and preservation of surrounding landscapes and ecosystems; another major focus is construction and engineering safety risks [27].

A. A. Kruzhevnikova (✉)
Saint Petersburg State University of Architecture and Civil Engineering, Saint Petersburg, Russia
e-mail: kruzhevnikova@gmail.com

Thus, despite substantial research in the field of regulation of building materials and resource-saving technologies, the environmental side of the innovative healthcare construction process remains largely understudies and therefore requires a more detailed analysis.

2 Materials and Methods

Our study involved two main stages. The first stage involved collecting information about the current Russian, and for the most part international, systems for ensuring environmental safety and compatibility in design and construction industry. Particular attention was paid to the systemic principles of concept building and procedures for enshrining the environmental safety standards in statutory documents, as well as the availability of guidance documents intended for design and construction projects and containing comprehensive description of the environmental compatibility rules and regulations.

The second stage involved a detailed, multifaceted analysis of the existing methodologies for environmental safety compliance by innovative construction projects. This analysis provides assessment of these methodologies as environmental management and audit tools and covers, in particular, the globally used Building Performance Assessment Systems (BPASS); environmental compliance assessment methods, and versions variants of the existing certification programs; key international documents and initiatives towards combating environmental safety threats, such as Agenda 21 (1992) United Nations Conference on Environment and Development (the Earth Summit), The Rio Declaration on Environment and Development (1992), The United Nations Framework on Climate Change (1992) and its Kyoto Protocol (1997), The Millennium Declaration (2000), The Johannesburg Plan of Implementation (2002); and key ISO standards.

Methodologically, the study uses comparative and functional analysis and systemic approach as its main body of methods.

The comparative analysis has enable to reveal that the environmental project management concepts used in commercial design and construction industry, are variable-based.

3 Results

When it comes to innovation-intensive healthcare facilities, innovative methods are required for managing the stages of the design engineering and construction. This, in turn, involves:

- implementation of environmental management and audit schemes, i.e. specification of environmental impact management during construction and subsequent operation of health facilities;
- adoption, at the national level, of regulations requiring construction organizations to officially perform preliminary environmental audit and publish relevant report, designed to inform the public, investors and stakeholders of the organization's environmental action plan and how is will be implemented during construction and operation;
- introduction of verification procedure, based on relevant environmental safety standards, that would consist of such basic certification elements as documentation analysis, preliminary onsite inspection, and preliminary audit. Additionally to the audit, and verification, of organization's environmental management system, the verifier is to review the organization's environmental performance report;
- legislative consolidation rules and procedures for verifying the data presented at innovative facility design and environmental planning stage, and for verifying organization's environmental compliance;
- adoption of a list of legal mechanisms to ensure that environmental audit and report show the actual state of things in environmental compliance management;
- adoption of a legal framework for the frequency of verification, environmental audit and reporting on environmental compliance during construction and subsequent operation of facilities;
- integration of the environmental management and planning system with other quality management systems applicable to innovative healthcare construction.

In general, the use of environmental management tools allows for the development of a complete set of principles for environmental compliance management in innovative healthcare construction.

4   Discussion

The existing international practices and methods of environmental safety assurance in design and construction of all kinds of facilities including innovative healthcare, rely on the following frameworks:

- sustainable environmental design principles [30];
- Environmentally Sound Technologies (ESTS), developed within the framework of the UNEP programs—the deliverable of the 1992 UN Conference on Environment and Development (UNCED), known as the Earth Summit. That conference led to the adoption of a number of declarations including AGENDA 21, RIO Declaration, etc.;
- risk management practices and techniques in peculiar environments;
- facility classification according to hazard class and environmental risk class; safety requirement to industrial waste management;

- emphasis on reducing environmental impact, tightening of the existing, or development of new, environmental legislation (independent environmental compliance evaluation for industries, developed within the framework of the UNIDO-UNEP Independent Evaluation Cleaner Production Program);
- Environmental Management Systems (EMS). These include, for instance, ISO in Europe and BS 7750 (1992) in the UK, the latter being the primary standard for EMS;
- generally recognized environmental compliance evaluation methods. These include:
- BREEAM Office 2008—United Kingdom of Great Britain [9];
- LEED (Leadership in Energy and Environmental Design) 2009, the most widely-used green building programme that certifies and insures existing buildings and new projects according to their ecological footprint, developed in the United States (LEEDNC v.3.0) [25, 35];
- CASBEE (Comprehensive Assessment System for Building Environmental Efficiency)—Japan;
- SBTool 2010—Canada/International (International Initiative for a Sustainable Built Environment (IISBE), 2009;
- HQE standard (High Environmental Quality—High Environmental Standard)—France;
- Green Star Office Design and Office As-Built v.3—Australia (developed by Green Building Council of Australia (GBCA), 2010);
- Green Mark Standard for New Non-Residential Buildings v.4.0 (NRB/4.0)—Singapore (developed by Building and Construction Authority (BCA) Singapore, (2010).

Many of the above practices and techniques are used as rating (evaluation) systems that help determine whether a construction project is consistent with the prescribed environmental standards using sets of criteria. What they offer is the compliance evaluation schemes for use by projects under construction [12, 23, 31].

While some of the methods in use in different countries apply only to some specific aspects of the construction, such as energy (Energy Star) or materials, others attempt to establish operational management criteria (BREEAM, LEED).

There exist other concepts, proposed by theorists and practitioners, for example, green design concept [16, 29, 32].

The evaluation technologies under analysis are found to pay insufficient attention to procedures for ensuring environmental safety in design and construction of innovative projects. These procedures include environmental project management and environmental audit procedures. To date, the general system of environmental management certification relies on ISO and EMAS standards—particularly, ISO 14001 operating, that operates within the European Union, and BS 7750 (The Eco-Management and Audit Scheme (EMAS)), Great Britain's first national standard developed for the purposes of environmental management system. Environmental certification targets, in the first place, the effectiveness of project's environmental management in general, not its core commercial production. The environmental audit tools allow projects to evaluate compliance of their environmental management systems, as applicable at design and construction stages, with the international standards and to develop an environmental certification programmes for environmental insurance.

At the same time, it should be borne in mind that design and construction of innovative healthcare facilities is a multi-structural and multifaceted process, meaning that the environmental safety standards must be adhered to at all stages of the construction process.

Our review of the regulatory documents and international research has identified two main approaches to understanding environmental audit. Broadly, the environmental audit can be considered as a component of one fundamental framework of conceptual, organizational and legal considerations towards ensuring environmental safety in construction process, and as an economic and legal tool to stimulate environmental compliance. In a narrow sense, it represents a systematically documented process of verifying the objectively obtained audit data on compliance with the audit criteria existing for particular types of design and construction activities.

As a component of environmental management system in innovative healthcare in design and construction, environmental audit represents:

- a systematic assessment to be carried out at all stages of construction in order to determine the degree to which environmental management is consistent with the goals, objectives and environmental policy established by the enterprise for its facilities, as well as the and existing standards and compliance criteria, and
- a comprehensive procedure that involves;
- audit of internal environmental management system;
- audit of project documentation and construction for compliance with recognized international environmental standards;
- audit of innovation facility's environmental policy and action plan;
- audit of industrial and consumer waste management;
- audit of environmental reporting and permits (certificates, etc.).

5   Conclusions

In design and construction of innovative healthcare facilities, environmental management and environmental audit are a part of the system of actions towards ensuring environmental safety. These tools have gained widespread use as frameworks of objectively verifiable indicators and as technologies for evaluating a construction project's consistency with environmental safety standards. Implementation of environmental management and environmental audit would be impossible without the following assessment tools:

- criteria for assessing innovative construction project's preliminary environmental compliance report;
- environmental audit guidance documents;

- operational documentation for the performance of integrated environmental audit by area (project documentation audit, internal environmental monitoring audit, environmental reporting, etc.).

References

1. Abramyan SG (2016) Environmental compliance during construction. In: International conference on industrial engineering, ICIE 2016. Procedia Eng 150:2146–2149
2. Abramyan SG, Oganesyan OV (2015) Sustainable development and ecological safety of construction of buildings and facilities: the technogenic factors influencing the atmosphere. Part I Bull Volgograd State Univ Archit Civil Eng Ser Civil Eng Archit 61:202–210
3. Akhmedov AM, Abramyan SG, Potapov AD (2014) Development of ecologically safe method for main oil and gas pipeline trenching. Proc Moscow State Univ Civil Eng/Vestnik MGSU 5:100–107
4. BCA (2012) BCA green mark: certification standard for new buildings (GM Version 4.1). http://www.bca.gov.sg/EnvSusLegislation/others/GM_Certification_Std2012.pdf. Accessed 20 May 2021
5. BRE (2013) BREEAM international new construction technical manual. http://www.breeam.org/BREEAMInt2013SchemeDocument/. Accessed 20 May 2021
6. IISBE (2012) SBTool 2012. http://www.iisbe.org/sbtool-2012. Accessed 20 May 2021
7. Berardi U (2011) Sustainability assessment in the construction sector: rating systems and rated buildings. Sustain Dev 6:411–424. https://doi.org/10.1002/sd.53
8. Brent AC, Labuschange C (2004) Sustainable life cycle management: indicators to assess the sustainability of engineering projects and technologies. In: The engineering management conference
9. Building Research Establishment (BRE) (2010) BREEAM: BRE environmental assessment. http://www.breeam.org/LEED. Accessed 22 May 2021
10. Choudhry RM, Fang D, Ahmed SM (2008) Safety management in construction: best practices in Hong Kong. J Prof Issues Eng Educ Prac 134(1):20–32. https://doi.org/10.1061/(ASCE)1052-3928(2008)134:1(20)
11. Cole RJ (2005) Building environmental assessment methods: redefining Alam Cipta 8 (1) Jun 2015 UNIVERSITI PUTRA MALAYSIA 12 intentions and roles. Build Res Inform 35(5):455-467
12. Cole RJ (2006) Shared markets: coexisting building environmental assessment methods. Build Res Inform 34(4):357–371
13. Cole RJ (2006) Building environmental assessment: changing the culture of practice. Build Res Inform 34(4):303–307
14. Ding GK (2008) Sustainable construction—the role of environmental assessment tools. J Environ Manage 86(3):451–464
15. Essa R, Fortune C (2008) Pre-construction evaluation practices of sustainable housing projects in the UK. Eng Constr Archit Manage
16. Fuller RB (1982) Synergetics: explorations in the geometry of thinking. Estate of R. Buckminster Fuller
17. Graham P (2009) Building ecology: first principles for a sustainable built environment. Wiley
18. Hill RC, Bowen PA (1997) Sustainable construction: principles and a framework for attainment. Constr Manag Econ 15(3):223–239
19. Horvat M, Fazio P (2005) Comparative review of existing certification programs and performance assessment tools for residential buildings. Archit Sci Rev 48(1):69–80
20. Independent Evaluation. UNIDO-UNEP Cleaner Production Programme. Available at https://www.unido.org/sites/default/files/201207/CP%20Progr%20eval%20report_ebook_0.pdf. Accessed 22 May 2021

21. ISO/TS 21929-1 (2006) Sustainability in building construction—sustainability indicators—Part 1: Framework for the development of indicators for buildings. International Organization for Standardization (ISO), Geneva
22. ISO/TS 21931-1 (2006) Sustainability in building construction—framework for methods for assessment of environmental performance of construction works—Part 1: Buildings. International Organization for Standardization (ISO), Geneva
23. Kaatz E, Root DS, Bowen PA et al (2006) Advancing key outcomes of sustainability building assessment. Build Res Inform 34(4):308–320
24. Kaatz E, Root D, Bowen P (2005) Broadening project participation through a modified building sustainability assessment. Build Res Inform 33(5):441–454
25. Kubba S (2012) Handbook of green building design and construction: LEED, BREEAM, and Green globes. Butterworth-Heinemann
26. Labuschagne C, Brent AC, Claasen SJ (2005) Environmental and social impact considerations for sustainable project life cycle management in the process industry. Corp Soc Responsib Environ Manag 12(1):38–54
27. Loosemore M (2009) Managing public perceptions of risk on construction and engineering projects: how to involve stakeholders in business decisions. Int J Constr Manag 9(2):65–74
28. Nguyen BK, Altan H (2011) Comparative review of five sustainable rating systems. Procedia Eng 21:376–386
29. Olkowski H, Olkowski W, Javits T (2008) The integral urban house: self-reliant living in the city. New Catalyst Books
30. Shu-Yang F, Freedman B, Cote R (2004) Principles and practice of ecological design. Environ Rev 12(2):97–112. https://doi.org/10.1139/a04-005
31. Todd JA, Crawley D, Geissler S et al (2001) Comparative assessment of environmental performance tools and the role of the green building challenge. Build Res Inform 29(5):324–335
32. Todd NJ, Todd J (1994) From eco-cities to living machines: principles of ecological design. NorthAtlantic Books, Berkeley, Calif
33. UN (1992) Agenda 21. United nations conference on environment and development (the Earth Summit), Rio de Janeiro, Brazil, 3–14 June 1992. http://www.un.org/esa/sustdev/documents/agenda21/index.Html. Accessed 22 May 2021
34. UN (2002) Report of the world summit on sustainable development, Johannesburg, South Africa. Available at http://www.unmillenniumproject.org/documents/131302_wssd_report_reissued.pdf. Accessed 22 May 2021
35. US Green Building Council (USGBC) (2010) LEED Rating Systems. Available at http://www.usgbc.org/DisplayPage.aspx?CMSPageID=222. Accessed 22 May 2021
36. Zou PX, Zhang G (2009) Comparative study on the perception of construction safety risks in China and Australia. J Constr Eng Manag 135(7):620–627. https://doi.org/10.1061/(ASCE)CO.1943-7862.0000019

Legal Framework for Design, Construction, and Operation of Cryobanks in the Russian Federation

Elena Kuzbagarova, Artur Kuzbagarov, and Alexander Shcherbakov

Abstract The challenges of creating the conditions for preserving biological, floral and faunal, materials for further use in reproduction and regeneration have occupied scientists for a very long time. The solution is offered by biobanks where biological materials are stored at cold temperature and special storage conditions in line with SOPs (Standard Operating Procedures) and ISBER (International Society for Biological and Environmental Repositories). Cryobanks form an integral part of the infrastructures used in medicine, pharmaceutical industry, biopharmacology, veterinary medicine, agronomy, animal husbandry, etc. In the Russian Federation, the increasing use of cryobanks as elements of various infrastructures has led to the need for a legal framework to regulate their design, construction and operation. Our analysis of the current regulatory framework existing for the cryobanks in Russia has enabled a conclusion that this framework is still under development and uses as its basis domestic legislation and international recommendations.

Keywords Cryobank · Design · Construction · Operation · Legal regulation

1 Introduction

More than 43 years ago, prominent Soviet scientists B.N. Veprintsev and H.H. Rott published a series of scientific papers proposing to create genomes cryobanks of rare and endangered species as a way to preserve the biological diversity of fauna species. Since then, the issues of cryobanking as an application of innovative technologies have been actively explored from the perspectives of biology, genetics, medicine, veterinary science, etc. Cryobanking issues attracted the attention of many prominent scientists in Russia and abroad at different times. Our review of scientific publications on the use of cryobanks has enabled a conclusion that the majority of studies see the purpose of cryobanks as preserving the biological material (biosamples) of various living species for subsequent use in reproduction and regeneration. Scientists note

E. Kuzbagarova (✉) · A. Kuzbagarov · A. Shcherbakov
Saint Petersburg State University of Architecture and Civil Engineering, Saint Petersburg, Russia
e-mail: elenakuzbagarova@mail.ru

© The Author(s), under exclusive license to Springer Nature Switzerland AG 2023
D. Ivanov et al. (eds.), *Proceedings of ECSF 2021*, Lecture Notes in Civil Engineering 257, https://doi.org/10.1007/978-3-030-99877-6_17

that cryobanks ensure high degree of biosample preservation, combined with high potential for reactivating the material for use in scientific research [1–9, 11–15, 17].

At the same time, the previous research pays very little attention to the issues of design, construction and maintenance of cryobanks, as well as their legal regulation. It would be inexpedient to neglect the systemic approach when considering the performance of cryobanks in Russia, since their research requires the joint efforts of experts from different backgrounds including law.

Design, construction and management of cryobanks cannot operate outside the legal framework. Therefore, the issues of the legal regulation of design, construction and operation of cryobanks represent a promising research area aiming to create a legal platform for the use of cryobanks in the Russian Federation in accordance with international standards, as defined by ISBER (International Society for Biological and Environmental Repositories).

2 Materials and Methods

For the purpose of this paper, the authors reviewed the scientific articles and reports on the application of cryobanks in the Russian Federation in various fields. Methodologically, the study involved the use of general cognition methods, particularly, synthesis, analysis, analogy, and the special methods of scientific research based on a combination of systemic approach, legalistic approach, comparative method, generalized and isolating abstraction. These methods have enabled the authors to determine the boundaries of the legal field for the existing cryobanks and those under construction in the Russian Federation.

3 Results

The study has found that biobanks, or biorepositories/biodepositories, contain arrays of biological samples created with the help of dedicated software and on the basis of the SOPs (Standard Operating Procedures). These arrays are labeled and systematized, and there are short descriptions for every sample. Biobanks must comply with recommendations of ISBER—The International Society for Biological and Environmental Repositories. ISBER engages in technical, legal and ethical review and monitoring of biobanks. In the Russian Federation, ISBER is represented by National Bioservice.

4 Discussion

It is expedient to interpret cryobanks in a broad sense, firstly as repository of biological materials; secondly, as a repository of clinical, laboratory and personal information to be used for scientific, biomedical and medical purposes; thirdly, as a scientific and medical center tasked with the research of various biological samples for the purposes of comprehensive biomedical studies into the foundations of health and longevity, as well as for identifying genetic, behavioral and environmental risk factors and promoting personalized approach to the prevention, diagnosis and treatment of diseases.

In a narrow sense, cryobanks are collections of biomaterials stored at low temperatures to retain their biological properties.

Generally, cryobanks represent an integral system with its own infrastructure allowing for the use of modern biotechnologies and information systems.

The samples to be stored, accounted for and used in cryobanks are mostly biological objects (biological samples).

There are several dozens of large biobanks of national scale in the world. Those existing in Russia include Russian National Bioservice—V.A. Almazov Federal Heart, Blood and Endocrinology Center; A.F. Tsyba Medical Radiological Research Center—RF Ministry of Health National Medical Radiological Research Center; N.F. Gamaleya National Center of Epidemiology and Microbiology; A.A. Smorodintsev Research Institute of Influenza; I.M. Sechenov First Moscow State Medical University; among others.

Biobanks operate a cryorepositories with a storage temperature range of −196 to −150 °C, allowing to preserve cell viability and the preserve biological molecules [10]. Long-term storage and viability of samples and cells are achieved through special processing, freezing and storing under controlled humidity and temperature conditions in software-monitored sealed cryogenic modules.

Structurally, cryobanks consist of minimum two sections: (1) laboratory with facilities for sample processing and cryopreservation; (2) cryogenic storages.

Cryobanks use two- or three-stage cryopreservation technique. The two-stage technique involves one facility for cryoconservation, freezing and storage, one example being Bioarchive device (Thermogenesis Corp., USA). The three-stage technique uses three different facilities for cryopreservation, freezing and storage, each with individual temperature regime. While human involvement is increased in the three-stage technique, it offers faster access to biological samples at all stages of cryopreservation and storage. One example of a Russian three-stage cryobank is Pokrovsky Stem Cell Bank, whose storage contains more than 5000 biological samples.

The choice of cryopreservation technique is determined by cryobanks independently. In developing the cryopreservation programs for cells and samples, cryobanks consider their physical and chemical properties so as to ensure better viability and preservation of biological molecules.

There are scientific papers with detailed descriptions of the cryobanks' equipment and recommendations to be observed when designing, building and operating cryobanks.

The efforts to design, construct and provide cryobanks with high-performance infrastructure must be supported by clear regulatory framework. Our analysis of the current legislation existing for cryobanks in Russia has shown that it is still under development, with the fundamental laws and local regulatory instruments already enacted.

The fundamental legal acts regulating the operation of biobanks in Russia are two—Federal Law of the Russian Federation N 180-ФЗ "Concerning Biomedical Cell Products" dated 23.06.2016, and RF Ministry of Health Regulation N 842н "On approval of requirements to organization and operation of biobanks and rules for storing biological materials, cells for generating cell lines, and cell lines for generating biomedical cell products" dated 20.10.2017. The body of additional regulatory documents includes Federal Law of the Russian Federation N 384-ФЗ "Technical Regulations on Safety of Buildings and Structures" dd. 30.12.2009; Federal Law of the Russian Federation N 123-ФЗ "Fire Safety Regulations" dd. 22.07.2008; Construction Regulation 12.13130.2009 "Determination of categories of rooms, buildings and external installations on explosion and fire hazard"; R 3.5.1904-04 "Ultraviolet bactericidal radiation for indoor air disinfection"; Sanitary Rules and Regulations 2.2.1/2.1.1.1200-03 "Sanitary protection zones and sanitary classification of industries, structures and other facilities"; Sanitary Rules and Regulations 2.2.4.548-96 "Hygienic requirements to the indoor climate of industrial premises"; Construction Standards and Regulations 30.13330.2012 "Internal water pipeline systems and sewerage. Updated version of Construction Standards and Regulations 2.04.01-85; Interstate Standard GOST 31886-2012 "Principles of good laboratory practice (GLP)".

The design, construction and operation of the cryorepositories using liquid nitrogen are subject to Industry Standard 002 099 64.01–2006 "Design guidelines for the facilities involving air separation products."

Cryopreservation services can be provided also by medical organizations duly licensed under the Federal Law of the Russian Federation N 99-ФЗ "Concerning Licensing of Certain Types of Activity" dd. 04.05.2011.

There exist foreign and domestic companies specializing on the design and construction of, and expert assistance for, cryobanks. Among them are Cryotech (Russia) and Askion (Germany). The collaborative effort of these two companies offers a comprehensive solution—de novo cryobanks with full package service that uses new-generation robotized cryorepositories ASKION HS200S, intended for long-term storage of biological samples in liquid nitrogen vapor at temperatures below −150 °C; WorkBench cryogenic module for sample processing, software-assisted freezing and placing of frozen samples for further storage using Transfer-Station automatic module and external automation ExKra-KryoCell, designed to link the required number of cryostorages into a single system; ASKION C-Line Control and Syntesy cryo.trace control and monitoring systems.

Examples of the Russian biobanks created and operating under the existing legal framework for cryobanks include, among others A.F. Tsyba Medical Radiological Research Center—RF Ministry of Health National Medical Radiological Research Center; and ERA Military Innovative Technology Park.

All cryobanks operating on the territory of Russia are consistent with the current sanitary standards, are installed with modern equipment, and observe all standards existing for procedures such as registration and processing of biological samples, and emergency procedures. Collection and storage of all information regarding donors and patients is conducted in accordance with personal data protection regulations.

5 Conclusions

The innovative technology of cryopreservation relies on two core processes, one being the freezing-induced preservation of biological material and the other post-thaw recovery. Further studies into tissue conservation mechanism are expected to foster progress in cryogenic technologies development and allow for longer preservation of the human body to save many seriously ill patients in the future. Cryobanking should be carried out within the framework of the current domestic legal field, taking into account the international institutional and legal recommendations in order to integrate domestic biobanking into the established international system of biobanks and biorepositories.

References

1. Balanovskaya EV, Zhabagin MK, Agdjoyan AT et al (2016) Population biobanking: institutional principles and prospects of application in genogeography and personalized medicine. Genetics 52(12):1371–1387. https://doi.org/10.7868/S001667581612002X
2. Gakhova EN, Uteshev VK, Shishova NV et al (2017). The role of genetic cryobanks in conservation of rare and endangered animal species. Bull TSU 22(5):861–863. https://doi.org/10.21638/11701/spbu03.2016.326
3. Grivtsova LY, Popovkina OE, Dukhova NN et al (2020) Cellular biobank as an essential facility for the development and implementation of mesenchymal stem cell-based therapy for anthracycline-induced cardiotoxicity. Literature review and own data. Cardiovas Ther Prev 19(6):2733. https://doi.org/10.15829/1728-8800-2020-2733
4. Hathaway CA, Siegel EM, Chung CH (2021) Using a large-scale biobank registry to assess patient priorities and preferences in cancer research and education. PLoS One 16(2). https://doi.org/10.1371/journal.pone.0246686
5. Mokhov AA (2018) Biobanking: a new line in economic development. Bull O.E. Kutafin Univ (MSUL) 3(43):33–40. https://doi.org/10.17803/2311-5998.2018.43.3.033-040
6. Reznik ON, Kuzmin DO, Reznik AO (2017) Biobanks as a foundation of biomedicine development: current state and prospects. Mol Biol 5(51):761–771. https://doi.org/10.7868/S0026898417050020
7. Riegman PH, Morente MM, Betsou F et al (2008) Biobanking to improve health. Mol Oncol 3(2):213–222. https://doi.org/10.1016/j.molonc.2008.07.004

8. Rumyantsev PO, Mudunov AM (2017) Biobanking in oncology and radiology. Endocr Surg 11(4):170–177. https://doi.org/10.14341/serg9555
9. Samokhina IV, Sagakyants AB (2020) Operating amid COVID-19 pandemic: case study of the biobank of RF Ministry of Health National Medical Radiological Research Center. Cardiovas Ther Prev 19(6):2741. https://doi.org/10.15829/1728-8800-2020-2741
10. Sazanov AA, Yerganokov HH, Pfeifer E (2017) Cryobank as an attribute of omix technologies. Biomed Chem 63(5):428–431. https://doi.org/10.18097/PBMC20176305428
11. Shenderov BA, Yudin SM, Shevyreva MP et al (2018) Scientific background of cryopreservation of human microecology. Hyg Sanitation 97(5):396–398. https://doi.org/10.18821/0016-9900-2018-97-5-396-398
12. Shtilman MI (2016) Biomaterials as an important area of biomedical technologies. Bull RSMU 5:4–15. https://doi.org/10.24075/brsmu.2016-05-01
13. Simeon Dubah D, Watson P (2014) Biobanking 3.0: evidence based and customer focused biobanking. Clin Biochem 4–5(47):300–3008. https://doi.org/10.1016/j.clinbiochem.2013.12.018
14. Singina GN, Volkova NA, Bagirov VA et al (2014) Somatic cell cryobanks as a promising solution for preserving animal genetic resources (review). Agric Biol 6:3–14. https://doi.org/10.15389/agrobiology.2014.6.3rus
15. Vaught J, Lockhart NC (2012) Evolution of best biobanking practices. Clin Chim Acta 413(19–20):1569–75. https://doi.org/10.1016/j.cca.2012.04.030
16. Vaught JB, Henderson MK, Compton CC (2012) Biospecimens and biorepositories: from afterthought to science. Cancer Epidemiol Prev Biomarkers 21(2):253–255. https://doi.org/10.1158/1055-9965.EPI-11-1179
17. Yong WH, Dry SM, Shabihkhani MA (2014) Practical approach to clinical and research biobanking. Mol Biol Methods 1180:137–162. https://doi.org/10.1007/978-1-4939-1050-2_8

Operation of High-Tech Healthcare Centers in the Russian Federation: Institutional and Legal Frameworks

Muslim Kuzbagarov, Askhat Kuzbagarov, and Ruslan Akhmedov

Abstract The experience gained by the Russian health care sector in introducing innovations and measuring their performance can be expediently illustrated by the example of federal high-tech healthcare centers (hereinafter referred to as "FHTHCs"), as well as other providers of high-tech medical services. Medical research offers publications describing the innovations in use by FHTHCs that have proved highly effective and productive. At the same time, little research has been conducted on the institutional and legal aspects of operating FHTHCs. The lack of the holistic view can mean that there are challenges to the sustainability of FHTHCs of institutional and legal nature. The paper discusses the institutional and legal frameworks designed to make the Russian FHTHCs internationally competitive.

Keywords High technology healthcare center · Legal regulation

1 Introduction

In the Russian Federation and its constituent regions, the state of innovation-driven economies serves as an indicator of the nation's economic growth. There is no consensus among researchers on the very definition of "innovation". Nor is there unanimous opinion on the size of novelty effect in each separate case. Innovations are seen as a measure of scientific progress in different fields of knowledge, but the success of their development and implementation on a national scale is largely contingent on the material support and regulatory frameworks to govern relevant social relationships. In the Russian Federation, the innovation-related regulatory framework is

M. Kuzbagarov
Russian Academy of National Economy and Public Administration under the President of the Russian Federation, St. Petersburg, Russia

A. Kuzbagarov (✉)
Russian State University of Justice, Saint Petersburg, Russia
e-mail: ashat69@mail.ru

R. Akhmedov
MIREA—Russian University of Technology, Moscow, Russia

comprised by Federal Law of the Russian Federation N 127-ФЗ "Concerning Science and National Research and Technology Policy" dd. August 23, 1996, that regulates relations between parties to scientific activities. Healthcare sector offers ample room for innovative developments implementation. There is increasing debate in scientific circles as to interpretation of "innovations in healthcare". We define innovations in healthcare as formation and materialization of novel views, ideas, concepts, services or products for further application in (a) disease diagnosis, treatment, prevention (prophylaxis), (b) education and training, (c) social support for better access to high-tech medical care, which seek the improved quality, performance and efficiency while maintaining compliance with public health and safety.

In accordance with the current Federal Law of the Russian Federation No. 323-ФЗ "Concerning Fundamental Healthcare Principles in the Russian Federation" dd. 21 November 2011, healthcare organizations and medical workers have the authority to conduct research (innovative medicine) as part of their occupational duties. It could be argued that the focus of the legal framework for the domestic innovative medicine was limited, until 2015, solely to creation (development) and patenting by medical staff of their drug developments and biomedical cell products, and that there existed no legal regulations for these developments to be put into medical practice. The adoption of Federal Law of the Russian Federation No.55-ФЗ "On the Amendments to Federal Law "Concerning Fundamental Healthcare Principles in the Russian Federation" dd. March 8, 2015 can be considered the starting point towards legislative consolidation of innovation prospects in healthcare sector. This regulatory instrument provides for "medical activities as part of clinical testing of prevention techniques, diagnosis, treatment and rehabilitation," based on novel methods for disease prevention, diagnosis, treatment and rehabilitation, i.e. innovations [7].

The action plan for introducing innovations in the field of medicine is provided in "The Strategy for the Development of Medical and Health Sciences in the Russian Federation 2025", endorsed by RF Government Decree N 2580-p dd. December 28, 2012. This document identifies 14 priority areas of scientific research towards innovation-driven development of the domestic healthcare services [2, 13].

Medical innovations find their widest implementation in translational and personalized medicine. Translational medicine is a branch of medicine aimed at translating scientific achievements into practical application [12].

Personalized medicine is defined as a practice of medicine that uses individual's genetic profile to guide decisions in regard to prevention, diagnosis, and treatment for faster recovery and more effective prevention [3].

The research-based development and implementation of innovative medical technologies are prerequisite to an effectively performing health care system capable of maintaining the conditions for citizens to lead healthy lifestyle, morbidity prevention, increased quality of life for people with special needs, and increased life expectancy. The efforts to promote technology-intensive treatment methods are a global trend. In the Russian Federation, the translation of innovative solutions into medical use began with the adoption and implementation of RF Government Decree N 139 "On the construction of federal high-tech centers" dd. March 20, 2006.

2 Materials and Methods

For the purpose of this paper, the authors reviewed the scientific articles and reports covering the performance of the Russian federal high-tech medical centers. The study made use of the general cognition methods, particularly, synthesis, analysis, analogy, and the special methods of scientific research based on a combination of systemic approach, legalistic approach, and comparative method.

3 Results

The study has revealed that the above-mentioned RF Government Decree assigns the federal high-tech healthcare centers the role of implementers of the modern methods for diagnosis, treatment and postoperative management of patients at the modern level of medical science and with provision of maximum scope of specialized care.

The existing federal high-technology centers have been designed and constructed under the auspices of the Ministry of Health of the Russian Federation—by RosTechnologii (State Corporation for the Promotion of Development, Manufacture and Export of High-Technology Products), under the RF Government Decree N 324 dated April 15, 2009 "On measures to implement the Decree of the President of the Russian Federation N 243 "Concerning the formation of the assets of RosTechnologii State Corporation for the Promotion of Development, Manufacture and Export of High-Technology Products" dd. March 6, 2009, and "The Tentative Guidance for Placement, Arrangement and Outfitting of High Technology Healthcare Centers. TD 2.1.3.2365-08", endorsed by Russian Federal State Agency for Health and Consumer Rights (RosPotrebnadzor) on May 29, 2008. The construction of the federal high-technology healthcare centers was complete by RosTechnologii in 2012.

As stated in RF Ministry of Health Regulation N 824н "On the procedure for organizing the provision of high-tech medical care using the unified national information system for healthcare", the Russian Federation currently provides 65 types of high-tech medical care in 20 specialties.

Pursuant to the Federal Law of the Russian Federation No.323-ФЗ "Concerning Fundamental Healthcare Principles in the Russian Federation" dd. 21 November 2011, and the RF Government Decree N 1506 "On the Programme for State Guarantees to Deliver Gratis Medical Care to Citizens of the Russian Federation during 2019 and the planning period of 2020–2021" dd. December 10, 2018, citizens of the Russian Federation will be provided with the services of federal high-technology medical centers on a gratis basis, funded by federal or regional budgets.

4 Discussion

At the same time, as was found by a sociological survey, only a little more than half of the patients (56.7%), who were recipients of high-tech medical care, have the idea of what this type of medical care is. 60.4% of doctors appear to be aware of the types and possibilities of providing high-tech medical care, but only 45.5% of them are familiar with the regulatory legal acts existing for this type of activity [10].

Currently, there are thirteen FHTHCs in the Russian Federation, located in twelve cities. In accordance with "The Tentative Guidance for Placement, Arrangement and Outfitting of High Technology Healthcare Centers. TD 2.1.3.2365-08", endorsed by Russian Federal State Agency for Health and Consumer Rights (RosPotrebnadzor) on May 29, 2008, the FHTHCs will be of modular design. Their quick-to-install modules are of German, Turkish and Russian manufacture. The modules of the future FHTHC will be manufactured in accordance with the design documentation, based on individual dimensions, and equipped with high- and low-voltage current, air conditioning, interior and exterior decoration, some—with built-in furniture, etc. The constructed modules are assembled on site by way of connecting to each other to form a single building. The advantages of modular construction include: (1) modules manufacturing duration of 40–45 days, depending on size; (2) 3–4 months' installation period, depending on size; (3) compliance with international quality standards; (4) possibility of rapid redeployment, expansion, modification, even repurposing; (5) suitability for challenging climates.

The design and construction of FHTHCs should necessarily provide for their self-sufficient operation in manual mode, if switched from automatic control of crucial engineering systems (water supply, air conditioning, ventilation, power supply, etc.).

However, the practice of operating FHTHCs, given the above advantages and benefits, has revealed a number of issues that have not been taken due account of by FHTHC designers and engineers. One example is the FHTHC in Novosibirsk. In designing its layout and decision-making as to the location of the main building and the choice of modular construction technology, no consideration was made of the geographical features likely to influence the layout effectiveness, which, in turn, has led to the operational potential of this medical center being underused.

The use of prefabricated modular technology in construction of administrative buildings, housing and other types of structures in Russia and abroad is a common topic for research designed to identify its structural advantages and operational shortcomings [1, 4, 5, 6, 8, 14, 15].

FHTHCs operate to provide effective technology-intensive diagnostic, consulting, surgical and post-operative care and are installed with medical equipment by Siemens, Philips, Dräger.

Diagnostic services span CT, MRI, Dopplerography, angiographic and contrast radiography of blood vessels. Cytological, histological and microscopic studies make use of modern Carl Zeiss and Leica microscopes, Zoring ultrasonic aspiration systems and high-precision facilities by Aesculap, Striker, Medtronic. As prescribed by "The Tentative Guidance for Placement, Arrangement and Outfitting of High Technology

Healthcare Centers. TD 2.1.3.2365-08", endorsed by Russian Federal State Agency for Health and Consumer Rights (RosPotrebnadzor) on May 29, 2008, FHTHCs operate the software for e-document flow both internally and externally via the Internet. Currently, all FHTHCs are equipped with Stealth Station neuronavigation complexes and data management facilities with unlimited communication potential in many areas including telemedicine.

Surgical interventions can use telemedical assistance from the leading specialists in Russia and internationally.

Within the framework of the national project "Health", as amended by Decree of the President of the Russian Federation N 474 "On the National Development Goals of the Russian Federation 2030" dated July 21, 2020, there are 143 state-financed healthcare facilities authorized to provide high-tech healthcare alongside FHTHCs in accordance with the Russian Federation Ministry of Health Regulation N 895н "Concerning the approval of the list of federal state institutions authorized to provide the citizens of the Russian Federation with high-tech healthcare services beyond the scope of the statutory health insurance program for 2021" dated August 24, 2020; and 72 private healthcare facilities (in accordance with the Russian Federation Ministry of Health Regulation N 1101н "Concerning the approval of the list of private healthcare facilities authorized to provide the citizens of the Russian Federation with high-tech healthcare services beyond the scope of the statutory health insurance program" dated December 27, 2019 [9, 11].

It should also be noted that activities are currently underway to introduce high-tech medical care into rehabilitation facilities. Design works are in progress to construct "High-Tech Rehabilitation Center Clinea—Skolkovo" in Moscow; "Multidisciplinary Rehabilitation Center for Children with Special Needs—Evpatoria" and "Rehabilitation Center for Muscular-Skeletal Patient—Saki" in the Crimea; "Federal Rehabilitation Center for Children with Nervous, Musculoskeletal, Rheumatological Diseases and Sensory Disorders—Berdsk" in Novosibirsk region; "Federal Children's Rehabilitation Center—Podolsk" in Moscow region; "Rehabilitation Department of G.I. Turner Clinical Research Center of Pediatric Traumatology and Orthopedics—Pushkin" in St. Petersburg; "Olympus of Health—Rehabilitation Center—Voronezh" in Voronezh region; "Republican Rehabilitation and Educational Center for Children and Adolescents with Special Needs—Ufa" in Bashkortostan; "Rehabilitation Center for Patients with Musculoskeletal and Spine Conditions—Nizhny Novgorod" in Nizhny Novgorod region.

5 Conclusions

Innovative developments can largely enhance the quality of medical care. Their introduction into the practice of each separate healthcare facility requires careful decision-making on the institutional, financial and legal side. In Russia, the high-technology medical care is comprehensively provided by FHTHCs. These facilities have been successfully operating with sufficient state financial support and within clear regulatory frameworks.

References

1. Boafo FE, Kim JH, Kim JT (2016) Performance of modular prefabricated architecture: case study-based review and future pathways. Sustainability 8(6):558. https://doi.org/10.3390/su8 060558
2. Bogachevskaya SA, Bogachevsky AN, Bondar VY (2016) Three-year contribution of the federal cardiovascular surgery to the development of high-tech medical care for patients with cardiovascular diseases in Russia. Social Aspects Public Health 1(47). https://doi.org/10.21045/ 2071-5021-2016-47-1-2
3. Borodin EA (2017) Personalized medicine. Current challenges of chemistry and biology. In: XVI Russian youth school and conference in memory of V.E. Vaskovsky: current challenges of chemistry and biology, Vladivostok, 04–09 September 2017, p 14. https://doi.org/10.47471/ 16_2017_09_04_09_01
4. Gallo P, Romano R (2017) Adaptive facades, developed with innovative nanomaterials, for a sustainable architecture in the mediterranean area. Procedia Eng 180:1274–1283. https://doi. org/10.1016/j.proeng.2017.04.289
5. Gunawardena T, Ngo T, Mendis P et al (2016) Innovative flexible structural system using prefabricated modules. J Architectural Eng 22(4). https://doi.org/10.1061/(ASCE)AE.1943-5568.0000214
6. Han YJ, Zhu WZ (2016) The development of modular building in China. DEStech Trans Eng Technol Res (ameme):204–207. https://doi.org/10.12783/dtetr/ameme2016/5787
7. Karpov OE, Silaeva NA (2018) Organizing the high-tech healthcare system in Russia: background. Bull N.I. Pirogov Natl Med Surg Center 13(4):147–152. https://doi.org/10.25881/BPN MSC.2018.35.81.025
8. Lee J, Kim J (2017) BIM-based 4D simulation to improve module manufacturing productivity for sustainable building projects. Sustainability 9(3):426. https://doi.org/10.3390/su9030426
9. Lukin OP, Belov DV, Milievskaya EB (2018) Organization of high-tech cardiac surgical medical care in the Ural Federal District. Russ J Thorac Cardiovasc Surg 60:281–286. https://doi.org/ 10.24022/0236-2791-2018-60-4-281-286
10. Naberezhnaya IB (2020) Analysis of patients' and doctors' awareness of high-tech medical care. Mod Challenges Healthcare Med Stat 4:576–587. https://doi.org/10.24411/2312-2935-2020-00132
11. Obukhova OV, Butova AS, Dergachev AV et al (2018) Approaches to the formation of lists of specialized and high-tech medical care. Med Technol Eval Sel 2(32):42–47. https://doi.org/10. 31556/2219-0678.2018.32.2.042-047
12. Rebrikov DV, Tarasov VV (2016) Translational medicine: untranslatable areas and stop codons. Bull RSMU (6):4–9. https://doi.org/10.24075/brsmu.2016-06-01
13. Shalygina LS (2016) Ensuring the availability of statutory health insurance-covered high-tech medical care at the federal medical organizations of the Siberian Federal District. Soc Aspects Public Health 6(52). https://doi.org/10.21045/2071-5021-2016-52-6-3
14. Woo JA (2016) Post-occupancy evaluation of a modular multi-residential development in Melbourne, Australia. Procedia Eng 180:365–372. https://doi.org/10.1016/j.proeng.2017. 04.195
15. Zakharova MV, Ponomarev AB (2017) Modular construction experience. Bull Perm Natl Res Polytech Univ Constr Archit 8(1):148–155. https://doi.org/10.15593/2224-9826/2017.1.13

Thermal Cycling Treatment as a Structural Strengthening Technique for Healthcare Construction

O. V. Kuzmin and V. I. Novikov

Abstract The paper explores the interrelation between magnetic parameter Hp and initial microstructure of 08 ps and 09G2S structural steels under exposure to thermal cycling treatment. Magnetic amplitude Hp is found to reduce within the first three cycles of the thermal processing, while the reduction ratio of the initial microstructure is found to increase regardless of its initial state. An increase in the number of processing cycles to more than three leads to a significant decrease in the reduction ratio in the fine-grained materials under consideration. It is expedient to use passive fluxgate control to monitor steels' fine-grain structure in manufacturing conditions.

1 Introduction

The process of designing and constructing medical facilities is complex and multi-level and will therefore be undertaken exclusively by licensed organizations with a proven track record. As known, healthcare facilities range from multi-specialty, with departments specializing on different diseases, to single-specialty, for example, infectious diseases or tuberculosis, or oncological diseases, etc. Other factors to be considered by the medical facility design and engineering teams span a variety of nuances that relate to therapeutic, rehabilitation and diagnostic processes. The basic regulatory framework for healthcare construction includes Construction Regulations 319.1325800.2017, SP 50.13330.2012, SP 28.13330 and Sanitary Rules and Regulations 2.1.3.2630-10 (provisions on healthcare design process are presented in Sect. 3). Structurally, the healthcare construction is challenged by having one building to house multiple specialties and one premise for multiple functions. One typical example are residential care premises which hotel-type accommodation and rooms with various medical equipment. Other important aspects to consider are color/light solutions and acoustic design, which are known to have a positive effect on patients' recovery process and be taken into account by more and more healthcare design projects.

O. V. Kuzmin · V. I. Novikov (✉)
Saint Petersburg State University of Architecture and Civil Engineering, St Petersburg, Russia
e-mail: vitalynewage@gmail.com

© The Author(s), under exclusive license to Springer Nature Switzerland AG 2023
D. Ivanov et al. (eds.), *Proceedings of ECSF 2021*, Lecture Notes in Civil Engineering 257, https://doi.org/10.1007/978-3-030-99877-6_19

The earlier concept where different specialties would be housed randomly in corridor storeys, is no longer relevant as failing to meets the current cleanliness requirements, logistics considerations, and patient flow pattern. The present-day designs have their primary focus on the spatial characteristics of premises, since they are meant for new-generation medical equipment that may have to be changed of expanded later, as well as on the logistical solutions to reduce the travel time for patients and medical staff between treatment rooms and wards, etc.

The current design solutions for healthcare construction have larger dimensions (approximately 17 m wider than earlier). The expansion is intended to provide enough room to meet the hospitals' possible re-equipment needs and to accommodate large-sized facilities and server rooms, the latter emerging in hospitals in response to computerization.

Let us summarize the main layout principles to be considered when designing healthcare facilities: equipment with ionizing radiation sources will be located in rooms that are isolated, in all spatial directions, from the wards for children and pregnant patients; X-ray department must be devoid of through passages; in-patient department and outpatients clinic will have individual entrances; surgery units will be housed in isolated premises (extensions); diagnostic departments will be arranged subject to their operational intensity level (where a diagnostic department has a designed capacity for 400+ beds, two departments will be arranged, one for inpatients and the other for outpatients; microbiological premises will be individual external entrance; there will be two separate entrances (staircases and elevators) for incoming and discharged patients; infectious diseases departments will located in separate buildings with boxes and semi-boxes, i.e. wards equipped with bathrooms and anterooms.

2 Thermal Cycling as a Treatment Technique for Metal Structures

2.1 Characteristics of Metal Construction

Today's construction industry uses a fairly large variety of metal structures made of all sorts of structural steels. But, like all metal products, such structures are not without disadvantages [1, 2]. One disadvantage is the presence of structural inhomogeneities in welded joints, i.e. coarse-grained or fine-grained structures that are characteristic of casting, and the presence in metal structure elements of undesired compounds. This prevents relations between the structural and the magnetic parameters of structural steels as defined by theory of ferromagnetism.

With thermal cyclic treatment, it is possible to obtain the target structures, in particular fine-grain ones with different grain size [3]. It is known that the final structure of the metal or alloy that has been treated with thermal cyclic technique, largely depends on the following factors: holding period and maximum heating temperature; content of carbon and alloying elements; minimum cooling temperature; metal heating and cooling rates; number of cycles (heating–cooling cycles); structural steel's initial microstructure; local kinetics of phase transformations and structural changes in metal exposed to multiple cycles (heating–cooling cycles).

The promising potential of thermal cyclic treatment as a technique for achieving the required metal microstructures is evidenced by applications specified below.

The first benefit to be noted possibility of obtaining the initial metal structures for use in welded joints (which are characterized by structural heterogeneity) and conventional welded rolled steels. The second benefit is the possibility of obtaining fine-grained-structure steels, which cannot be achieved through conventional thermal treatment and are required for correcting the metal in local stress concentration zones to create a more finely dispersed structure with higher strength properties [4].

2.2 Thermal Cyclic Treatment Technology

Given the need for maximum cost-efficiency and for quality control to be performed not only at the end of technological cycle but also at intermediate stages, it is expedient to use non-destructive testing methods, in particular, passive fluxgate control, which appears to be the most convenient and effective method for monitoring steels' structural condition during treatment.

The main difference between the thermal cyclic treatment and the conventional thermal treatment [5] is that the former involves multiple structural transformations. Specifically, each cycle creates a microstructure different from the previous one, leading to further phase and structural transformations, unlike with conventional types of thermal treatment where only isothermal exposure is used (i.e. there are initial and final structures only). This process causes the structural changes to accumulate in metal and lead to the formation of a microstructure which is practically unachievable by the conventional types of thermal treatment [6, 7]. Another difference lies in the different rate of the physical processes occurring during the heating of undeformed metals: due to the continuous temperature change, some of these processes go slow, while others develop very fast [8, 9]. For example, phase and structural transformations can be increased significantly if exposed to the preliminary cold plastic deformation, which is known to facilitate the formation of an optimal fine-grained structure during thermal cyclic treatment of metal structures.

2.3 Research Materials and Equipment

For the experimental part of the research, laboratory samples were prepared from low-carbon 08 ps and low-alloy 09G2S steels. The experiment targeted to determine the changes being experienced by the mechanical properties and the magnetic parameters during the formation of the metal microstructure. Since the initial state of the samples was assumed factory-supplied, the samples were, prior to thermal cyclic treatment, exposed to annealing (heated up to 900 °C) and cold plastic deformation (deformation degree 50%). During the thermal cyclic treatment, the samples were heated to 770 °C and immediately cooled in the open air.

The structural changes occurring in the material under treatment were recorded using microstructural analysis and passive fluxgate control. The latter uses the relationship between magnetic, structural and mechanical parameters of ferromagnetic materials if they concurrently depend on the same factors, be it microstructure, chemical composition, micro- or macro-stresses or processing modes. In case of low-alloy and low-carbon structural steels, these factors will definitely have an effect on their mechanical and magnetic properties, which makes passive fluxgate control an effective tool for use during thermal cyclic treatment [10].

Magnetic parameter Hp was monitored by ICNM 2-FP device with a two-channel converter. The microstructural studies made use of the following equipment: Brilliant 220 precision cutting machine (with laser marking and automatic feeding along three axes) and automatic Opal 460 press—for sampling the above materials; Saphir 560 facility—for sample preparation; metallographic microscopes Carl Zeiss Axio Observer and Leica DMI. Thermal cyclic treatment was carried out in programmable muffle furnaces with SNOL 8.2/1100 microprocessor control.

3   Results

The study has revealed the relationship between the magnetic parameter Hp for 08 ps and 09G2S steels and the number of thermal treatment cycles (Figs. 1 and 2). For

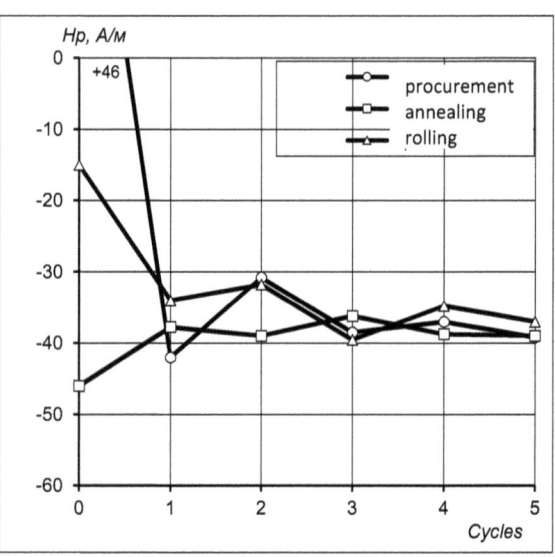

Fig. 1 Hp parameter versus number of cycles for 08 ps steel

Fig. 2 Hp parameter versus number of cycles for 09G2S steel

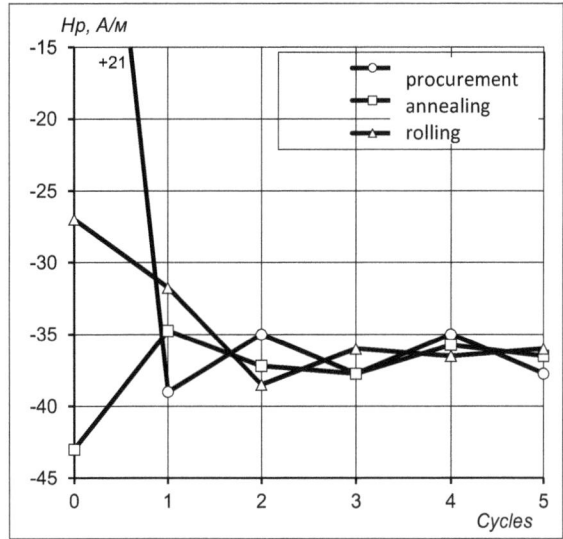

the materials with different initial states, the most significant changes in magnetic parameter Hp occur during the first two cycles, following which Hp achieves stability.

The study of the samples taken from the materials in factory-supplied state (Fig. 3) has shown that after the first cycle their Hp parameter not only experiences a significant change but reverses and remains reversed. After the fifth cycle of thermal cyclic treatment, the Hp amplitude remains less than 4 A/m, which indicates proportionality of the average grain size of the steels under analysis.

Fig. 3 Hp parameter versus number of cycles for materials in factory-delivered condition

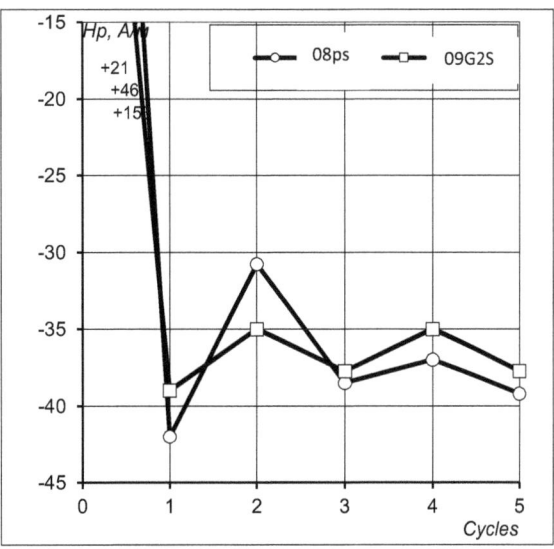

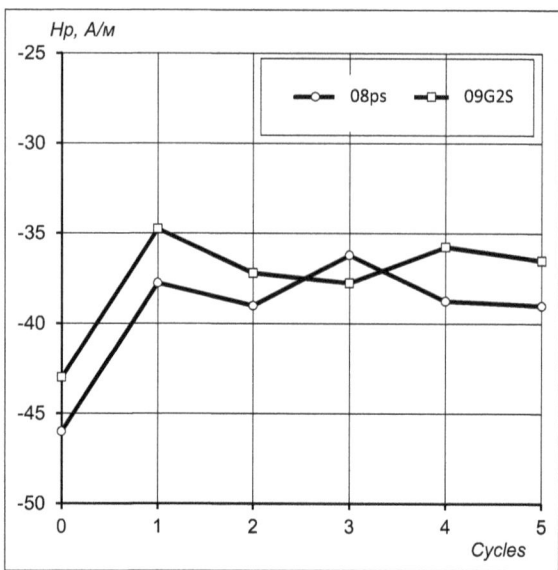

Fig. 4 Hp parameter versus number of processing cycles for materials in factory-delivered state and annealed at 900 °C

The coarse-grained structure (annealed at 900 °C) has shown less significant changes in Hp parameter starting from the first cycle, compared to finer-grained (factory-supply condition) structure (Fig. 4). This suggests that the occurrence of significant changes depends on a particular structural state of steels: the steels exposed to preliminary cold plastic deformation show higher magnetic parameter Hp after the second and third cycles (Fig. 5).

Thus, the repeated exposure of metal to heating and cooling cycles has shown a significant change in its magnetic parameter Hp towards a fine-grained structure with different grain sizes, induced by the phase transformations and structural changes, which means that the structural changes can be monitored and controlled in steels at different stages of production process.

4 Conclusions

The interrelation is shown between the magnetic parameter Hp and the initial microstructure in 08 ps and 09G2S steels under exposure to thermal cyclic treatment.

The first three processing cycles are found to cause the magnetic amplitude of Hp to decrease and the reduction ratio of the initial microstructure to increase regardless of the initial state. An increase in the number of cycles to more than three will lead to a significant decrease in the reduction ratio of the fine-grained structure of the materials under study.

The passive fluxgate control is found expedient for monitoring steels' fine-grain structure in manufacturing conditions.

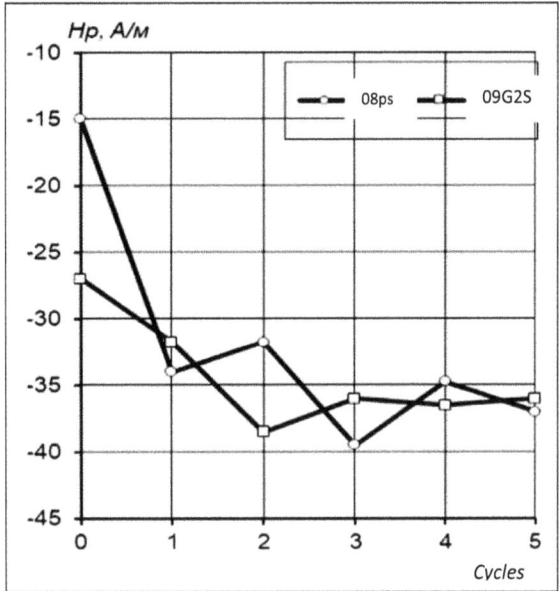

Fig. 5 Hp parameter versus number of cycles for materials in factory-delivered state and with subsequent 50% deformation

Summarizing the above results, we conclude that thermal cyclic treatment represents a highly expedient method for changing the steel structure towards improved mechanical properties and, consequently, improved bearing capacity of metal structures.

References

1. Gordienko VE, Abrosimova AA, Kuzmin OV et al (2018) The influence of thermal and thermal cyclic treatment on the mechanical properties of structural steels. Bull Civ Eng 1(66):128–133. https://doi.org/10.23968/1999-5571-2018-15-1-128-133
2. Gordienko VE, Abrosimova AA, Trunova EV et al (2017) Some features of verifying the reliability of mathematical models chosen for design calculation of welded metal structures with corrosion damage. Bull Civ Eng 1(60):210–213
3. Gordienko VE, Ivanov IA, Abrosimova AA et al (2018) On influence of the initial state of welded parts on the structure and properties of welding joints. Bull Civ Eng 4(69):150–155. https://doi.org/10.23968/1999-5571-2018-15-4-150-155
4. Gordienko V, Chernykh A, Repin S (2020) Influence of recrystallization annealing regime on the formation of a fine-grained structure in structural steels. In: E3S web of conferences. https://doi.org/10.1051/e3sconf/202016408021
5. Novikov AE, Motorin VA, Lamskova MI et al (2018) Composition and tribological properties of hardened cutting blades of tillage machines under abrasive deterioration. J Friction Wear 2(39):158–163. https://doi.org/10.3103/S1068366618020137
6. Prudnikov AN, Prudnikov VA (2015) Influence of thermal-cyclic deformation and hardening heat treatment on the structure and properties of steel 10. Appl Mech Mater 788:187–193

7. Prudnikov AN, Prudnikov VA (2016) Hardening low carbon steel 10 by using of thermalcyclic deformation and subseauent heat treatment. Mater Sci. Noneguilibrium Pahse Transform 4:10–13
8. Scherbakov A, Monastyreva D, Smirnov V (2019) Passive fluxgate control of structural transformations in structural steels during thermal cycling. Paper presented at the E3S web of conferences 135. https://doi.org/10.1051/e3sconf/201913503022
9. Zinchenko SA (2020) Thermocyclic treatment (TCT) as a method to decrease carbide segregation of hypereutectoid steels. Eur Phys J Spec Topics 229:459–465. https://doi.org/10.1140/epjst/e2019-900107-0
10. Zinchenko SA, Zolotarev NA, Vasilyeva NN (2020) Reducing carbide segregation of hypereutectoid steels by thermal cyclic treatment. Ferrous Metallurgy. Bull Sci Tech Econ Inf 76(5):477–481. https://doi.org/10.32339/0135-5910-2020-5-477-481

Hospital Campuses in the Urban Green Frame of Sevastopol

Elena Ovsyannikova and Nikolai Vassiliev

Abstract Built after the Great Patriotic War, the hospitals of Sevastopol were designed as architectural, landscaped expressions of the "Victory style". Their pavilion structure with low-rise buildings and gardens contributing to patients' rehabilitation is in harmony with the warm climate of the seaside city. This paper is the first to present the archival materials featuring the original design layouts of Sevastopol hospitals and the names of their developers.

1 Introduction

Unlike pre-revolutionary hospitals, housed in historic estates with large parks, the hospitals of the Soviet era and their landscape elements have received very little research attention. There were no publications about Sevastopol's healthcare facilities of the Soviet period. Nor was this restricted-access city with a naval base dedicated any detailed descriptions of its architecture. The existing books about Sevastopol were written by local historian E. V. Venikeyev, who was a witness to the post-war reconstruction of Sevastopol, but its healthcare institutions did not attract his much of his attention [8] Nor were the therapeutic gardens in this part of the Crimea covered in any research publications.

E. Ovsyannikova
Moscow Architectural Institute (State Academy), Moscow, Russia

Sevastopol State University, Sevastopol, Russia

N. Vassiliev (✉)
V.I. Surikov Moscow State Academic Art Institute, Moscow, Russia
e-mail: nikolai@vassiliev.net

Moscow State Research University of Civil Engineering, Moscow, Russia

2 Materials and Methods

As doctors were fighting against tuberculosis, landscaping became part of the general climatology concept. 1914 marked the opening of Romanov Institute of Physical Medicine, known since 1921 as I. M. Sechenov State Institute for Clinical Medicine. After the Great Patriotic War, this institute was moved to Yalta to become part of the Institute for Tuberculosis Climatotherapy as climate started to be seen as an important part of convalescence strategy.

Landscaping efforts started to be contributed by scientists, and trees and shrubs would be selected based on their competent rationales. Biologist N. A. Milchakova refers to V. A. Vodyanitsky, "a prominent hydrobiologist and public figure, elected, on December 21, 1947, to the Sevastopol City Council in the first post-war elections …" [2].

N. A. Milchakova also emphasizes that "as a biologist by education, he was appointed chair of the Council's green construction and landscaping committee, which he headed for almost 20 years." Importantly, this highly respected specialist "was promoting research-based landscaping and facilitated a significant increase in Sevastopol's green construction: the number of parks, boulevards and squares had increased almost 20-fold…" In 1953, V. A. Vodyanitsky summarized his land-scaping research in "Expanding Urban Greenspace through Research", published in Slava Sevastopolya (Glory of Sevastopol) newspaper [2]. Vodyanitsky emphasized the importance of seedling nurseries and planting with perrenial drought-resistant plants. In 1947, Sevastopol Committee of the Communist Party of the Soviet Union issued an important order towards "landscaping of each newly built housing project… and planting, over the course of 10 years, of 500,000 trees and about 2 million shrubs" [2]. The implementors of Sevastopol's landscaping strategy were its public organi-zations, school students and residents, there existed a society for green construction promotion, led by the Zelenkhoz Trust. The seedlings for Sevastopol were grown in the forestry blocks of Balaklava, Bakhchisaray, Kuibyshev, Sudak and Yalta, and Nikitsky Botanical Garden, while ornamental flowering plants would be sourced from Inkerman and the farm called "Ornamental Plants". In his papers, Vodyanitsky emphasized futility of "… the unjustified pursuit of the south-coast, hygrophilous vegetation as unsuitable for planting in dry air and soil conditions of Sevastopol …, and proposed to "focus mainly on the drought-resistant vegetation, the vegetation of dry, not humid, subtropics and of semi-deserts …" [2].

Following Vodyanitsky's recommendations, the areas around Sevastopol started to be planted with Pallas pine and those in the center with "large batches of trees and shrubs typical of the southern zone (bitter almonds, plane trees, deodar cedar, Atlantic cedar, Lebanon cedar, Lenkoran acacia, ball-shaped acacia, pyramidal acacia, maple, ash, yew, magnolia, cherry laurel, boxwood, birch bark, etc.), sourced from the Crimea (Yalta and Nizhnegorsk), Odessa, Melitopol, Mariupol, Dnepropetrovkaya, Sochi, Adler and Nalchik; roses and lawn grass seeds would be brought from Donetsk and vigorous grapes from Moldova. In 1954, 300 plane trees were received from Odessa forest nursery for planting on the boulevards, Nakhimov Avenue and Lenin

Street. By 1958, the number of trees planted in Sevastopol had doubled the figure for 1944–1950" [2]. Thus, the emphasis was placed on the wild growing species of the Crimea, as well as on the established plants introduced from the southern regions of the USSR [1]. These were the plants and species selected by dendrologists for Ushakova Balka park, developed by the Leningrad GosInzhProjekt [3, 4].

3   Results and Discussions

As can be seen from general urban planning schemes of the 1940s–1960s, the green zones with low-density hospital campuses were being integrated in the green framework of Sevastopol (Fig. 1).

City Hospital No. 1

N. I. Pirogov City Hospital No.1 is known for its representative architecture. Built in 1868 as a county hospital, it was revived in the 1920s and rebuilt after the Great

Fig. 1 The general plan of Sevastopol. Fragment. Red circles are the green areas with hospital campuses. G. I. Barkhin, 1946

Fig. 2 The hospital in Vosstavshikh Square. Bird's-eye view. I. A. Braude, V. P. Petropavlovsky, 1949

Patriotic War to receive a large park. The hospital received a new building of its polyclinic that faced the Vosstavshikh Square. Its exquisite Corinthian portico had turned the square into a landmark place. All other buildings of the hospital campus are covered with lush greenery and cannot be seen from outside. The architectural ensemble of Hospital No. 1 can be clearly seen in the previously unpublished layout below (Fig. 2).

As can be seen from this layout, landscaping formed an essential part of the restored and expanded hospital. The diversity of plants has survived until our days, creating a special microclimate on previously was bare hill, which began to be planted with the start of the construction. This hospital campus is home to some of the oldest trees in the city.

Vosstavshikh Square faces the central hill with post-war blocks of housing for senior officials—the architectural face of Sevastopol. Surrounding these upland blocks is the circle of central streets with Nakhimov Square at Grafskaya Pier, Lazarev Square with Chernomorets Central Design Bureau of the Black Sea Fleet, Ushakov Square with Sailors Club, which together form Sevatopol's architectural dominant. It was only logical that the old hospital on the neighboring hill entered the list of high-priority renovation projects. It became one of the city-forming architectural designs and urban infrastructure facility, while the territory behind Vosstavshikh Square was built with new residential buildings.

Tuberculosis Hospital

The second post-war hospital, Tuberculosis Hospital (currently Infectious Diseases Hospital No. 8), is located much further from the center of Sevastopol than Hospital No. 1—behind the historic cemetery adjacent to Pozharov Street. Designed as a greenscaped area, it cannot be seen from the distance. Topographically, this hospital lies deep in the ravine descending to Karantinnaya Bay. With fertile soils, ravines are perfect places for planting large tree species. Over one hundred and fifty years, this ravine and its cemetery became one of the greenest places of the city (urban development started here shortly before the 1940s). Like City Hospital No. 1, the Tuberculosis Hospital is of pavilion design but is much more modest architecturally, owing its originality to the use of natural stone. The name of its designer is now known—A. N. Sibiryakov, a Leningrad-based architect known for a number of residential projects (developed in cooperation with A. A. Junger and N. N. Lebedev).

Hospital on the Northern Side

The third hospital to be built in the post-war period is City Hospital No. 4, located on the Northern Side. Its architect is also known today. His surname is Rogovenko (given names unknown). It is noteworthy that Rogovenko was also the architect who designed the large-scale residential development adjacent to the hospital on the Northern Side. Architecturally, the hospital campus is a pavilion-type cluster of country estates. Amply landscaped, its trees and shrubs do not, however, seem to be key element of its appearance, as is the case with Hospitals No.1 and Tuberculosis Hospital. This is due to the then general concept of the Northern Side as a garden city.

Ushakova Balka

The last to be built during the Soviet period in Sevastopol is Dasha Sevastopolskaya Hospital (currently City Hospital No. 3). Originally intended, in the late 1950s, as a hospital to service S. Ordzhonikidze Marine Shipyard on Korabelnaya Side, this hospital was built in the 1970s to reflect the pragmatics of the Soviet architecture of its time. With simplified forms very different from the "hand-made" architecture of the 1950s, the hospital didn't receive many greenspaces due to its compact size. We attribute this also to the fact that it abuts Ushakova Balka (currently a nature reserve), overgrown with trees and shrubs. The original design of the 1950s did envisage elaborated landscaping, which, however, was not implemented by the team of Leningrad-based GosInzhProjekt to the full extent, probably because of the availability of the adjacent landscapes of the Ushakova Balka [3] (Fig. 3).

   Given that Sevastopol lies in an arid area with limited water supply sources, landscaping was of special importance in its planning Each neighborhood was to have a garden or flower beds. This can be seen from the previously unpublished general plan of Sevastopol, developed by V. M. Artyukhov and Yu. A. Trautman and approved in 1949. Shown in this plan drawing are also Hospital No. 1, Infectious Diseases Hospital, and the hospital on the Northern Side. These three hospitals appear also on the general plan of 1965. Dasha Sevastopolskaya Hospital was yet to be built. It these drawings, the hospitals are shown as green areas [5–7] (Fig. 4).

Fig. 3 The general plan of Sevastopol. Fragment. Green circles are hospital campuses. The area near Ushakova Balka is yet to be built up. V. M. Artyukhov and Yu. A. Trautman, 1949

4 Conclusion

To which extent were those large-scale landscaping plans for Sevastopol inclusive of the therapeutic effect? It is difficult to give the answer today. The fact is that there existed different opinions as to the health-improving effect of, for example, cypress species that were brought to the Crimea back in antiquity. Here is what dendrologist L. N. Zgurovskaya wrote in her 1984 "Book of the Crimean Trees": "Unfortunately, the opinion spread in the Crimea more than twenty-five years ago that cypresses contributed to multiplication of mosquitoes and tuberculosis bacillus. To think that of the Crimea, the only hope and "holy grail" of so many pulmonary patients… Unthinkable. Over a short time, 75,000 mature cypresses were cut on the Southern Coast. After they did some serious research, they realized what a big mistake they had done and planted them anew." [1]. This indicates a lack of scientific research, which was hard to re-start after the war. At the same time, there was a general understanding of landscaping as having a beneficial effect on the climate of any city. This led to some of the Soviet cities planting their green spaces with incompetently selected

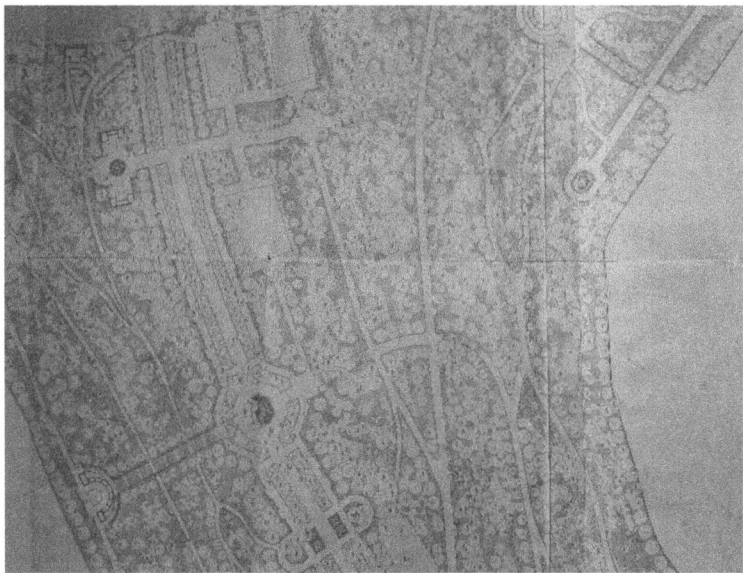

Fig. 4 The fragment of the layout of the park in Ushakova Balka with a grand staircase near the future Dasha Sevastopolskaya Hospital. GosInzhProjekt, Leningrad, 1951–1952. Published for the first time

species and, consequently, the loss of their architectural and spatial composition due to landmark buildings being obscured by trees (Fig. 5).

As known, the concept of climatotherapy has its key focus on natural factors and took shape in the nineteenth century first in England and then in France, Spain and other countries. The garden city concept had gained spread as an alternative to industrial cities, where air pollution was taking on horrific dimensions. Later, when scientists turned to plant-derived phytoncides, there occurred a real landscaping boom in the USSR in the 1920s and 1930s. As a result, many Russian regions, and especially health resorts in the Crimea, received more beautiful appearance with "man-made" forests, parks and gardens. The regulations that existed for the hospital construction in the 1950s provided for a certain percentage of hospital's territory to be occupied by greenscapes. This percentage was taken due account of by, for instance, V. M. Artyukhov, the technical assignments developer for the original hospital projects referred to in this paper.

Our study has enabled the following conclusions:

1. the pavilion design of hospitals, that took shape in the nineteenth century and was developed in the 20th, has converted the hospital campuses into parks;
2. landscaping of hospitals was incorporated into the building codes and regulations (SNIPs) thanks to the study of the health benefits of phytoncides;
3. the design concept for Sevastopol's post-war hospitals had its major focus on improved climate and more comfortable environment for patients and visitors of healthcare facilities;

Fig. 5 The general plan of Sevastopol. Fragment. Green circles are the hospital campuses. V. M. Artyukhov, 1965

4. the concept of therapeutic garden was yet emerging at that time but its fundamental principles had already been formed;
5. landscaping of hospitals was part of the 1949 and 1965 general urban development plans for Sevastopol.

In conclusion, the above principles have not lost their relevance also in the context of healthcare and rehabilitation challenged by the pandemic phenomena such as the COVID-19.

Acknowledgements The paper has been prepared with the grant support from Sevastopol State University—"Landscaping concept of Sevastopol healthcare facilities". Identification number: 29/06-31.

References

1. Zgurovskaya LN (1984) Tales of the Crimean trees. Tavria, Simferopol
2. Milchakova NA (2019) Vodyanitsky's contribution to restoration of the Sevastopol biological station and beautification of Sevastopol after the Great Patriotic War. Hist Biol Res 11(3):7–29

3. Ovsyannikova EB (2020) The 1952 Ushakova Balka Park layout in the historical and architectural context, comfortable environment—healthy environment. Creating therapeutic gardens in urban green frames. In: International scientific conference, 26–27 Dec 2020. Sevastopol, pp 63–70
4. Vasiliev N, Ovsyannikova E (2020) Role of Leningrad architects in last-recovery repair of center of Sevastopol and problem of preserving their heritage. In: E3S web of conferences, vol 164. EDP Sciences, p 05016
5. Vasiliev NY, Ovsyannikova EB (2019a) Architect V.M. Artyukhov and his role in shaping Sevastopol's post-war image. Archit Mod Inf Technol 2(47):28–42
6. Vassiliev N, Ovsyannikova E (2019b) Postwar Sevastopol architectural heritage: discoveries and preservation concerns. In: 2019 International conference on architecture: heritage, traditions and innovations (AHTI 2019). Atlantis Press, pp 330–334
7. Vasiliev NY, Ovsyannikova EB (2018) The post-war architecture of Sevastopol downtown. Archit Mod Inf Technol 4(45):135–144
8. Venikeev EV (1983) The architecture of Sevastopol. Tavria, Simferopol

Clustering the Territories of St. Petersburg by Strategic Priorities for the Construction of Healthcare Facilities

Tamara N. Orlovskaya

Abstract The modern problems of providing the population of the megalopolis with healthcare facilities are considered. The development of a methodology for assessing the provision and accessibility of healthcare facilities for the population ensures the balanced development of social infrastructure. The proposed clustering methodology is based on the analysis of providing the population with the most important health-care facilities—polyclinics, hospitals, ambulance stations. The research was carried out according to the actual data posted on the Digital Master Plan of St. Petersburg. Based on the results of the density analysis, there were identified four groups of the districts of St. Petersburg—the highest density, high density, moderately low and low. A combined analysis of the indicators made it possible to identify balanced and unbalanced areas of the city in terms of the degree of providing the population with healthcare facilities in order to establish strategic priorities for construction.

1 Introduction

At the present stage of development, the focus on improving the quality of life, creating decent conditions for preserving and maintaining the health of citizens is becoming a priority of government policy. Medical assistance to the population in megalopolises is largely provided by the availability of healthcare facilities, primarily policlinics, hospitals, and emergency medical stations. Such a system forms the framework of the quality of the urban environment in the field of health, providing a decent standard of living.

The study of gaps in the level of provision of the population with healthcare facilities in the territory of one megalopolis allows, according to Stadelbauer and Pacione [6, 9], smooth out conflict situations. Monitoring the need for social infras-tructure, including healthcare infrastructure, is an extremely important area of urban economics and sociology, the foundations of which were laid in the works of Bout and Zimmel [12, 15]. The feasibility of analytical work on the provision of health

T. N. Orlovskaya (✉)
Saint Petersburg State University of Architecture and Civil Engineering, St. Petersburg, Russia
e-mail: e-tamara@mail.ru

© The Author(s), under exclusive license to Springer Nature Switzerland AG 2023
D. Ivanov et al. (eds.), *Proceedings of ECSF 2021*, Lecture Notes in Civil
Engineering 257, https://doi.org/10.1007/978-3-030-99877-6_21

facilities in the city districts is an urgent task for the development of a strategy for socio-economic and spatial development. The results obtained allow us to compare the development of healthcare infrastructure in different megalopolises, identify "underdeveloped" territories, and identify the reasons for differences in the provision of reference needs of the population not only in different cities, but also within the territories of one megalopolis.

The study of the development strategy of St. Petersburg in terms of the uniformity of the provision of healthcare facilities to the citizens has both scientific and practical significance, allowing us to form an effective and balanced program for the construction of healthcare facilities.

2   Materials and Methods

The scientific basis of the research is the works of Russian and foreign scientists that research of the quality of life in global cities [1, 4, 14], the information and statistical data on St. Petersburg [2, 5], research materials about Russian and foreign cities [7, 9, 10, 13], UN-HABITAT and PwC research on the development of healthy lifestyles in the world cities [7, 8, 11].

Assessing the availability of public healthcare facilities in each particular district of St. Petersburg, the author used the data posted on the Digital Master Plan of St. Petersburg and systematized the data on the placement of healthcare facilities (polyclinics, hospitals, ambulance stations) by the districts of St. Petersburg.

Clustering the districts of St. Petersburg and identifying the level of provision of the population with healthcare facilities, taking into account statistical data on demography, the author calculated the indicators—the services of policlinics, hospitals and ambulance stations per 1000 residents. The analysis of the services of health facilities for the citizens was carried out for all the administrative districts of St. Petersburg.

The analysis of the local services of healthcare facilities in various territories of St. Petersburg was carried out on the basis of normalized indicators by reducing the size scale to a dimensionless one.

The work uses general scientific methods of cognition, methods of comparative and logical analysis, expert methods of evaluation.

3   Results

The provision of reference healthcare needs is one of the most important in the period of post-pandemic coronavirus infection. Moreover, this indicator, according to [11, 13], is one of the priorities in assessing the quality of life. Identifying the extent of the gap between the provision of services in different parts of the city the main methodological problem is to assess the degree of proximity of areas to identify priority areas for construction and placement of healthcare facilities. For the transformation and normalization, the use of the most significant absolute indicators of the population's provision with health facilities was proposed as input indicators for assessing the population's provision. These are the indicators of the population's provision with:

- capacity of policlinics, visits per shift (X_1);
- capacity of hospitals, beds (X_2);
- capacity of ambulance stations, ambulance vehicle (X_3).

To transform and normalize as input indicators of the density of healthcare infrastructure placement by district, it is proposed to use the relative indicators of the population's availability of policlinics, hospitals and ambulance stations per 1000 residents in various districts of St. Petersburg:

- provision with policlinics, visits per shift per 1000 residents (X_4);
- provision with hospitals, beds per 1000 residents (X_5);
- provision with an ambulance, ambulance vehicle per 1000 residents (X_6).

During the conversion, the dimensionless value of the input exponent $X\_i^{\wedge\prime}$ changes from 0 at $X_i = X_{min}$ to 1 at $X_i = X_{max}$. The indices for each territory are calculated on a cumulative basis.

The author calculated the index of the population's provision with healthcare facilities (IPF) for each district of St. Petersburg by summing normalized dimensionless values of the input indicator for each group of indicators. There was used the actual data posted on the Digital Master Plan of St. Petersburg. The result is shown in Fig. 1.

The assessment of the healthcare infrastructure development, taking into account the demographic component, was carried out by each district of St. Petersburg on the base of the relative values of the population's provision with healthcare facilities— the placement density index for the healthcare facilities (IPD). For this calculation, the author also summed normalized dimensionless values of the input indicator by each group of indicators. The result is shown in Fig. 2.

The results of the combined analysis of the two groups of indicators are shown in Table 1.

The study of the population's provision with healthcare facilities in megalopolises, taking into account the territorial and demographic aspects, involves the study of various groups of indices. There is a need of the group of relative indicators to carry out the comparative analysis of provision indicators in the context of demographic and territorial components. The group of absolute indicators reflects the objective component of the provision level, showing how much a particular area of the city is sufficiently saturated with healthcare facilities.

The combined analysis of the two groups of indicators allows us to build an adequate model of the territory development in terms of healthcare infrastructure development, to prioritize construction in time and space.

In general, the results of the study allow us to draw a conclusion about the direction and priorities of the strategy for construction of healthcare facilities by the districts of St. Petersburg in the future.

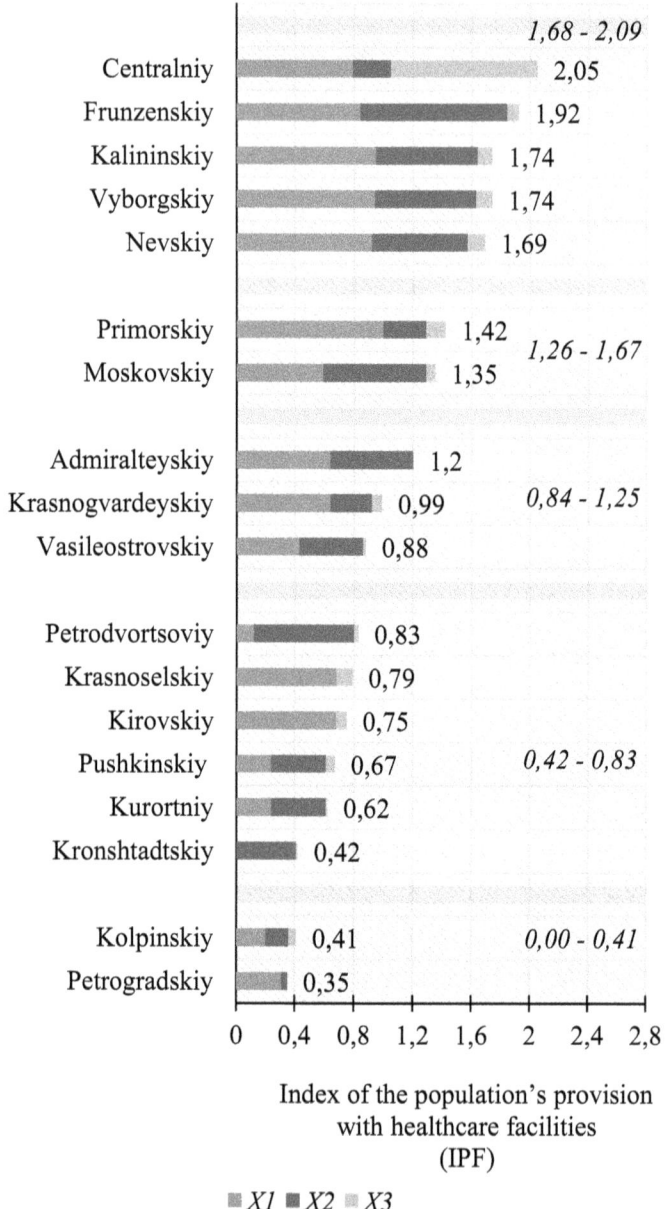

Fig. 1 Cumulative assessment of the population's provision with healthcare facilities by the administrative districts of St. Petersburg

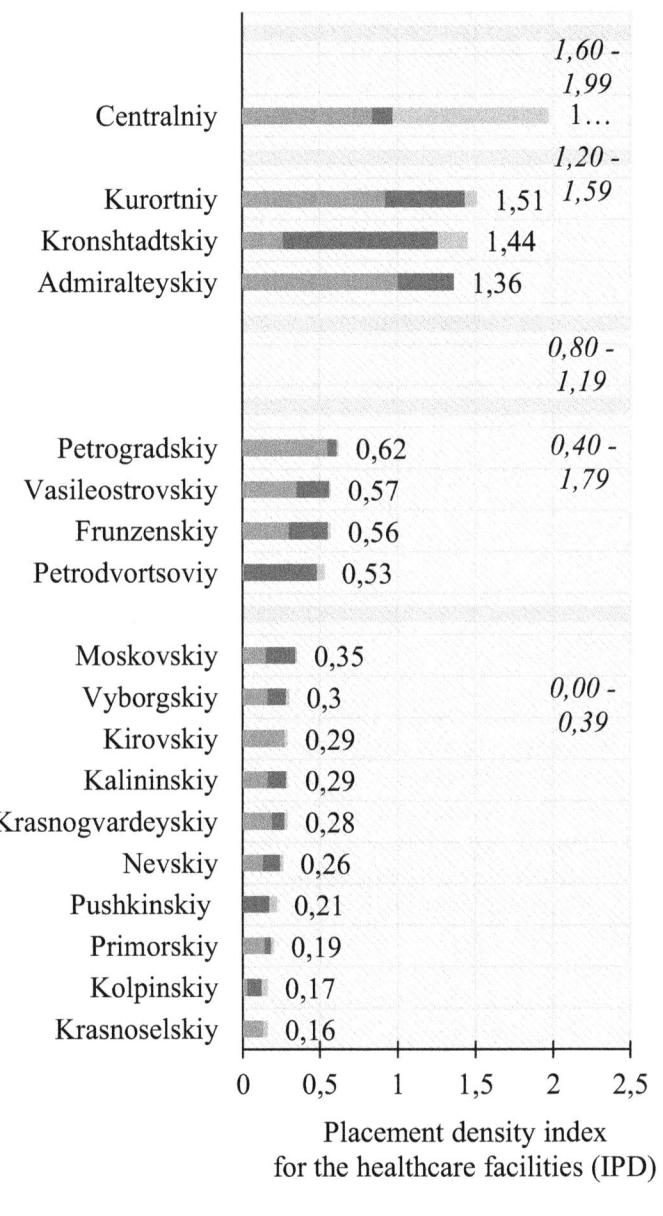

Fig. 2 Cumulative assessment of healthcare facilities density by the administrative districts of St. Petersburg

Table 1 Differentiation of metropolitan areas by the level of population's provision with healthcare facilities (for example, St. Petersburg)*

Placement density index for the healthcare facilities (IPD)	Index of the population's provision with healthcare facilities (IPF)				
	0.00/0.41	0.42/0.83	0.84/1.25	1.26/1.67	1.68/2.09
0.00/0.39	Kolpinskiy	Kirovskiy, Krasnoselskiy, Pushkinskiy	Krasnogvardeyskiy	Moskovskiy, Primorskiy	Vyborgskiy, Kalininskiy, Nevskiy
0.40/0.79	Petrogradskiy	Petrodvortcoviy	Vasileostrovskiy		Frunzenskiy
0.80/1.19					
1.20/1.59		Kronshtadtskiy, Kurortniy	Admiralteyskiy		
1.60/1.99					Centralniy

*Compiled by the author

4 Discussion

The harmonization of urban space, reducing the level of conflict and ensuring the safety of life is achieved, by the organization of effective placement of healthcare facilities. The formation of balanced plans for the construction of healthcare facilities become not only important, but priority areas of socio-economic and spatial development of the city, the assessment of its creativity. Today, there is a change in the role of healthcare in the development plans of urban management [3, 7]. Monitoring the location of healthcare facilities on the urban territory is one of the most relevant areas of effective policy for the formation of a comfortable and safe urban environment and the life quality.

The methodology and methods for assessing the location, accessibility and provision of healthcare facilities remain the objects of close attention of scientists [7, 10]. The methodology for assessing the quality of life and the quality of the urban environment today should be focused on new priorities related to the harmonization of public space, ensuring equality and accessibility of healthcare facilities for the citizens.

The main task of the study was to cluster the districts of St. Petersburg according to the level of accessibility and availability of healthcare to the citizens in order to develop a strategy for the construction of healthcare facilities.

5 Conclusion

According to the results of the analysis of the healthcare facilities development by the districts of St. Petersburg, five groups of territories were identified:

1. the most affluent districts are Vyborgskiy, Kalininskiy, Nevskiy, Frunzenskiy, and Centralniy (IPF = 1.68/2.09);
2. the districts with a large number of the healthcare facilities: Moskovskiy and Primorskiy (IPF = 1.26/1.67);
3. the districts with an average number of the healthcare facilities: Admiralteyskiy, Vasileostrovsky, and Krasnogvardeysky (IPF = 0.84/1.25);
4. the districts with moderately a little number of the healthcare facilities: Kirovskiy, Krasnoselskiy, Kronshtadtskiy, Kurortniy, Petrodvortsoviy, and Pushkinskiy (IPF = 0.42/0.83);
5. the districts with the less number of the healthcare facilities: Kolpinskiy and Petrogradskiy (IPF = 0.00/0.41).

According to the results of the analysis of healthcare facilities density, four groups of territories were identified:

1. there is the highest density in most districts of the city: Vyborgskiy, Kalininskiy, Kirovskiy, Kolpinskiy, Krasnogvardeyskiy, Krasnoselskiy, Moskovskiy, Nevskiy, Primorskiy, and Pushkinskiy (IPD = 0/0.39);
2. the districts with a high density in Vasileostrovsky, Petrogradskiy, Petrodvortsoviy, and Frunzenskiy (IPD = 0.40/0.79);
3. the districts with moderately low density: Kronshtadtskiy, Kurortniy, and Admiralteyskiy (IPD = 0.80/1.59);
4. the district with the lowest density: Centralniy (IPD = 1.60/1.99).

According to the results of the combined analysis of the two groups of indicators there is a balance of indicators in districts: Admiralteyskiy, Kronshtadtskiy, Kurortniy (average/average-low number and low density); Vasileostrovsky and Frunzenskiy (large/average number and relatively high density).

There is an unbalance in the districts: Kirovskiy, Kolpinskiy, Krasnoselskiy, Petrogradskiy, Petrodvortsoviy, Pushkinskiy (the less/moderately a little number and high/highest density); Moscovskiy, Primorskyy, Vyborgskiy, Kalininskiy, Krasnogvardeyskiy, Nevskiy, (large/average number and highest density); and Central (the largest number and the lowest density).

References

1. Andersson DE et al (2001) The gateway to the global economy. Fazis, Moscow
2. Ershova S, Orlovskaya T (2019) Differentiation of Russian megacities by level of investment in comprehensive residential development. In: IOP conference series: materials science and engineering, vol 687, no 5. IOP Publishing, p 055069. https://doi.org/10.1088/1757-899X/687/5/055069

3. International Telecommunication Union. (2020) Measuring digital development. Facts and figures. Geneva. https://www.itu.int/en/ITU-D/Statistics/Documents/facts/FactsFigures2020.pdf. Accessed 17 May 2021
4. Okrepilov VV, Gagulina NL (2019) Development of estimating quality of life of regional population. J Econ Econ 16(3):318–330. https://doi.org/10.31063/2073-6517/2019.16-3.1
5. Orlovskaya T, Ershova S (2019) The study of the sociospatial aspect of green building in St. Petersburg: theoretical and practical aspects. E3S Web Conf 135. https://doi.org/10.1051/e3s conf/201913503054
6. Pacione M (2001) Urban geography: a global perspective. Routledge, London, New York
7. PwC (2018a) Healthcare efficiency study. https://www.pwc.ru/ru/publications/health-research/issledovanie-effectivnosti-zdravoohraneniya-v-gorodah-mira.pdf. Accessed 15 May 2021
8. PwC (2018b) Megapolis of the future. Space for living. Summary of the study. Retrieved from https://www.pwc.ru/ru/publications/megapolis-future-rus.pdf. Accessed 17 May 2021
9. Stadelbauer J (2007) Mega-cities as a conflict of space global city: theory and reality. Avanglion, Moscow
10. Tulicheva LD (2007) The use of comparative studies to determine the strategic priorities of regional development. SUAE, St. Petersburg
11. United Nations human settlements programme (UN-HABITAT) (2009) Planning for sustainable cities: global report on human settlements 2009. Earthscan, London & Sterling, Virginia. http://unhabitat.ru/assets/files/publication/GRHS_2009.pdf. Accessed 17 May 2021
12. Volosnikova EA (2012) Evolution of sociological views on the city within a classical metaparadigm. Bull Tomsk State Univ Philos Sociol Polit Sci 1(17):107–113
13. Vysokovskiy A et al (2014) The struggle for the citizen: human potential and urban environment. https://mosurbanforum.ru/upload/iblock/791/7914595f72b603a31eafa8185ee07909.pdf. Accessed 15 May 2021
14. Yavon SV (2018) Research practices in urban space. Azimuth Sci Res Econ Adm 72(23):376–380
15. Zimmel G (2002) Big cities and spiritual life. Logos (3–4). http://magazines.russ.ru/logos/2002/3/zim.html. Accessed 15 May 2021

Therapeutic Garden Design: Aesthetics and Semiotics

M. D. Popkova and E. E. Krasilnikova

Abstract The paper discusses the basic semiotic codes of garden landscapes. The problem of aesthetic concept choice for hospital's therapeutic garden is solved in favor of the "living picture" based on pastoral idyll and paradise garden model. The principles of landscape evolutionist aesthetics are critically analyzed. The semiotic analysis of therapeutic landscapes has shown the abandonment of radical forms of Naturgarden style and deconstructivism, as dehumanized geographic models inconsistent with the local cultural landscape and the tasks of therapeutic effect.

Keywords Semiotics of gardens · Aesthetic harmonization of environment · Healing garden · "Living picture" · Anthropocentric code · Architectural and landscape environment of hospital complexes

1 Introduction

In conditions of Covid-19, the design of therapeutic landscapes is gaining relevance as an element of urban environment. Of particular importance in this context is landscaping of healthcare facilities. We all have experienced isolation first-hand. For in-patients, isolation is one of the many trials associated with the disease. Aggravated by stress and limitations associated with hospital treatment, the disharmony in mental and physical state can be alleviated through aesthetically harmonization of hospital premises with the world of nature. From this perspective, the issues of therapeutic garden design acquire special importance.

With many Sevastopol's hospitals, the pavilion-type layout has been a tradition enabling to convert their territory into therapeutic landscapes. In domestic and international healthcare construction and landscaping, the design principles of centralized and mixed layouts gained strong foothold in the second half of the twentieth century (see, for example, Gradov [12]). The hospitals of Sevastopol are of pavilion design and therefore represent a unique case whose territorial resources should be

M. D. Popkova · E. E. Krasilnikova (✉)
Department of Architecture and Design, Institute of Urban Development, Sevastopol State University, Sevastopol, Russia
e-mail: landurbanizm@gmail.com

© The Author(s), under exclusive license to Springer Nature Switzerland AG 2023
D. Ivanov et al. (eds.), *Proceedings of ECSF 2021*, Lecture Notes in Civil Engineering 257, https://doi.org/10.1007/978-3-030-99877-6_22

189

approached in a creative way while taking into account the benefits of the unique southern climate.

The theorists and practitioners of the Sevastopol architectural and landscape school has as one of their fundamental tasks the development of the design code for the local healthcare facilities and their therapeutic gardens.

This task requires exhaustive analysis of the aesthetic principles and semiotic codes relevant in the design of therapeutic gardens. Based on its results, it will be possible to identify stylistic solutions that would fit the local architectural and landscape features best.

The connection of value and meaning, described by Lotman [18] and linking semiotics with axiology, became the starting point for the hypothesis of this study. The symbolic nature of our perception is the reason why we are looking for the symbolic meanings to code our design systems and perception of the spatial environment.

2 Materials and Methods

Methodologically, the study relies on comprehensive analysis of international and domestic landscape design practice, paradigmatic models of garden design and their relevance for therapeutic gardens. Our decoding of the semantics of the visual representation of the architectural and landscape environment uses the results of historical, cultural, semiotic, axiological and aesthetic interpretations.

While developing the conceptual model of therapeutic garden design, we relied mostly on visual-compositional analysis as the visual perception dominates all other forms of perceptual interaction with the therapeutic landscapes (due to limitations imposed by hospital regime and condition of patients).

The studies into the actual effect of hospitals' therapeutic landscapes show that they are used mostly for viewing. Shan Jiang, Kirsten Staloch, and Sofija Kaljevic argue that "however, most gardens are primarily used for viewing purposes, and complicated factors contribute to a lower physical use of the spaces" [15]. There are, indeed, a lot of factors (of individual somatic, sanitary and epidemiological nature) that hinder the physical use of therapeutic gardens by patients. Accordingly, when designing therapeutic gardens, of all the sensory aspects of their perception, the visual aspect remains the leading one. Therefore, it is the principle of "living picture", proposed by landscape park theorist H. Repton, that turns out to be the fundamental one in the design of hospitals' therapeutic gardens.

So, what are the principles to be observed when creating these "living pictures"?

3 Results

The study of the typology of garden landscapes reveals that man, as animal symbolicum [6], encodes and decodes visual series in accordance with symbolic series. Eco points out the "the architectural and landscape environment is represented by a set of signs and symbols denoting the utilitarian functions of the structure and connoting its symbolic meaning" [10]. Therefore, the design of a hospital's therapeutic garden should be preceded by decoding of semantics of the basic semiotic models of a garden design.

In modern culture, the **positive axiological accentuation of biophilic semantics**, when *Nature* is seen as an ultimate value, is associated with such axiological constants as *Life* and its endless cycle of rebirths knows as *Eternity*. While *Life-Eternity* concept forms the value vertical, *Harmony and Joy* form the horizontal relating to the emotional experience of *Nature*.

Perceptions harmony. Ideal landscape

At the same time, the studies into the history and theory of garden design show that the descriptions of ideal harmony of landscape and its physical expressions are conditioned by the cultural and historical situation and human self-consciousness. Based on the ideas of harmony and ideal landscape, alternative garden design models have appeared, shaping the formal garden and informal landscape design styles.

While the European and Middle Eastern cultures originally adopted as a model the formal garden, the Far Eastern ones (China, Japan) practiced the informal landscape design. It wasn't until the eighteenth century that the European culture adopted landscape design, but one it did, not did it become the aesthetic norm in garden and park designing, it ousted the formal garden as a devalued model.

Why all the classical gardens of Egypt, Babylon, Persia, Rome, as well as the monastic gardens of Europe, the Moorish gardens of Spain, the Renaissance Italian gardens, the Baroque and classical parks of Europe were built according to formal garden design?

The answer obviously lies in the general semantics of the garden as an ideal landscape—the paradise garden embodying the image of Eternity and sacred world as opposed to the mundane, secular reality. That also explains why exotic plants would be so popular in formal gardens, marking the space as unworldly and supermundane.

In Europe, the **paradise garden model**, which was characteristic of Persian gardens, was further translated into Late Antiquity and the Middle Ages. The presence in the garden layout of a regular pattern embodied Cosmos and defeat of Chaos. The mundane world of regular forms started to acquire many positive connotations. This model remained dominant until modern age.

However, the formal paradise garden is not the only model of ideal landscape.

Landscape design

In the Far Eastern cultural region, the image of the Absolute, embodied, according to the Taoist tradition, in the natural landscape, is to be preserved, not transformed. This attitude is reflected in Sharawadgi—the style of landscape gardening or architecture that originates from China and avoids rigid lines and symmetry to give the scenery an organic, naturalistic look. "Nature is their model," said Chambers of the Chinese gardeners, and their goal is to emulate it with all its beautiful irregularities" [17]. The irregularity of natural forms, however, was skillfully modeled in the gardens

for the "wild" landscape is not always expressive enough and may not be composi-
tionally complete. The selection and simulation of expressive landscape compositions—often with symbolic connotations (frightening, laughing, idyllic landscapes
in the Chinese gardens)—is in favor of ideal models, not nature as it is.

The idealization of wildlife marks also the European culture, rooted in paganism
and "sacred groves". "Primordial nature is sinless," says Gregory of Nyssa in his
"Tale of the Primordial Beauty of the World" which gained popularity in Russia
during the spread of hermitry [16]. However, like in the European cultural milieu,
the decorative gardens of Russia followed formal design and carried the semantic
connotations associated with their symbolic load (the cruciform plan of cloisters, the
compositions having tree of life or fountain as their central element; etc.).

Everything changed dramatically with the onset of urbanism.

In general, the historical victory of landscape style over formal garden is connected
with rapid urbanization. The first to experience its negative consequences was the
industry-led England. It is indicant that the European the model of idyllic pastoral
landscape, like "paradise lost" paradigm, takes its origin in the bucolics by ancient
Roman poet Virgil: it was in Ancient Rome—the first city with a population of one
million—that the fashion for pastoral rural idyll appeared.

It is known that the spread of landscape style in the eighteenth-century England
was supported by the ideology of liberalism as opposed to French absolutism. These
nationality-specific features, however, have important semantic connotations. It is,
indeed, the sensation of freedom that we draw from contemplating the nature. And the
desire to abandon formal garden with its regular geometric layout as inconsistent with
experience of freedom becomes all the more evident as the power of technology, with
its rational, regular forms and rigid system of determination, comes upon us—and
this power is only growing, both in the industrial and post-industrial era.

The British school had adjusted the Far Eastern model to its own idea of the ideal,
supported by bucolic literature. Arrangement remained quite speculative. Artist H.
Repton, the theorist and practitioner of the English landscape park and the author of
the "landscape theory", believed that landscapes should have distinctive foreground,
middle ground and background. The foreground represents a trimmable frame with
decorative elements (including plants with formal geometry), the middle ground is
a park (modeled according to L. Brown's idyllic landscapes), and the background is
"wilderness" [16]. Notably, the "wilderness" remains largely formal, since the ideal
continues to be dominant in the arrangement of landscape parks.

Wild landscape and Naturgarden

A response to the increase in urban civilization takes the form of fashion for wild
landscape. Disturbed by uncontrollably progressing urbanization and technology
advance, the technogenic landscapes seem to be giving way to the positively perceived
"wild field".

A new wave in landscape design, Naturgarden is radical in its naturalness and
has become the face of our era. Its prototype traces back to the famous garden of
Pete Udolf, that had thistle as its central elements and won the first place in Chelsea
in 2000. Naturgarden uses meadow lawns instead of parterre lawns, and its flower

compositions built on cereals and perennials. Rustic style has gained wide use in the design of small architectural forms, with impromptu means and robotariums currently being a design trend.

Environmental aesthetics versus anthropocentrism

As modern man gets more and more dependent on, and neuroticized by, the technology, in an effort of repulsion he obviously wants to escape into the wild and, as suggested by anthropological theories, see himself as a natural biological species, leaving out both civilization and his own symbolic nature, reducing himself to the animal level. Roger Ulrich, a leading researcher of therapeutic gardens, writes: "humans have an unlearned predisposition to pay attention and respond positively to natural content (e.g. vegetation, water) and to configurations characteristic of settings that were favorable to survival or ongoing well-being during evolution" [25]. In his theory of habitat, D. Appleton argues that an animal's habitat selection is determined by "to see, but not be seen" principle, which largely explains why our aesthetic choices are driven solely by biological needs [2].

The currently trending aesthetics of environmentalism [4, 5] goes so far in its interpretation of the value of nature that it denies anthropocentrism, devaluing man and his aesthetic endeavors as opposed to nature [1, 23].

Biological interpretations of aesthetic choices reduce aesthetics to physiology, whereas landscape gardening has as its original purpose satisfying the aesthetic, not pragmatic, needs, it functions being primarily symbolic. As applied to therapeutic landscapes, the theory of "prospect-refuge" [2] explains the "inwards to outwards" principle in the design of Persian paradise palaces and parks, and the idyllic "living pictures" of the cosmic whole.

Similarly controversial is the interpretation of the current landscape design trend originating from "**savanna hypothesis**" [20] as a basic landscape model. Since real savannas are lifeless dried-up plains, they cannot serve as images of the ideal landscape. The archetype of the savanna, which has a common place in the semantic motivation for the choice of landscape, isn't indisputable. It is more a question of oasis as a form of the ideal landscape than of savanna as it is, with barren sun-scorched land [11].

Contextually determined style choice

Style classifies as a systemic quality and is associated, at functional level, with reflection of the value system of activity [14]. Any axiological interpretation of style is contextually determined.

The analysis of landscape design practice shows that the ideal landscape models build on the natural context of environment. In the European paradigm, garden landscape is inversely proportional to natural one: poor soils with scanty vegetation (Persia or Spain) explained the abundance of flowers and fruits in landscaping.

Similarly, in an urbanized environment with its rigid, over-rationalized functional architecture and abundance of hard surfaces, the "wild" landscape (rough and raw as it may seem) is often perceived as an authentic embodiment of natural harmony. It is this "wild landscape" that many modern urban parks and garden landscapes have as their key motif.

However, in the domestic practice, this global trend is not entirely appropriate. In Russia with its vast territory, where there are more undisturbed landscapes than in Europe, where anthropogenic rural landscapes are sporadic and occasional, represented by mostly garden associations or temporary tourist sites devoid of any stylistic cohesion, the rustic style (especially in its rawest form) often seems relevant.

It seems illogical to introduce rutariums (landscape composition of driftwood) into the garden space that neighbors overgrown and unkempt trees and bushes. The impression will be that of chaos and ugly neglect, not of natural beauty and harmony, much to the frustration of the viewer.

We believe that in a milieu with dominant influence of technogenic advance, the fashion for "unconventional forms" should not be used uncritically for landscapes of therapeutic purpose.

Blooming garden. Color code

As we noted earlier, the problem of the modern science is that it tends to reduce human response to purely physiological needs. This explains why some of the conclusions made by Martin Hůla and Yaroslav Flegr in their fascinating study "Habitat selection and human aesthetic responses to flowers" [13] may sound perplexing to some. The authors put to test the hypothesis by Hirwagen and Orians (1993) that people have an aesthetical response to flowers because they signal presence of food, but the choice between the flowers and the fruits is always in favor of the latter. The logic behind this theory suggests is that fruits signal food availability much more directly than flowers. Therefore, fruits are believed to evoke a stronger aesthetic response than flowers. However, the experiments conducted by Hůla and Flegr have provide this theory wrong.

Thus, the aesthetic response to flowers cannot be reduced to a utilitarian level. "The question remains open as to why flowers are preferred at the perceptual level," the authors write [13]. The answer become clear if we move the theory from physiological to symbolic level: flowering is perceived as a symbol of the beginning of new life (oriental cherry blossom in China and Japan as a harbinger of spring, spring as an image of Resurrection in the Christian tradition [16]. The paradigm of "eternal spring" mark eternal life in nearly all cultures.

The problem of the **color code** of therapeutic garden is also connected with the image of "eternal spring". The garden landscape is naturally based on green and its varieties. Green is the color of life, a signal to our psyche that the fertile forces of the earth are in their prime. As a basic design model for therapeutic gardens, the **paradise garden model** has its framework consisting of the traditional Crimean evergreens unpretentious in these climatic conditions—*Cupressus, Juniperus communis, Pinus pallasiana, Ilex aquifolium, Prunus laurocerasus, Buxus sempervirens, Taxus baccata*, etc. The studies conducted on the territory of Dasha Sevastopolskaya Hospital have shown that coniferous species account for 21% of the total number of plantings, and it is planned to increase their number two-fold (given the beneficial effect of coniferous trees' phytoncidal properties) and supplement the garden's green frame with hedges of laurel (*Laurus nobilis*).

At the same time, the paradigm of the Garden of Eden involves the widespread use of flowering plants. The study by Martin Hůla and Yaroslav Flegr provides valuable statistical materials on psychological response to flowering plants, describing yellow as the least preferred color [13]. However, field studies have shown that the nature of color perception is influenced by seasonality: in the climatic of Sevastopol, yellow may not be a good coloristic solution in the summer season, whereas in winter and autumn the compositions of yellow violas, yellow bulbous plants and chrysanthemums with yellow tones are perceived as enlivening and embodying sunny summer. In addition, yellow shades of roses, perceived as exotic, are welcomed by the majority of respondents throughout the year as perfectly fitting the paradigm of paradise garden.

In general, the flower compositions using roses, as hypoallergenic, everblossoming plants, are the main decorative solution with many hospital therapeutic gardens. Rarer varieties of complex shades can be used as compositional dominants. When designing flower arrangements, it should be borne in mind that in the south, when the sun is bright, human eye distinguishes fewer shades, so additional colors should be added to the flower arrangements for mutual reinforcement. The process of selecting rose varieties for therapeutic rosaries is led by L. I. Uleyskaya, leading researcher at Sevastopol State University's Department of Architecture and Design and lead expert at Nikitsky Botanical Garden. Aromas and color solutions are combined in series of experiments to achieve maximum therapeutic effect.

Notably, the coloristic diversity of plants can be used to model spaces according to the laws of light and air perspective and has proved a highly effective tool in small hospital garden landscaping. The colder shades of green in the foliage, the flower beds of blue hues, the soft silhouette of crowns contribute to the effect of caerulescence [3]. These coloristic principles are taken into account when developing the "living pictures" of hospital gardens.

The visual perception of the landscape is, however, not without limitations, as there is a risk in modern visual culture of gradual atrophy of all human perceptual capabilities, except vision [19], and of related neuroticism. This risk leads therapeutic landscapes specialists to design sensory gardens, which are, however, rare in the practice of hospital garden landscaping. Nevertheless, the creation in hospital gardens of "green offices" as zones of climatotherapy and recreation represents a promising task, especially for pavilion- and mixed-type hospital campuses, that will allow to bring the aesthetic perception from solely visual to multisensory level and thus enhance human involvement in the "living picture".

The fact that the projected scenery will have, along with visual impact, smell (with its own therapeutic effect) has generally been taken into account by designers of the therapeutic landscapes. The analysis and compilation of aromatic compositions is one of the tasks being effectively solved in the landscaping of the health facilities in Sevastopol.

4 Discussion

Postclassical cultures, inspired by I. Prigozhine's "order out of chaos" concept [22] and postmodern philosophers [7, 8], often chooses as their underlying basis chaos [24], contrasting it with cosmographic and anthropocentric classics. A geographic paradigm marks "wild" landscapes and deconstructivist architectural techniques (see, for example, the public space concept for Fifth XiangYa Hospital [21]) in hospital landscapes. The study of the basic semiotic principles of shaping enables a conclusion that the paradigm of chaos with its decomposition, imagelessness, amorphousness and irrationality due to dehumanization, is unacceptable for therapeutic purposes. The chaos paradigm questions academic assumptions, leading to the need for a more critical approach to the postclassical trends (deconstructivism and parametrism) among hospital garden architects and designers. "No metaphorical paradise here, but discomfort and the unbalancing of expectations," writes Bernard Tschumi (quoted by Dobritsyn [9]). This paradigm is simply inconsistent with the therapeutic function of hospital architecture and landscape and should be reviewed critically from the standpoint of anthropocentrism as contradicting the stylistic context of the Sevastopol classicism—the historically dominant style in Sevastopol's urban environment.

5 Conclusion

The analysis of functioning and axiological determination of hospitals' therapeutic landscapes has identified anthropocentric design code and the "living picture" model as most suitable.

Along with positive distraction being created by viewing of a therapeutic garden, the harmony of the aesthetically perfect system can enhance spiritual, mental and physical potential. The model of a pastoral idyll with elements of paradise garden is chosen as a conceptual basis for therapeutic garden designing based on the analysis of psychological responses in patients. The semiotic analysis of therapeutic landscapes has revealed irrelevance of naturgarden style and deconstructivism in the local cultural landscape.

References

1. Environmental Aesthetics (2019) Stanford encyclopedia of philosophy. In: Edward N. Zalta (ed). https://plato.stanford.edu/entries/environmental-aesthetics/#EnviAestEnvi
2. Appleton J (1975) Behaviour and environment: prospect-refuge theory. In: The experience of landscape
3. Bogovaya IO, Fursova LM (1988) Landscape art. Agropromizdat, Moscow
4. Carlson A (2010a) Contemporary environmental aesthetics and the requirements of environmentalism. Environ Values 19(3):289–314

5. Carlson A (2010b) Nature and landscape. An introduction to environmental aesthetics. Columbia University Press, Nueva York. Enrahonar, 45, 177. https://doi.org/10.5565/rev/enr ahonar.231
6. Cassirer E (1998) An essay on man. Gardarika, Moscow
7. Deleuze G, Guattari F (1983) Anti-oedipus: capitalism and schizophrenia (trans. Robert Hurley, Mike Seem, and Helen R. Lane. London: Athlone, 1984)
8. Derrida J (2000) On grammatology. Ad Marginem, Moscow
9. Dobritsyna IA (2004) From postmodernism to nonlinear architecture: architecture in the context of modern philosophy and science. Progress-Tradition, Moscow
10. Eco U (2006) Otsutstvuyushchaya struktura: Vvedenie v semiologiyu [Missing structure: Introduction to semiology]. Saint-Petersburg
11. Falk JH, Balling JD (2010) Evolutionary influence on human landscape preference. Environ Behav 42(4):479–493
12. Gradov GA (1953) On integrated design of hospitals and polyclinics. The Architecture of the USSR, 3. https://tatlin.ru/articles/bolnichnoe_delo. Accessed 10 May 2021
13. Hůla M, Flegr J (2021) Habitat selection and human aesthetic responses to flowers. Evol Hum Sci 3(E5). https://doi.org/10.1017/ehs.2020.66
14. Ikonnikov AV, Kagan M, Pilipenko V (1990) Aesthetic values of the subject-spatial environment. Stroyizdat, Moscow
15. Jiang S, Staloch K, Kaljevic S (2018) Opportunities and barriers to using hospital gardens: comparative post occupancy evaluations of healthcare landscape environments. J Ther Hortic 28(2):23–56
16. Likhachev DS (1998) The poetry of gardens: on semantics of garden and park styles. Garden as a text. 3rd rev. Consent: JSC "Type "News", Moscow
17. Liu Y (2008) Seeds of a different Eden: Chinese Gardening ideas and a New English aesthetic ideal. Univ of South Carolina Press
18. Lotman YM (2001) Semiosphere. SPb: Iskusstvo-SPb
19. Mitchell WJ (2002) Showing seeing: a critique of visual culture. J Vis Cult 1(2):165–181
20. Orians GH, Heerwagen JH (1993) Humans, habitats and aesthetics. The Biophilia Hypothesis, Washington, DC
21. Payette (2015) Fifth XiangYa Hospital. New construction. 2015 AIA—National Awards. https://www.architectmagazine.com/project-gallery/fifth-xiangya-hospital_o. Accessed 11 May 2021
22. Prigogine I, Stengers I (2018) Order out of chaos: Man's new dialogue with nature. Verso Books
23. Shapshay S, Tenen L, Carlson A (2018) Environmental aesthetics, ethics, and ecoaesthetics. J Aesthet Art Critic 76(4):399–410
24. Tschumi B (1996) Architecture and disjunction. MIT Press
25. Ulrich RS (1984) View through a window may influence recovery from surgery. Science 224(4647):420–421

Training System for Medical Facility Designers

Semen Prokopchuk and Dmitrii Zhivotov

Abstract There has been a significant advance in the development of medical technologies over the past few decades, posing healthcare engineering designers to more challenging tasks. These tasks are becoming more complex from the perspective of projects themselves and their functions, and in terms of internal and external infrastructure.

Keywords Medical facilities · Designers · Medical technologies · Training

1 Introduction

As medical facilities are becoming more and more technology-intensive, to the foreground comes the need for specialists capable of designing, engineering and implementing healthcare construction projects.

Today, the medical facility design is led by healthcare engineers as professionals determining a facility's functional scope. They are the key decision-makers for the project's technological solutions. Design specialists without background in healthcare engineering may often have only a vague understanding of the specifics of the medical facilities. Consequently, the resultant project documentation can be of poor quality and incapable of achieving the main medical tasks set directly by doctors.

This problem is largely due to civil engineering designers being generalists, not specialists. In order to improve the quality of healthcare project documentation and its consistency with regulatory standards and medical goals, it is necessary to create a special training system for medical facility designers, TS2MF.

S. Prokopchuk
Director General at M-Investproekt LLC, Saint Petersburg, Russia

D. Zhivotov (✉)
Saint Petersburg State University of Architecture and Civil Engineering, Saint Petersburg, Russia
e-mail: d.zhivotov@mail.com

D. Ivanov et al. (eds.), *Proceedings of ECSF 2021*, Lecture Notes in Civil Engineering 257, https://doi.org/10.1007/978-3-030-99877-6_23

2 Methods

The project documentation quality became an issue during the collapse of the Soviet Union. What followed was disintegration of design institutions in healthcare services sector (and not only healthcare), regulatory frameworks for documentation preparation, training systems.

If we turn to practices used in the Soviet Union, we will see that medical facilities, for all their complexity, would be designed by specialists with no background in healthcare construction, but the quality of their project documentation was high-level. The institution responsible for the design of medical facilities at that time was GrazhdanProekt Research Institute and its regional offices.

Consequently, it was something else, not narrow specialists, that determined the success of the Soviet approach to healthcare construction.

Our analysis of the regulatory documents of that time has revealed a multi-level documentation system. Its simplified version looks as follows (Fig. 1).

Each of the documents in the above categories would be developed by relevant research institute using research- and statistical data-based approach.

The second component of success was the understanding that designer is a generalist and therefore needs to be provided with the set of documents with instructions to guarantee the quality of the design documentation. This means that design engineers were developing their projects strictly according to the standards set out in the

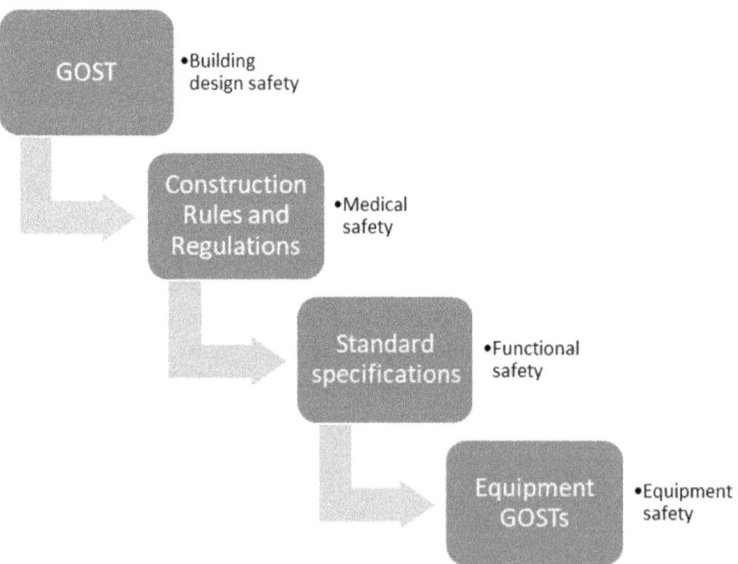

Fig. 1 Multi-level documentation system

regulatory documents. The echoes of this approach are still heard at project documentation review stage, when experts assume a principled stance on compliance with regulatory frameworks currently in force [11].

Another important component was the centralized nature of healthcare design development. The state acted—to use a word of the day—as a single customer with uniform standards for project documentation, training and regulatory documentation. The state had a training system that trained doctors for work in this given medical facility, design engineers for designing this given medical facility, and builders for building this given medical facility, and for each group of these specialists the occupational rules existed in the form of a regulatory framework. Whenever there emerged a novel medical technique, which proved effective and entailed general changes to the facility, the state, as a single customer, would update the regulatory documentation and provide relevant training to dedicated groups of specialists so as to ensure that its healthcare system can maintain high level of performance without going too deep into details [2].

On the one hand, such system assured high performance of its medical facilities, but on the other it had one significant drawback—slow pace of introduction of know-hows. In this passive, inert system, scientific development was slow, requiring lengthy investigation and collection of evidence, following which relevant regulatory document had to be executed, which was a time-consuming process, too. Only then could a know-how be put into practical use by a facility. But at the same time, once tested and proven, solutions were fast to scale. The effectiveness of such a system can be judged today through the prism of the pandemic of 2020. Countries such as China and Russia used a centralized approach and were quick to repurpose their hospitals and retain doctors. As practice has shown, inert systems can prove highly efficient in crisis situations, as evidenced by new coronavirus infection statistics for these countries [3].

To create such a system today would be technically impossible, and there is little need to do it. Even if this passive, inert system still existed, it would not allow healthcare facilities to operate at required level.

3 Results

The purpose of TS2MF is to change the approach, not reformat the system.

Its primary goal is to set a frame of reference for all specialists to follow and to give them understanding of why the system should operate the way it does [17].

In the classical approach, the patient load is distributed mainly at macro-level (acute patients, elective patients, outpatients), which leads to different interpretation of patient flows by different specialists and, consequently, errors. For further differentiation of the patient flows (elective surgical patient, elective therapy patient, etc.), TS2MF operates the concept of "flow phase": there are phases in the patient flows that require certain conditions to be met as to premises, equipment, and interaction with other flows. Therefore, due procedures and principles should be in place for patient flows management.

Further, TS2MF introduces the notion of "condition of premises", which is subject to specific phases of the flows and their interaction.

When two or more flows interact and their phases change, a transition occurs. In some of the departments, this transition must take place in premises with required clean room class and safety levels.

A medical premise's safety level is determined by the safety class of power supply, air purity class, ISO, water supply safety class, among other standards.

Let us take the example of an operating room. There is a statutory level of safety to be maintained in operating rooms as they experience a large number of transitions and flow phase changes: sterile equipment becomes dirty, textile products become wastes, patient status changes to restricted mobility, etc. These and other factors determine the requirements to operating rooms and become redundant once the operating room is closed for maintenance.

4 Discussion

This approach allows to build facility's technological matrix and a mathematical model of its process flows. This model will have numerical characteristics describing the effectiveness of the solutions being laid at the design stage. These numerical characteristics are a tool for comparing design solutions and linking them with external statistical data. Using mathematical modeling technique, we can identify the optimal solutions for specific situations and performance targets.

It is important to note that one and the same process can be approached differently by different specialists. This is especially noticeable in doctor-engineer interaction, when a design engineer teams up with a doctor to discuss nuances. Since the doctor and the engineer may have completely different visions, it is important that there is a frame of reference to allow them to speak common language and make communication more productive [19].

5 Conclusion

Given this perspective, TS2MF is designed to:

- set a uniform frame of reference for medical facilities design process-by introducing basic terminology, concepts, principles and approaches to be shared by all stakeholders in facility's life cycle;
- Identify best practices and solutions related to medical facilities;
- make medical facility design measurable and quantifiable with respect to process flows;

- be evidence-based (similarly to evidence-based medicine);
- operate a performance evaluation system applicable to individual processes and medical facility at large.

References

1. Asdrubali F, Ferracuti B, Lombardi L et al (2017) A review of structural, thermo-physical, acoustical, and environmental properties of wooden materials for building applications. Build Environ 114:307–332
2. Chilton J (2010) Tensile structures–textiles for architecture and design. In: Textiles, polymers and composites for buildings. Woodhead Publishing, pp. 229–257
3. Guan Y, Virgin LN, Helm D (2018) Structural behavior of shallow geodesic lattice domes. Int J Solids Struct 155:225–239
4. Gusev EL, Ivanova MA, Chernykh VD (2019) The development and application of generalized models long-term forecasting of the residual resource composite materials and structures at the impact of extreme climate factors of the Arctic zone. Procedia Struct Integrity 20:294–299
5. Halliwell S (2008) Ageing of composites in the construction industry. In: Ageing of composites. Woodhead Publishing, pp. 401–420
6. Ibrahim Y, Kempers R, Amirfazli A (2019) 3D printed electro-thermal anti-or de-icing system for composite panels. Cold Regions Sci Technol 166. https://doi.org/10.1016/j.coldregions.2019.102844
7. Kasyanov NV (2016) The evolution of architectural morphogenesis at the beginning of XXI century in the context of scientific advances. Procedia Eng 153:266–270
8. Kurta I, Zemlyansky V (2016) Preconditions for technological development of the construction industry of the north for the arrangement of the mineral complex of the Russian Arctic. Procedia Eng 165:1542–1546
9. Kuzmin SA, Egorova AD, Krasilnikov DA et al (2019) Durability of construction materials modified by polymeric additives. Procedia Struct Integrity 20:278–283
10. Lebedev MP, Startsev OV, Kychkin AK (2019) Development of climatic tests of polymer materials for extreme operating conditions. Procedia Struct Integrity 20:81–86
11. Pastukh OA (2021) Transformation and development of coastal industrial zones with using large-span construction structures. European experience of 1990–2000s. In: Contemporary problems of architecture and construction. CRC Press, pp. 58–61
12. Porta-Gándara MA, Gómez-Muñoz V (2005) Solar performance of an electrochromic geodesic dome roof. Energy 30(13):2474–2486
13. Shen ZY, Li YQ, Luo YF (2005) Instability research of single layer reticulated shells at Tongji University. In: Fourth international conference on advances in steel structures. Elsevier Science Ltd, pp. 83–92
14. Wei JP, Tian LM, Hao JP (2018) Improving the progressive collapse resistance of long-span single-layer spatial grid structures. Constr Build Mater 171:96–108
15. Wu Y, Takatsuka M (2006) Spherical self-organizing map using efficient indexed geodesic data structure. Neural Netw 19(6–7):900–910
16. Zhivotov DA, Pastukh OA, Tilinin YI (2021) Architectural and spatial planning solutions of spherical shape buildings. In: Contemporary problems of architecture and construction. CRC Press, pp. 91–96
17. Zhivotov DA, Pastukh OA, Panin A et al (2021) Application of engineering composites for elements of architectural and construction structures operated in Arctic zone. In: Contemporary problems of architecture and construction. CRC Press, pp. 292–296

18. Zhivotov D, Pastukh O (2020) Construction of geodesic domes made of wood and composite materials during restoration and conservation of cultural heritage objects. In: E3S web of conferences, vol 164. EDP Sciences, p. 02020
19. Zhivotov D, Tilinin Y (2020) Experimental studies of the strength of nodal joints of geodesic domes made of wood and fiberglass made on a 3D printer for the Arctic and Northern territories. Publ Ser LAB Univ Appl Sci 2:57–65

Innovative Medicine: What Challenges Does It Pose to Designers and Developers?

Semen Prokopchuk, Dmitrii Zhivotov, Olga Pastukh, and Aleksandr Panin

Abstract The pandemic has increased the role of healthcare services as a sector pivotal to many processes. The experience of different countries has shown that the availability of sufficient funding may not always guarantee excellent performance in healthcare sector. The natural evolution and technology advance have their effect also on healthcare services. This paper discusses the trends existing globally in the development of medical industry and the reasons for their formation. General principles have been identified concerning the design and construction of healthcare facilities.

Keywords Healthcare facilities · Designers · Medical technologies · Training · Innovation · Technological effectiveness

1 Introduction

From the healthcare development perspective, the evolutionary trends are basically two (Table 1). One trend is towards diversification and implies implementation of non-medical developments such as IT, computer-assisted services, robotics, etc., and the other is towards specialization that implies modifications to the treatment process and unfolds through hybridization, conversion, introduction of minimally invasive and high-tech treatment methods, etc.

S. Prokopchuk
M-Investproekt LLC, 28 Marshal Novikov Str, Saint Petersburg, Russia
e-mail: info@m-investproekt.ru

D. Zhivotov (✉) · O. Pastukh · A. Panin
Saint Petersburg State University of Architecture and Civil Engineering, Saint Petersburg, Russia
e-mail: d.zhivotov@mail.com

Table 1 Healthcare development perspective	Diversification	Specialization
	General trends beyond the scope of medicive	Modification of healthcare process
	Widespread introduction of IT	Hybridization of treatment processes
	Computer-assisted services	Conversion potential
	Robotics	Minimally invasive and
	etc	high-tech treatment methods
		etc

2 Methods

2.1 Healthcare Development Trends Beyond Medicine

2.1.1 Information Technologies

The development of IT has had both direct and indirect impact on the evolution of healthcare services. In the first case, the introduction of advanced information technologies into medical processes has led to high-tech care and revised approaches to internal and external processes. Prominent examples include medical information systems—laboratory information system (LIS), radiological information system (RIS), PACS, telemedicine, medical integration systems, among others.

The indirect influence of the IT advance manifests itself in the medical processes being modified through the use of gadgets and wearables to monitor health conditions, the development of Internet and computer-assisted services, which entails the natural evolution of medical and engineering equipment as a whole.

2.1.2 Automation and Dispatching

As a general trend, automation and dispatching penetrate all industries and are designed to replace humans in performance of mechanical tasks. Examples include the artificial intelligence, big data systems, utility networks management, bots and similar programs, etc.

2.1.3 Robotic Applications

Robots find active application in medicine. There are specialized and non-specialized medical robots. Examples of robots for specialized medical applications include Da Vinchi robotic surgery, robotic biopsy, robotic laser eye surgery, disinfection robots, among others.

Non-medical robotic applications in medicine include automatic cargo delivery systems, pharmacy robots offering high-density storage and rapid retrieval, assistant robots and bots, etc.

2.2 Specialized Modification Trends in Healthcare Development and Evolution

2.2.1 Hybridization of Medical Technologies

The advance in IT sector has triggered changes in the very approach to medical care. While earlier patient management followed a multiple-stage procedure, today it can be provided as single-stage, using the "golden hour" rule implying the provision of a wide range of specialized medical within one hour's time. Classical approaches to medical care are giving way to combined approaches as better suiting the current stage of healthcare development. One example can be found in hybrid operating room, a surgical workspace that integrates imaging devices with a multifunctional surgical table for managing difficult cases.

2.2.2 Development of High-Tech Healthcare

The above trends lead to healthcare services becoming a technology-intensive sector. One more associated trend is towards wider application of technologies in polyclinic and home care through telemedicine.

As healthcare facilities become more technology-assisted, they extend their coverage. This poses the need for narrowly specialized facilities with ample high-tech equipment.

2.2.3 Conversion in Healthcare

Increasing healthcare facilities' potential to be adapted to current needs is one of the latest trends. It gained special relevance during the pandemic, when hospitals had to be adapted for use as COVID facilities and installed with relevant equipment.

Repurposing is not new, but has come to the foreground during the pandemic. Other factors that necessitate repurposing include:

migration and population growth. Examples include large cities where population growth outruns infrastructure growth; and smaller cities with rapidly decreasing population density. Both cases require repurposing of their existing healthcare services;

major emergencies and man-made disasters. These events require hospitals to be repurposed for a shorter period;

introduction of the latest medical techniques, designed to upgrade facility's existing technological solutions.

2.2.4   Changes in Approaches to Healthcare Provision

In addition to the technological advance, healthcare facilities experience changes of purely medical nature:

- natural "aging" and "rejuvenation" of diseases;
- natural "death" of diseases;
- natural "birth" of diseases;
- changing approaches to treatment, diagnosis, and prevention;
- emergence of new drug and non-drug treatment methods, including methods of surgical interventional and their abandonment in classical cases;
- change population's attitude to their health and promotion of healthy aging;
- focus shift from therapy to prevention and healthy lifestyle.

All these trends have had a great impact on the tasks faced by healthcare facilities. While fifteen years ago hosting by healthcare facilities of schools for healthy living or diabetes were rather the exception than the normal practice, today they are must-haves and a new function that has already altered healthcare facilities design and construction.

2.2.5   Combining Training and Therapeutic Processes

The availability of onsite training allows doctors to perfect their skills without interrupting their duties and young professionals and students to start applying their skills early in their careers.

Training and advanced training are necessitated firstly by the high requirements existing for the level of skill expected of the medical personnel.

Secondly, given the rate at which new medical techniques are being introduced into medical practice, retraining and organization of training centers become a vital necessity.

The nuances of the medical training process should be taken due account of when designing healthcare facilities.

3   Results

In view of these trends, the classical, proven approaches to healthcare design and construction cease to be relevant [8].

Firstly, every time a new technology is put into operation, it renders previous approach obsolete and no longer meeting the requirements of medical personnel.

Secondly, the modernization potential of healthcare facilities is rather low, especially when it comes to state-funded healthcare construction.

Thirdly, with many healthcare facilities, the existing infrastructure and utilities often appear insufficient to meet modernization purposes.

Fourthly, the availability to healthcare facilities of IT infrastructure remains low.

These challenges are difficult to meet and would be impossible to overcome without changing the design and construction concept.

4   Discussion

In view of the foregoing, we have identified the following issues to be considered by new healthcare construction projects at the design stage:

- potential for future modernization;
- prospects of urban/district/local development [11];
- infrastructure. The project's infrastructure should be designed with account of the surrounding facilities—district infrastructure, approach ways, public transport, etc.;
- use of standard solutions, not standard designs [1, 4, 9];
- use prefabricated modular technologies [2, 3];
- use of "light" modernization technologies [5];
- "medical tasks first" principle.

All other standards and requirements should be subordinate to and aimed at fullest achievement of medical tasks [10].

5   Conclusion

Healthcare design and construction represent a niche area of expertise [7, 15]. The experience of building innovative healthcare projects shows that they should be designed by adequately trained teams.

The European experience shows the expediency of placing the tasks of design, construction and, importantly, operation of an innovative healthcare project in the hands of one single contractor.

An update is needed in the approach to reviewing project documentation [12, 13]. The project documentation review is the stage where a focus shift occurs from medical tasks to norms and standards [6]. It is currently impossible to build a project that would be consistent with every single regulation while meeting its medical purpose, so a change is needed in the approach to regulatory framework. This change can be achieved by:

developing a mechanism for validating Project Specific Standards across all project sections or canceling those regulations that seem to be no longer relevant [14].

By providing the innovative healthcare projects with carefully considered infrastructure, it would be possible to extend their service life without having to introduce cardinal upgrades on the medical side.

References

1. Asdrubali F, Ferracuti B, Lombardi L et al (2017) A review of structural, thermo-physical, acoustical, and environmental properties of wooden materials for building applications. Build Environ 114:307–332
2. Chilton J (2010) Tensile structures–textiles for architecture and design. In: Textiles, polymers and composites for buildings. Woodhead Publishing, pp. 229–257
3. Fuller RB (1982) Geodesic dome. Art, Architecture and Engineering Library
4. Guan Y, Virgin LN, Helm D (2018) Structural behavior of shallow geodesic lattice domes. Int J Solids Struct 155:225–239
5. Gusev EL, Ivanova MA, Chernykh VD (2019) The development and application of generalized models long-term forecasting of the residual resource composite materials and structures at the impact of extreme climate factors of the Arctic zone. Procedia Struct Integrity 20:294–299
6. Ibrahim Y, Kempers R, Amirfazli A (2019) 3D printed electro-thermal anti-or de-icing system for composite panels. Cold Reg Sci Technol 166:102844. https://doi.org/10.1016/j.coldregions.2019.102844
7. Kasyanov NV (2016) The evolution of architectural morphogenesis at the beginning of XXI century in the context of scientific advances. Procedia Eng 153:266–270
8. Kurta I, Zemlyansky V (2016) Preconditions for technological development of the construction industry of the north for the arrangement of the mineral complex of the Russian Arctic. Procedia Eng 165:1542–1546
9. Kuzmin SA, Egorova AD, Krasilnikov DA et al (2019) Durability of construction materials modified by polymeric additives. Procedia Struct Integrity 20:278–283
10. Lebedev MP, Startsev OV, Kychkin AK (2019) Development of climatic tests of polymer materials for extreme operating conditions. Procedia Struct Integrity 20:81–86
11. Pastukh OA (2021) Transformation and development of coastal industrial zones with using large-span construction structures. European experience of 1990–2000s. In: Contemporary problems of architecture and construction. CRC Press, pp. 58–61
12. Porta-Gándara MA, Gómez-Muñoz V (2005) Solar performance of an electrochromic geodesic dome roof. Energy 30(13):2474–2486
13. Shen ZY, Li YQ, Luo YF (2005). Instability research of single layer reticulated shells at Tongji University. In: Fourth international conference on advances in steel structures. Elsevier Science Ltd, pp. 83–92
14. Wei JP, Tian LM, Hao JP (2018) Improving the progressive collapse resistance of long-span single-layer spatial grid structures. Constr Build Mater 171:96–108
15. Zhivotov DA, Pastukh OA, Tilinin YI (2021) Architectural and spatial planning solutions of spherical shape buildings. In: Contemporary problems of architecture and construction. CRC Press, pp. 91–96

The Algorithm for Evaluating the Performance Specifications of Automated Medical Measurement Instrumentation

V. V. Reznichenko

Abstract This paper presents the study that provides a method for identifying the performance specifications of automated medical measurement instruments. Evidence is provided of the expediency of the method that uses multi-criteria optimization. This method consists in identifying, among the range of options, of a compromise solution that fits the given technical and economic policy best.

1 Introduction

Over the last two decades, the advance in microprocessor technologies for medical instrumentation has led to a widespread use in medical practice of automation tools and integrated control systems designed for use by a limited group of personnel. Real-time systems vary in operational speed but need to form an integrated system [3]. This circumstance has its impact on hardware, reliability and viability assurance, resource allocation [4].

In this context, the process of integrating real-time systems into a single system to provide concurrent decision-making on a large number of heterogeneous tasks, can be challenging and has an impact of the parameters of real-time automated process control systems. The solution lies in microprocessor-based automation control technology [5]. Not only does this technology enable further automation of technical processes and integrated control of medical equipment, it offers room for the robotic applications and the use of conventional methods to deal with a wide range of tasks through:

- adaptive control in real operational conditions;
- normal and emergency control decision-making;
- etc.

V. V. Reznichenko (✉)
Saint Petersburg State University of Architecture and Civil Engineering, Saint-Petersburg, Russia
e-mail: 12356788@mail.ru

2 Materials and Methods

The functional compatibility of the sub-systems of automated control systems (ACS) enables mutual exchange of information while dealing with local control tasks, based on the key purpose of automation. This makes it possible for management processes to be integrated into composite systems and controlled by a reduced number of regulators.

ACS involves various types of recognition instruments in the form of video cameras, supersonic locators, etc. Integrated recognition uses measuring information and is supported by microprocessor-based software. A mobile automated control system is a combination of two subsystems, one measuring and the other recognition. For implementation of the method, there have been developed:

- specific optimality criteria calculation methods for identifying the specifications and optimal performance evaluation of ACS instrumentation;
- algorithm for evaluating and identifying optimal performance specifications of the prospective ACS instrumentation in various technical and economic concepts.

The primary aspects that have been considered while developing the method for identifying optimal performance specifications of ACS instrumentation, were the following:

- applicability to all types of automated control systems and at both the initial stage of life cycle and functions implementation stage;
- perception friendliness;
- consistency with tasks and objectives of automated control systems in various operating conditions and at all levels.

Generally, the process of identifying the optimal performance specifications of measuring instruments is preceded by:

- accurate description of the designated purpose of the device, for which the automated control system application is developed, intended for actual results dissemination and benchmarking against similar projects;
- accurate description of the study methodology for simplified perception of the results obtained;
- a set of unambiguous indicators to assess benchmarking performance at different levels;
- indication of suitability of the obtained results for statistical analysis.

Next to be identified are the evaluation tasks and quantifiable parameters. Special attention is paid to:

- security;
- efficiency;
- user-friendliness (quality) of software;
- unification (embeddability);
- accessibility.

Evaluation will be performed only when the system under analysis has an actual impact on the above parameters. The study targeted, firstly, the technical specifications of ACS measuring instruments, on the basis of which statistical data are analyzed and their effectiveness is evaluated; and, secondly, results transferability, i.e. the extent to which the results obtained in the evaluation process can be applied in similar systems; local, external and introduced features are noted that may have an impact on the results; if several evaluation reports have been compiled, the results section can serve the purposes of comparative analysis and decision-making on the transfer of results.

The automated control system uses the electronic systems allowing to determine healthcare dynamics and replace humans. Its use was necessitated by the need to continuously collect more data. The main advantage of automatic measurements is that they allow receipt of data in electronic form and boast higher accuracy, as compared to manual method, although sometimes an incorrectly installed and/or poorly calibrated device can produce incorrect results. To avoid it, it is recommended, at least at the initial stage, to supplement automatic data collection with manual collection. Another drawback of the automatic data collection systems is errors made during data analysis. Indeed, given the vastness of data and the lack of adequate automation tools, the huge amount of information can be a big obstacle to flawless processing. Therefore, the decision-making as to new technology should be preceded by the analysis of data processing performance. The main disadvantages of this system are high investment costs, and installation and operation expenditures.

The automatic measurement system involves 4–5 components:

- sensor;
- converter;
- registrar/log;
- processor;
- communication.

The choice of automated control systems should account of data collection mode. There exist:

- continuous recognition mode; and
- temporary recognition mode.

Continuous recognition refers to the data collection process, which requires placing the appropriate devices in pre-determined locations for comprehensive measurement of the object under observation.

Temporary recognition involves partial data collection within a limited period of time. In this case, sensors will be placed at certain points and removed upon completion of the recognition. Data collection lasts only a couple of hours. This measurement technique can also involve wearable sensors, but they are vulnerable to negative influences of weather. The existing technologies that use instrumentation that maintains one measurement mode throughout their operation, can be used for temporary data collection. The following should be taken into account:

- installation in interference-free (or minimal interference) zones;
- it is possible to use existing infrastructure facilities;
- easy installation and calibration. Since the time allotted for data collection is limited, it makes no sense to spend a lot of time on installation and calibration;
- devices must be capable of data storing and transmitting;
- (solar) battery supply where in the absence of power sources;
- cost. Frequent use of temporary data collection devices is an important factor to consider when analyzing costs. It might be expedient to use continuous data collection, if the use promises to be frequent.

There are more factors to be considered when choosing the most optimal device for your purposes, and among them:

- power supply;
- data transmission and archiving.

3 Results

To ensure the operability of your device, first of all, it is necessary to check the availability of power supply. If connection to power supply is problematic for some reason, alternative power sources should be provided. In areas with a large number of sunny days per year, it is possible to use solar energy, although in some cases this method can be too costly.

When benchmarking automated control systems, the following factors should be considered [1]:

- initial capital investments;
- installation costs;
- maintenance cost during the entire service life;
- other expenses.

For the most optimal choice, specific criteria will be considered—recognition efficiency, cost and risk involved in measuring instrument (MI) development. There exists a set of Pareto-optimal criteria that are based on the assumptions to be considered when identifying the optimal performance specification of ACS instrumentation for operation in technical and economic conditions [2].

The efficiency of recognition of MI objects N_i is probabilistic in nature and is represented by a set of probabilities for solving k-th recognition problem P_k over time ΔT. It is advisable to evaluate these probabilities using software.

The effectiveness of solving the k-th problem of recognizing the MI objects can be presented as the degree of correspondence between the obtained probabilistic value P_κ and the required P_κ^e:

$$n_i = P_k^e - P_k, \text{ if } P_k^e \geq P_k$$
$$n_i = 0, \text{ if } P_k^e < P_k \tag{1}$$

Then, the effectiveness of solving the MI recognition problem can be assessed as follows:

$$Ni = \frac{1}{\sum_{k=1}^{N} q_k} \text{if} \sum_{k=1}^{N} n_k \neq 0$$

$$N_i = 1, \text{if} \sum_{k=1}^{N} n_k = 0 \tag{2}$$

Thus, the effectiveness of MI object recognition problem is characterized by coefficient $Ni = 0…1$ and depends on the degree of correspondence between the probability of solving each recognition problem and the targets specified in technical design assignment.

4 Discussion

The figure below shows the sequence of steps towards determining the optimal performance specifications of the measuring instrument to be used with given recognition subsystem (Fig. 1).

One of the effective methods for assessing the risks of operating a complex technical system is the "fixed scales" method. This method can be represented as an estimate of the technical success rate that takes into account the specific contributing factors:

$$R(g) = \sum_{i=1}^{n} Bi * n \tag{3}$$

where $R(g)$ is final feasibility value, measured in points, Bi is i-th factor of feasibility; n_i is i-th factor contribution.

5 Conclusion

The proposed method allows to obtain a satisfactory result with minimal financial costs and technological risk. Two variants of technical and economic specifications are presented. Proper performance of MI is achieved with the first variant. In this variant, the specific criteria can have identical significance, i.e. $w_1 = w_2 = w_3$. In the second variant, the specific criteria of cost and risk have a greater significance than the efficiency criterion.

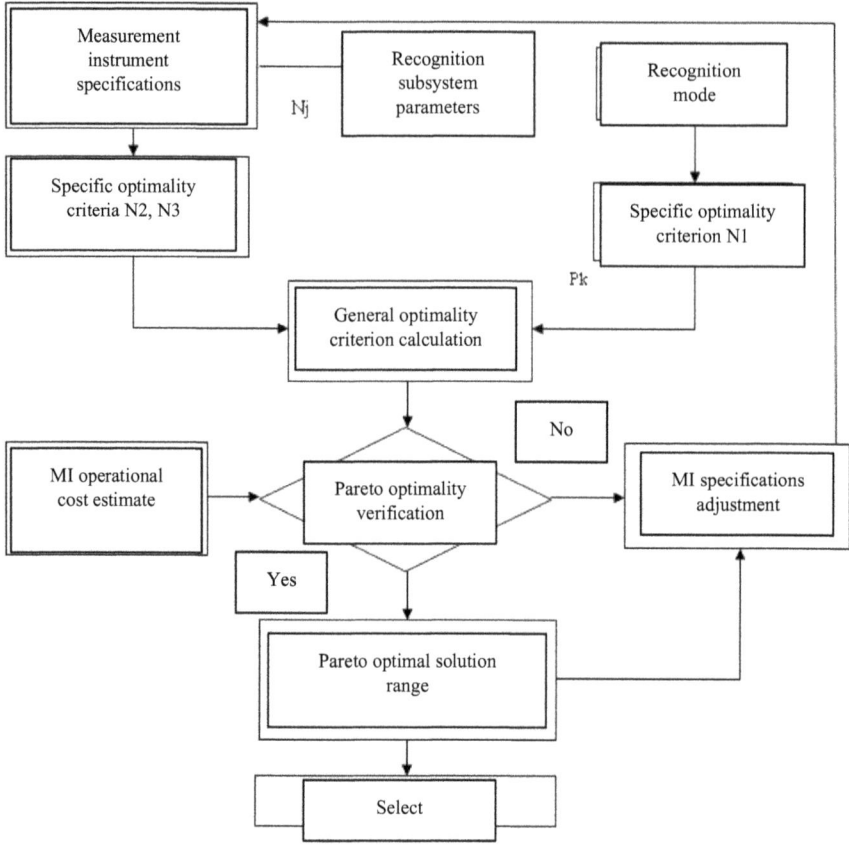

Fig. 1 The sequence of steps towards determining the optimal performance specifications of the measuring instrument

Errors in identifying optimal MI performance specifications can lead to low efficiency or unreasonable increase in the cost of the system. Our analysis of the published studies suggests that the problem of the common method for identifying the MI performance parameters at "measuring instruments—recognition" level has not been explored before.

The studies have found that the expedient method for identifying measurement instruments' performance requirements is the one that uses multi-criteria optimization, which consists in identifying, among the range of options, of a compromise solution that fits the given technical and economic policy best.

References

1. Di X, Liu HX, Ban XJ (2016) Second best toll pricing within the framework of bounded rationality. Transp Res Part B Methodol 83:74–90
2. Ivnitsky VA (2012) Non-observance risk identification of the prescribed average daily speed level by speeded-up container trains due to technical failures. VNIIZhT Bull 12:41–44
3. Safiullin RN, Afanasyev AS, Reznichenko VV (2019) The concept of development of monitoring systems and management of intelligent technical complexes. J Mining Inst 237:322–330. https://doi.org/10.31897/PMI.2019.3.322
4. Safiullin R, Marusin A, Safiullin R et al (2019) Methodical approaches for creation of intelligent management information systems by means of energy resources of technical facilities. In: E3S web of conferences, vol 140. EDP Sciences, p 10008
5. Safiullin RN, Reznichenko VV, Kalyuzhny AF (2019) Automation systems for motor traffic control. St. Petersburg: Lan'.

Legal Framework for the Design and Construction of COVID Hospitals in the Russian Federation

G. Z. Sitdikova, M. V. Lifanova, and O. V. Kornelyuk

Abstract The paper discusses the legal framework for the design and construction of modern high-tech prefabricated healthcare facilities. It highlights the challenges faced by design and construction works in the context of the coronavirus pandemic. With limited timeframes, healthcare construction projects are using prefabricated technologies that are supposed to meet the highest of construction standards and allow for operation of modern healthcare equipment. This task is being solved also at the legal level. All things considered, the issues of legal regulation of COVID hospitals design and construction in the Russian Federation relevant require due attention.

Keywords Prefabricated healthcare facilities · COVID hospital · Design and construction · Legal coverage

1 Introduction

Currently, the spread of infectious diseases requires a rapid response from the construction sector in form of modern infectious healthcare facilities. In the current epidemiological situation, prefabricated design is the only adequate solution to counter this novel threat. Notably, the construction industry has adapted to the new reality rather quickly, meeting the public demand in healthcare services.

For the purposes of the above task, the Ministry of Construction, Housing and Communal Services of the Russian Federation has prepared draft Amendment No.3 to Construction Regulations 158.13330.2014 "Buildings and premises of healthcare organizations. Design Rules", adopted on March 1, 2021.

According to experts, the amendments are aimed to reduce the risk nosocomial infection and improve the fire safety of healthcare facilities. A number of requirements concerning the availability and number of individual boxes for the reception and placement of infectious patients during the pandemic have been canceled so

G. Z. Sitdikova (✉) · M. V. Lifanova · O. V. Kornelyuk
Bashkir State University, Ufa, Russian Federation
e-mail: g40773@yandex.ru

© The Author(s), under exclusive license to Springer Nature Switzerland AG 2023
D. Ivanov et al. (eds.), *Proceedings of ECSF 2021*, Lecture Notes in Civil
Engineering 257, https://doi.org/10.1007/978-3-030-99877-6_26

as to allow healthcare facilities to increase their capacity within their allocated and existing land plots. Optimization covered also the minimum permissible distances between the buildings of a healthcare facility, and the permissible distance between the COVID hospital (other infectious diseases facility) and the buildings located outside the hospital territory (beyond the red line).

One of the innovations established in Amendment No.3 to Construction Regulations 158.13330.2014 concerns space-planning solutions for infectious diseases hospitals. Updated requirements have been established for the premises intended for reception and hospitalization of patients (emergency room, sanitary check-points, boxes for people with infectious diseases, treatment sections); laboratories of various types (depending on the level of biological hazard); preventive cleaning and disinfection; rules for managing patient flows and healthcare personnel flows.

Since prefabricated solutions have not been used widely in healthcare sector before, the above regulatory act specifies also the requirements to air exchange for radionuclide diagnostics and therapy departments, as well as stack rooms. A provision has been introduced on the minimum width of the corridors in such buildings, which allows for a lower width (compared with the width established for ordinary infectious diseases hospitals).

For the purposes of Order of the President of the Russian Federation V. V. Putin No.Пр-843 "On optimizing the regulatory and legal framework for civil construction", dated May 25, 2020, changes have been introduced into other regulatory documents applicable to the design and construction of healthcare facilities.

Concurrently with Amendment No.3 to Construction Regulations 158.13330.2014, there have been developed and duly endorsed Construction Regulations SP 60.13330.2020 "Heating, ventilation and air conditioning" and Construction Regulations 70 "Load-bearing and enclosing structures".

We presume that the multifaceted amendment of these regulations has led to higher energy efficiency of modern prefabricated healthcare facilities, cardinally reduced construction timeline, and simplified organizational procedures associated with prefab construction.

At the same time, the research-based interpretation of the challenges and the conceptual foundations for their solution lags behind the practice.

In the modern healthcare research, the COVID-19 pandemic is a widely covered topic. Among notable international researchers in this field are Marie Scully [15], Shimabukuro [16], Polack [9] and other scientists [17], Sadoff et al. [12], Pogue et al. 2020, [2].

In Russia, the research of healthcare during the pandemic is led by Mozgovoy et al. [7]; Salukhov et al. [14]. Melekhin [6], among other healthcare scientists, psychologists and psychotherapists.

The studies cover also some of the economic and political aspects of the challenges that have arisen in connection with the COVID-19 pandemic [1], Saad [11], Rau [10], [5]. One repeatedly raised issue is schooling and education during the pandemic [3].

At the same time, the legal aspects of countering the COVID-19 pandemic have not received sufficient attention. Comprehensive research is scarce, in particular, of the legal regulations for prefabricated design and construction in modern high-tech healthcare sector.

As we noted earlier, the comprehensive efforts to update the legal framework for prefabricated healthcare design and construction have enabled a solution to one key challenge in the fight against the pandemic—construction timeline. COVID hospitals can now be built within the record time of 30 to 60 days. The Russian and Chinese experience of deploying prefabricated healthcare buildings shows their expediency and good performance in terms of quality, energy efficiency and suitability for the treatment of patients with COVID-19.

2 Materials and Methods

Methodologically, the study relies on dialectical materialism as a universal way of understanding social and legal phenomena and involves comparative legal analysis, logical analysis, case study, comparative historical and statistical methods.

3 Results

The COVID-19 virus and its varieties have evolved into biogenic stimulators, disrupting the normal course of social relations. The pandemic has changed the plans and stability of people's lives by disrupting their independence in decision-making and individual's relative independence from society [13], Jiaqi [18]. All countries have concentrated their efforts on combating COVID-19 by taking measures to quickly create infrastructures for sanitary and epidemiological well-being of the population. The use of prefabricated design has become a relevant trend in healthcare construction.

The experience of rapid and efficient construction of COVID hospitals in Wuhan, Hubei Province has made a great contribution to solving the organizational, technical and infrastructure-related issues of prefabricated COVID hospitals in many countries.

The state policy of the Russian Federation in the field of healthcare construction underwent a serious revision during the pandemic, having caused relevant legal institutions to transform and turn to prefabricated design for healthcare buildings.

The regulatory framework exists for social relations at all stages of the construction process, up to commissioning, and is supported by consolidated standards in the field of prefabricated healthcare construction and state review of project documentation.

The safety of buildings and structures, as well as their design (including surveying), construction, installation, commissioning, operation and demolition, is ensured through compliance with Federal Law No.384-FZ "Technical Regulations on Safety of Buildings and Structures" dated December 30, 2009. This law is intended to protect the life and health of citizens, property of individuals or legal entities, state and municipal property, environment, fauna and flora; to prevent misleading actions; and to ensure energy efficiency of buildings and structures.

The initial stages of the legal regulation in healthcare span the study of natural conditions and anthropogenic factors likely to affect the rational and safe use of territories and land plots allocated for COVID hospitals; preparation of territorial planning rationales; site planning and architectural design.

Legal regulation of the design and construction of Clinical and Diagnostic Infectious Disease Centers (COVID hospitals) is intended to create the legal environment for safe construction and operation of prefabricated innovative healthcare buildings.

Legal regulation of the design and construction of prefabricated healthcare facilities, including COVID hospitals, involves several lines of audit.

(1) Legal audit:

- standards, regulations and rules for prefabricated healthcare construction;
- construction permits;
- permit for connecting the project to the energy and utility infrastructure.

As established in Resolution of Chief Public Health Officer of the Russian Federation No.58 "On approval of Sanitary Rules and Regulations 2.1.3.2630–10—Sanitary and Epidemiological Requirements to Healthcare Services Providers" (as amended on June 10, 2016), dated May 18, 2010, healthcare facilities will be located on the territory of residential development, in a green or suburban area at a distance from public, industrial, communal, economic and other organizations in accordance with the requirements for planning and development of urban and rural areas, as well as in accordance with hygienic requirements to sanitary buffer zones. Inpatient care facilities with round-the-clock supervision and treatment for infectious diseases will be located at a distance of at least 100 m from residential development. Infectious diseases hospitals with 1000 + beds will be located in suburban or green zones.

The body responsible for setting the National Standards, including in the field of construction, is the Federal Agency for Technical Regulation and Metrology (RosStandart). The National Standards serve as basis for the following Codes:

- "Public buildings and structures" (enacted by Regulation No. 822/пр. "On approval of Amendment No. 4 to Construction Regulations 118.13330.2012—Sanitary Rules and Regulations 31–06-2009" (as amended on February 22, 2020) issued by the RF Ministry of Construction on December 19, 2019);
- "Construction Climatology" (enacted by Regulation No.859/or. "On approval of Construction Regulations 131.13330.2020—Sanitary Rules and Regulations 23–01-99 "Construction Climatology" issued by the RF Ministry of Construction on December 24, 2020;
- "Design and construction of foundations of buildings and structures" (enacted by Resolution No.28 "On the Code "Design and construction of foundations of buildings and structures" issued by the RF State Committee for Construction, Housing and Utilities of March 9, 2004).

(2) Construction audit:

- review of project documentation;
- review engineering survey results;
- construction and technical review.

Construction audit is intended to ensure the compliance with requirements to safety of buildings and structures, as well as processes of design (including surveys), construction, installation, commissioning, operation and demolition:

- mechanical safety;
- safety in case of dangerous natural processes/phenomena and (or) man-made impacts;
- safe and healthy living conditions;
- safety of users of buildings and structures;
- accessibility of buildings and structures for people with disabilities and other groups with limited mobility;
- energy efficiency of buildings and structures.

The legal provisions on fire safety are established in Federal Law No.123-FZ "Technical Regulations on Fire Safety" (as amended on 30.04.2021) dd. 22.07.2008, Code "Fire protection systems. Limiting the spread of fire at protection facilities. Requirements to space-planning and design solutions", Code "Fire protection systems. Automatic fire alarm and fire extinguishing installations", endorsed by RF Ministry of Emergency Situations Regulation No.288 dd. 24.04.2013.

(3)   External infrastructure audit:

- planning of healthcare facility's transport accessibility;
- ensuring environmental safety;
- anti-terrorist protection.

(4)   Economic audit:

- pricing policy;
- construction materials suitable for use in special operating conditions;

(5)   Internal infrastructure audit:

- commissioning of pre-fabricated healthcare facilities;
- facility's connection to resource-supplying sources and information environment;
- commissioning facility's healthcare equipment and life support systems.

More than twenty COVID hospitals were built in different regions of the Russian Federation in a short time by military specialists of the RF Ministry of Defense.

Amply equipped modern healthcare centers operate in Moscow and Moscow Region (Odintsovo), St. Petersburg, Rostov-on-Don, Sevastopol, Kaliningrad, Ulan-Ude, Omsk, Podolsk, Nizhny Novgorod, Volgograd, Novosibirsk, Orenburg, Smolensk, Petropavlovsk-Kamchatsky, Khabarovsk, and Ussuriysk.

One more clinical and diagnostic infection center for patients with coronavirus appeared in Russia's Ufa (Zubovo village). Pre-fabricated and demountable, this center takes up 16,000 square meters and was built in 55 days.

The Ufa Clinical and Diagnostic Infectious Diseases Center is a one-story multi-colored complex shaped like kurai flower (Bashkir reed pipe). It comprises several independent units, one hundred boxes (82 for round-the-clock stay) and eighteen intensive care units, and has capacity for 484 additional beds in case of emergency.

Another facility for coronavirus patients has been built in the Republic of Tatarstan—The Republican Clinical Infectious Diseases Hospital. This hospital has 232 beds, including 12 intensive care beds, and the capacity for additional 504 beds.

The measures taken towards creating the infrastructure for sanitary and epidemiological well-being of the population made it possible to localize the pandemic wave.

4 Conclusion

In conclusion, the construction business, like any other business, seeks to minimize not only construction timelines but also expenditures. In this connection, the task of legal and technical regulation in order to ensure safety of constructed facilities is highly relevant. This task can be solved by providing a framework of clear standards and requirements to construction projects, process flows, and compliance monitoring.

References

1. Bashkova IV (2020) COVID-19: Russification of internationalism. The World of Science Culture Educ 6(85):689–692
2. Bazar S (2020) Coronavirus pandemic (COVID-19) and its aftermath on contemporary world affairs. Eurasianism and World 1:61–73
3. Dzhikiya MD (2020) Does COVID give the right to a discount: the legal subtleties of e-learning. Int J Humanit Natural Sci 11(4):75–78
4. Filho AS (2020) Neoliberalism and the Covid-19 pandemic: a political and economic analysis. Proc Free Econom Soc Russia 223(3):565–572
5. Gurieva SD et al (2020) Trust as a way to overcome the crisis. Modern Soc Stud 12(2):248–265
6. Melekhin AI (2021) Sleep disorders during the COVID-19 pandemic: Specifics, psychological examination and psychotherapy. Bulletin of the Udmurt University "Philosophy. Psychology. Pedagogy" series 31(1):27–38
7. Mozgovoy ED, Ochkolias MV (2020) Hyperbaric oxygenation therapy for treating complicated COVID-19: first experience. Extreme Med 3:64–67
8. Pogue JM, Lauring AS, Gandhi TN et al (2021). Monoclonal antibodies for early treatment of Covid-19 in a world of evolving SARS CoV-2 mutations and variants. In: Open forum infectious diseases
9. Polack FP, Thomas SJ, Kitchin N et al (2020) Safety and efficacy of the BNT162b2 mRNA Covid-19 vaccine. N Engl J Med 383:2603–2615. https://doi.org/10.1056/NEJMoa2034577
10. Rau J (2021) COVID-19 and its impact on international political relations in 2020. Modern Scientif Thought 1:116–127

11. Saad-Filho A (2020) From COVID-19 to the end of neoliberalism. El trimestre económico 87(348):1211–1229
12. Sadoff J, Le Gars M, Shukarev G et al (2021) Interim results of a phase 1–2a trial of Ad26. COV2. S Covid-19 vaccine. New England J Med 384(19):1824–1835
13. Salari N, Hosseinian-Far A, Jalali R et al (2020) Prevalence of stress, anxiety, depression among the general population during the COVID-19 pandemic: a systematic review and meta-analysis. Glob Health 16(1):1–11
14. Salukhov VV, Gulyaev NI, Dorokhina EV (2020) Assessment of systemic inflammatory reactions and coagulopathy against the background of hormonal therapy in covid-associated lung damage. Meditsinskiy Sovet 230–237
15. Scully M, Singh D, Lown R et al (2021) Pathologic antibodies to platelet factor 4 after ChAdOx1 nCoV-19 vaccination. N Engl J Med 384(23):2202–2211
16. Shimabukuro TT, Kim SY, Myers TR et al (2021) Preliminary findings of mRNA Covid-19 vaccine safety in pregnant persons. N Engl J Med 384(24):2273–2282
17. Shinde V, Bhikha S, Hoosain Z et al (2021) Efficacy of NVX-CoV2373 Covid-19 Vaccine against the B. 1.351 variant. New England J Med 384(20):1899–1909
18. Xiong J, Lipsitz O, Nasri F et al (2020) Impact of COVID-19 pandemic on mental health in the general population: a systematic review. J Affect Disord 277:55–64

GIS Based Infrastructure Support for the Preservation and Transfer of Biomaterials

L. A. Soprun, V. K. Averyanov, A. A. Melezhik, M. Yu. Demidionov, O. V. Mironenko, and E. A. Fedorova

Abstract The paper describes the alternative schemes for ensuring biological safety in the organization of the microbiological laboratories network on the territory of a megalopolis. For St. Petersburg, Russia, two options of work with biomaterial were considered—centralized and decentralized schemes. It is shown that continuous information support at all the stages of biomaterials life cycle is required to meet the quantitative and qualitative objectives. Such support is the most promising through an integrated approach based on the common data environment with the information modelling in GIS and the collaboration and tracking services.

1 Introduction

In 2020, by Decree of the Government of the Russian Federation No. 66, the new coronavirus infection was classified as a disease that poses a danger to others. The SARS-CoV-2 virus is a pathogenic group II virus so it is necessary to comply with the requirements for working with group II pathogenic of microorganisms [15]. Therefore both medical personal, patients and other urban dwellers must be protected from biological contamination.

L. A. Soprun (✉) · O. V. Mironenko
St. Petersburg State University, St. Petersburg, Russia
e-mail: lidas7@yandex.ru

V. K. Averyanov
Russian Academy of Architecture and Construction Sciences (RAACS), St. Petersburg, Russia

A. A. Melezhik
IEEE Association, St. Petersburg, Russia

M. Yu. Demidionov
The Herzen State Pedagogical University of Russia, Saint Petersburg, Russia

E. A. Fedorova
North-Western State Medical University Named After I.I. Mechnikov (NWSMU), St. Petersburg, Russia

To achieve a significant improvement in the prevention of nosocomial infections and to ensure sanitary-epidemiological and environmental well-being in St. Petersburg it's necessary to develop proposals for improving the system of ensuring the safety of work with biomaterials. Both correct choices of centralized and decentralized schemes of organization laboratory network and IT-platform should be integrated with GIS. So the correct cost-effective and environmentally friendly technology that helps prevent the spread of infections, is more likely to be selected.

2 Background

Despite the specifics of working with biomaterials, some analogies can be found with other types of activities. As the closest analogs in process digitalization two main activities can be determined: retail and construction. Practical implementation in these fields is based on data analysis with information modelling in GIS and BIM, and the integrated collaborative environment [2, 3, 5, 9, 10, 12].

Using location intelligence based on GIS, retailers analyze potential locations for single stores as well as groups of outlets across a region, and logistical arrangements [1, 7, 13, 14].

Information modelling integrated and collaborative working for multi-disciplinary team become the most important mechanism for construction industry. It helps to avoid collisions and to make quickly the right decisions (both in normal and emergency modes) [1, 11, 14].

On the basis of earlier studies [4, 6] the authors of the article investigate the use for the medical purposes both of the above IT-directions:

- the possibilities of using GIS to optimize work with biomaterials;
- the pilot operation of the integrated information system with GIS for medical purposes is planned to be started.

3 Materials and Methods

The following data and tools were used:

- The OpenStreetMap spatial data layer of the roads (ESRI Shapefile) and the address list of one of the major networks of medical centers in St. Petersburg.
- ZuluGIS and QGIS software with a network analysis library (QGIS network analysis library) and QNEAT3 module was used for preparing geodata and calculating the time of delivery of biomaterials.
- Datrics software integrated with GIS was used for organization an integrated collaborative environment.

The original road layer contained all the paths in the form of linear objects. Only roads suitable for traffic were selected for the analysis. The layer has been reprojected from a geographic coordinate system (WGS84 EPSG: 4326) to a rectangular one (Pseudo Mercator EPSG: 3857). The list of addresses was converted into point layer of the position of medical centers on the map.

The final processing and analysis of geodata was done in QGIS (version 3.18.2). Using the network analysis library, an operation was performed to build graphs, on the basis of which the calculation of the travel time between the given points was performed in the QNEAT3 module. In the calculations, the following were taken: average speed of motor transport—40 km/h; maximum travel time—6000 s (1 h 40 min). On the basis of the calculations performed, areas of accessibility of the territory were created depending on the time of delivery of the biomaterial by road [8].

Additionally, the following initial data were taken into account:

- In the medical centers accepted for analysis, polymerase chain reaction (hereinafter PCR) studies are performed to determine a new coronavirus infection caused by the SARS-CoV-2 virus, indicating the power, calculated coefficients of quantitative justification for this type of research.
- Passport of a medical organization (MO), a license for medical activities.
- Accounting and reporting documentation in the MO, in the laboratory for working with II–IV pathogenicity groups.
- Project for the location of the laboratory for work with II–IV pathogenicity groups.

4   Results

For the comparative analysis, the addresses of the medical centers are divided according to their functional characteristics into the following 2 types: (1) the biomaterials sampling points, and (2) the laboratories for collection and researching biomaterials.

The following options were considered:

Centralized scheme of work with biomaterials with one laboratory located in the south of the city on Pulkovskoe highway and ten biomaterials sampling points.

Decentralized scheme for working with biomaterials with two laboratories located in the south (as for option 1) Pulkovskoe highway and in the north of the city on Bogatyrsky prospect, and nine biomaterials sampling points.

The locations of the laboratories were taken as starting points, relative to which polygons and point objects were constructed by interpolation, with information about the required time.

The results of the calculations for both options are presented in Figs. 1 and 2. For the centralized scheme, the service area time ranged from a minimum of 0–10 min to a maximum of 100 min. At the same time, for the centralized scheme, the delivery time of biomaterials was reduced to 70 min, and the corresponding economic costs decreased by 42%.

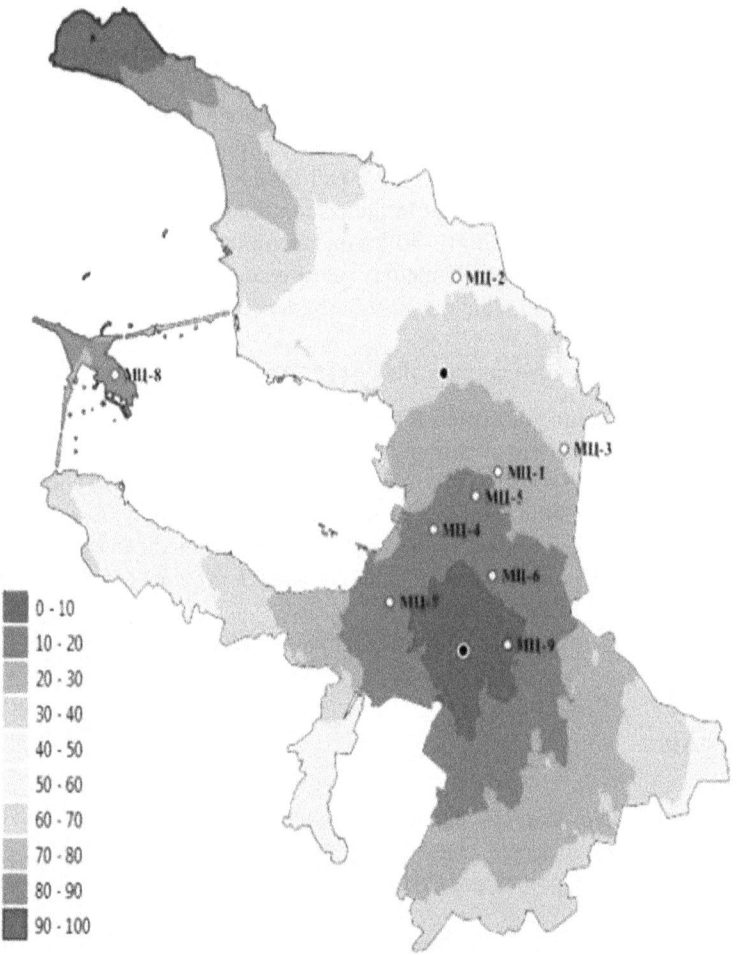

Fig. 1 Centralized scheme for working with biomaterials (with one laboratory)

Despite the clear advantage of a decentralized scheme in terms of delivery time for biomaterials, it is necessary to take into account the requirements for the placement of separate specialized equipment and adherence to the principles of flow.

The life cycle of work related to biomaterials includes both the physical processes of their collection, storage, transportation and analysis, and informational—the accumulation and use of relevant databases and knowledge for medical, economic and commercial purposes. The work is carried out in various conditions (mobile field, stationary, laboratory and clinical) with the participation of a large number of multi-disciplinary specialists. The quality of production, safety and efficiency of the use of biomaterials and information about them is ensured by strict adherence to regulations and their control, the use of special equipment and software. Thus, it is

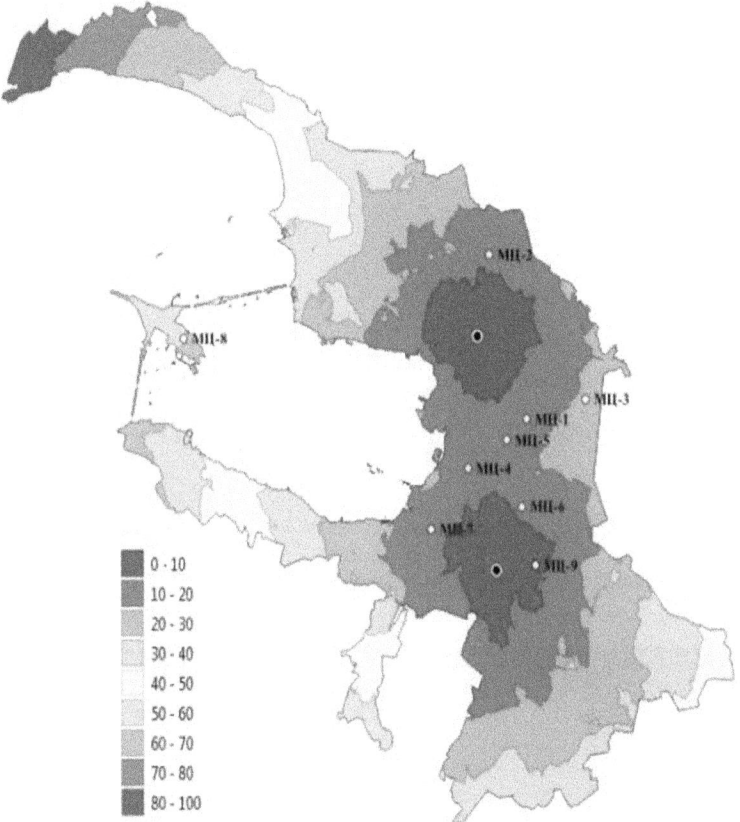

Fig. 2 Decentralized scheme for working with biomaterials (with two laboratories)

required to optimize planning processes, as well as organizational and management and production and technological processes for the effective support of work with biomaterials.

The use of GIS technologies allows maintenance of interconnected base of graphic and semantic information as well as usage of hyperlink tool to provide transitions to external resources. At the same time, the integrated approach to product lifecycle management implies the provision of more flexible and secure opportunities for documentation and communication between all stakeholders. Screenshots of interaction using the integrated environment are shown in Figs. 3 and 4.

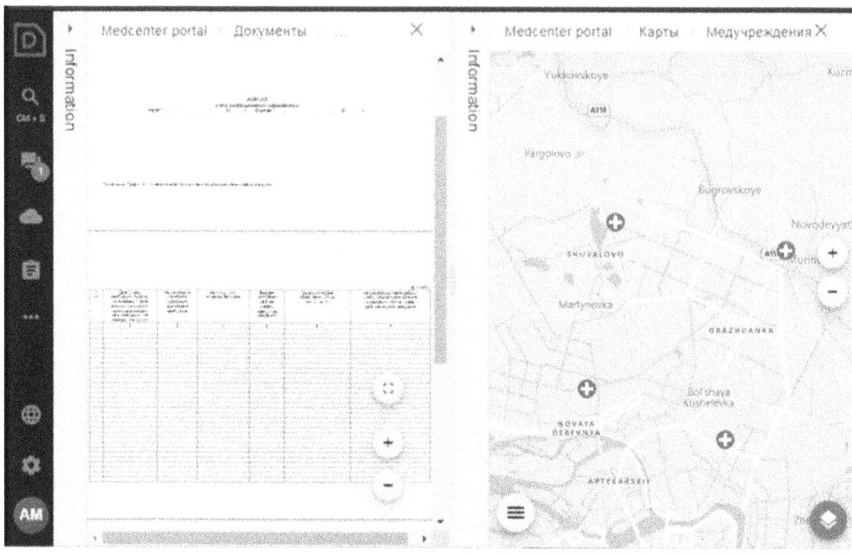

Fig. 3 Interaction of the duty service with the mobile brigade

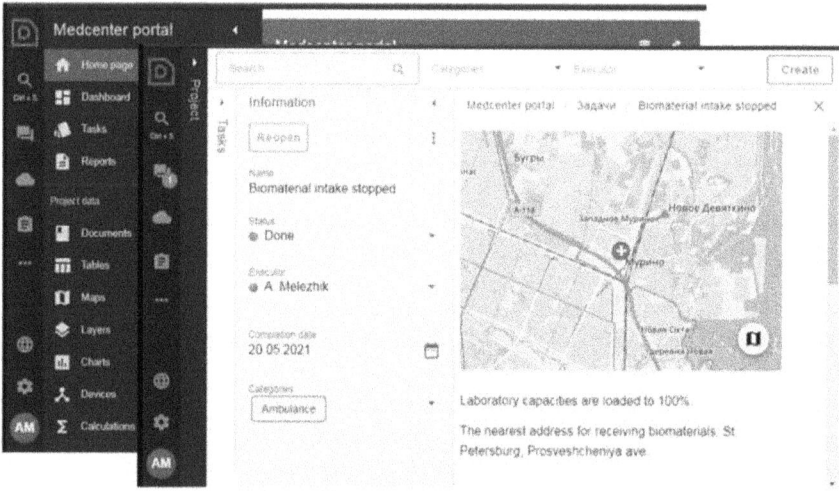

Fig. 4 An example of joint work with documents and maps for objects related to biomaterials

5 Discussion

The purpose of developing proposals for improving the system for ensuring work with biomaterials is to achieve a significant improvement in the prevention of nosocomial (hospital-acquired) infections and to ensure sanitary-pidemiological and environmental well-being in St. Petersburg through the introduction of centralized and decentralized systems for the placement of MCs for work with I-II pathogenicity groups for microorganisms, using the most advanced, cost-effective and environmentally friendly technologies that are guaranteed to prevent the spread of infections and ensure the impossibility of reuse of individual components of biomaterials, as well as a modern biomaterials management system in the region should include a unified system of accounting and reporting documentation and a system for tracking the movement of biomaterials at all stages.

Initial data required for the development of proposals for handling biomaterials for the use of GIS:

- A list of the region's MO with an indication of the capacity, the calculated coefficients of the quantitative substantiation of the formation of biomaterials.
- Passport of the Ministry of Defense, license for medical activities.
- Data on the technologies used in the MO or on the territory (a package of technical and permissive documentation).
- Accounting and reporting documentation in the MO, in the laboratory center.
- Regulatory documents in the field of activities of the Ministry of Defense and work with biomaterials.
- Project of the site for the location of the laboratory center.
- View of the premises of the site in 3D format.

6 Conclusions

1. New challenges and modern requirements for the epidemiological and environmental safety of infected biomaterials determine the need to create and use an appropriate infrastructure for collection, storage and transportation in conjunction with special technologies that ensure the safety of work.
2. When choosing the most reliable and economical layout of medical centers for working with biomaterials, it is effective to carry out spatial analysis based on GIS, taking into account the urban planning and transport conditions of the locations in question, the types and capacity of medical organizations for conducting microbiological laboratory studies.
3. The development of biomaterials management systems in regions could be based on decentralized, centralized or mixed principle of MC placement. The studies of the collection and transportation of biomaterials have shown that it can take less than 2 h (in hermetically sealed conditions, in compliance with all the principles of asepsis and antiseptic).

4. The use of an integrated environment (Datrics software united with GIS) allows users to manage the life cycle of work with biomaterials more effectively and economically, to save and accumulate the necessary medical knowledge, as well as contribute to improving the quality of communication of specialists of medical institutions.

References

1. Aytur T, Foley J, Anwar M et al (2006) A novel magnetic bead bioassay platform using a microchip-based sensor for infectious disease diagnosis. J Immunol Methods 314(1–2):21–29
2. Chakhalian D, Shultz RB, Miles CE et al (2020) Opportunities for biomaterials to address the challenges of COVID-19. J Biomed Mater Res Part A 108(10):1974–1990
3. Cromley EK (2003) GIS and disease. Annu Rev Public Health 24(1):7–24
4. Flahault, A (2003) SARS-CoV: 2. Modeling SARS epidemic. Medec Sci M/S 19(11):1161–1164
5. Kim J, Biondi MJ, Feld JJ et al (2016) Clinical validation of quantum dot barcode diagnostic technology. ACS Nano 10(4):4742–4753
6. Melezhik A, Petrovtsev P (2020) Web application for collaborative work with documents, GIS and monitoring "Datrics". Russian state registration certificate №2020619077. Moscow, FIPS
7. Mick P, Murphy R (2020) Aerosol-generating otolaryngology procedures and the need for enhanced PPE during the COVID-19 pandemic: a literature review. J Otolaryngol-Head and Neck Surg 49:1–10
8. Mironenko OV et al (2020) Hygienic assessment of pyrolytic incineration technologies for medical waste of the Russian Federation. Prevent Clin Med 4(77):46–56
9. Nhavoto JA, Grönlund Å (2014) Mobile technologies and geographic information systems to improve health care systems: a literature review. JMIR mHealth and uHealth, 2(2):e3216
10. Shaw N, McGuire S (2017) Understanding the use of geographical information systems (GIS) in health informatics research: a review. J Innov Health Inform 24(2):228–233
11. Shen W et al (2008). Systems integration and collaboration in construction: a review. In: Proceedings of the 12th international conference on computer supported cooperative work in design (CSCWD 2008), pp 11–22
12. Suh J, Kim SM, Yi H et al (2017) An overview of GIS-based modeling and assessment of mining-induced hazards: soil, water, and forest. Int J Environ Res Public Health 14(12):1463
13. Trubint N, Ostojić L, Bojović N (2016) Determining an optmal retail location by using GIS. Yugoslav J Oper Res 16(2):253–264. https://doi.org/10.2298/YJOR0602253T
14. Udugama B, Kadhiresan P, Kozlowski HN et al (2020) Diagnosing COVID-19: the disease and tools for detection. ACS Nano 14(4):3822–3835
15. Ye ZW, Jin DY (2020) Diagnosis, treatment, control and prevention of SARS-CoV-2 and coronavirus disease 2019: back to the future. Sheng wu gong cheng xue bao= Chinese journal of biotechnology 36(4):571–592

Architectural and Artistic Techniques for Humanization of Healthcare Spaces

V. M. Supranovich and N. M. Drizhapolova

Abstract The paper highlights the issue of architectural planning and artistic design of healthcare facilities with account the psychological and emotional status of patients. The evolution of healthcare architecture is traced in the context of the changing medical concepts. Characteristic examples are given of the international and domestic healthcare campuses. The study is limited to urban non-infectious and non-mobile complexes, aiming to explore only the architectural and design techniques contributing to the therapeutic effect of architectural environments. The search for architectural and artistic techniques makes use of graphical analysis and form making methods. Recommendations have been developed for consideration by healthcare design and construction teams at the level of architectural and space-planning design and artistic design solutions. The materials of this paper are intended for landscape architects, medical professionals, government bodies responsible for national healthcare policy, as well as students.

Keywords Architecture · Design · Architectural and artistic techniques · Healthcare facilities · Innovations

1 Introduction

In the modern world, the issues of re-shaping the image of medical and healthcare facilities are being widely discussed. As the development of innovative medicine offers more effective treatment, rehabilitation and quality care for seriously ill people, the length of patients' stay at hospitals and medical centers increases, acquiring, in some cases, a systematic nature. Therefore, the place of medical care ceases to be perceived solely in terms of its original function. The environment of patients' stay, both internal and external, is one of the factors affecting patients' psychological state and, consequently, recovery.

V. M. Supranovich (✉) · N. M. Drizhapolova
Saint Petersburg State University of Architecture and Civil Engineering, Saint Petersburg, Russia
e-mail: vmsupranovich@gmail.com

© The Author(s), under exclusive license to Springer Nature Switzerland AG 2023　　　235
D. Ivanov et al. (eds.), *Proceedings of ECSF 2021*, Lecture Notes in Civil
Engineering 257, https://doi.org/10.1007/978-3-030-99877-6_28

2 Materials and Methods

The history of healing originates in the ancient world [14], but the appearance of medical institutions as we know them today is associated with the Greco-Roman Period [15]. Those early hospitals were intended for short stay, but gradually, as the treatment concepts changed, the duration of stay increased.

In the 20th century, the main criterion in the design of medical buildings was functionality [12]. In the 21st century, evidence was found of the architecture and design having a greater influence on the condition of patients and the performance of doctors than was previously thought [12]. It is therefore important that modern medical facilities receive expressive and sustainable architectural solutions [12] for their layouts and facades.

The national project "Health", that has been implemented since 2006, reflects the tasks of healthcare design as established in Federal Law N 323-ФЗ "On fundamental healthcare principles in the Russian Federation", enacted on November 21, 2011. It is stated in Article 6, part 1, paragraph 5 of this Law that "design and location of healthcare facilities will be considered based on the sanitary and hygienic standards and the need to ensure comfortable conditions for patients" [12].

This study aims to identify and test the architectural and artistic methods for creating the modern image of healthcare facilities that take into account the psychological factor.

Achieving this purpose involves:

- identifying the main trends in the design of architectural and space-planning solutions for medical facilities (philosophies behind layouts; symbolism of facades);
- developing the shape options for elements of healthcare buildings; and
- establishing the relationship between the natural landscape and the healthcare facility.

In medical practice, the concept of "psychological status of patient" has gained a strong foothold. According to its definition, the recovery process depends not only on patient's physical condition and methods of treatment, but also on the moral status [7]. The environment of patient's stay, the surrounding space and the buildings are found to be able to influence patient's psychological state [11]. Their effect can be negative or positive [4]. To achieve its tasks, the study made use of the following methods:

- survey of the international experience and trends in healthcare construction. The example of Paimio tuberculosis sanatorium, designed by architect Alvar Aalto, has been studied [10]. Developed in 1928 and subsequently implemented between 1929 and 1933, the design concept for Paimio Sanatorium assigned it the role of a constituent element of landscape [6]. One special element of the design is sundecks (balconies and verandas with special furniture), allowing both patients and employees to stay in the open air [13] and divided into categories (for severe patients, for patients in moderately grave condition, and for employees),

- survey of the domestic experience of innovative medical facilities design. The example of the Regional Center for Reproductive Medicine has been studied, a custom design developed in 2015–2016 by Reinberg and Sharov Architectural Studio. According to its concept, the building was to be located in the urban environment, and its image and environment were meant to create a positive psychological effect on the patients. The philosophy of reproductive technologies has found reflection in both planning and architectural design. The key design criteria included orientation of the building and the welcoming interior, the latter enhanced by the central core having two functionally different zones on its sides. The design solution for the façade uses the "encrypted code" resembling the human DNA code;
- graphical analysis. Graphical analysis was applied by students of the Architectural Design Department as an association search tool (Figs.1, 2 and 3; photo courtesy of V.M. Supranovich; layout by V.M. Supranovich, 2021). This method uses graphic interpretations of various images and representations of natural textures

Fig. 1 The graphical analysis of pine grove image: from nature to the façade design solution (student A. Safronova, academic supervisor V.M. Supranovich)

Fig. 2 The graphical analysis of pine trunk: from natural texture to facade design solution (student A. Safronova, academic supervisor V.M. Supranovich)

Fig. 3 The graphical analysis of pine trunk: from nature to axial plan (student A. Safronova, academic supervisor V.M. Supranovich)

for application of the resultant ornaments and textures in the facade (Figs. 1 and 2) and space-planning solutions for medical facilities (Fig. 3). It is the living nature that unveils the laws of harmony between function and form [9],

- form making [8]. According to the available data, the appeal to nature in the architecture of medical buildings helps to soften patients' perception and stabilize psychological. Therefore, the use of natural components in shaping of objects offers a way to humanize medical facilities.

3 Results

The study has revealed a number of recommendations (prerequisites) to be considered when developing the healthcare design concepts at the level of architectural and space-planning design and artistic design tools.

For the design solution to achieve psychologically positive and emotionally comfortable effect on patient's mood, it is necessary to focus on the following architectural and space-planning aspects:

1. Maximized creative expression that inspires interest and distracts from negative thinking.
2. Customized expressive framework. Since the association with typical hospitals is that of painful treatment and can be negative with many patients, new hospital campuses should use more expressive design solutions.
3. Greater thematic expression of individual buildings within healthcare campus for easier orientation.
4. Clear space-planning solution is prerequisite to easy indoor and outdoor navigation within the healthcare campus. Buildings should create impression of

full-fledged public spaces [5]. Patients are not supposed to be in panic from feeling lost and late for treatment.

5. The availability of recreation and entertainment premises can decrease negative thinking in patients and distract them from everyday life (cafe, library, winter garden, billiard room, cinema, gym, family meeting place, etc.).
6. Green spaces as elements of outdoor and indoor environments of healthcare campus [1]—softscapes, water bodies, beautiful landscapes, etc.
7. Noise protection (integration of noise protection devices into architectural and space-planning solutions) to provide patients with quiet environment and sense of protection.

Further, the innovative healthcare projects can be humanized using a range of artistic design techniques. The following recommendations apply here:

1. Use of various textures, surface finishes or ornaments associated with natural motifs as elements of facades and interior design.
2. Integrated approach to interior shapes that avoids flat, angular forms and welcomes smooth curves.
3. Research-based use of color solutions for interior spaces. Bright and muted tones are recommended for spaces where patients need to be cheered up or calmed; cool and warmer colors are recommended for rooms with fever reducing treatment or to eliminate chilly sensations.
4. Noise protection solutions in design layout (partitions, screens, green areas, upholstered furniture, etc.).

Of special attention is the spiritual component of the treatment process. Healthcare facilities can be recommended to have premises where patients could practice their religion (elements of Orthodox church, synagogue, mosque).

Careful consideration of the above recommendations in the design of high-tech medical facilities will allow creating a modern therapeutic environment designed to facilitate recovery and further social adaptation. Humanized healthcare environments offer better working conditions for the medical staff and their communication with patients can become more relaxed.

4 Discussion

The proposed architectural and space-planning recommendations and artistic design approaches have been tested as part of theses delivered by undergraduate and master's students of Architectural Design Department, Saint Petersburg State University of Architecture and Civil Engineering:

- draft design solution for rehabilitation center for obesity treatment, developed by Arina Safronova, student of Architectural Design Department, Saint Petersburg State University of Architecture and Civil Engineering (academic supervisor: V.M. Supranovich).

The design brief describes this rehabilitation center as located in the town of Roshchino, Leningrad Region. Its building is subject to reconstruction and repurposing. The design concept involves modernization of the building's exterior based on the results of graphical analysis and intended to integrate the building into the existing natural environment. The design brief prescribes a symmetrical axial composition of the building layout and construction of a catering and sports building. The exterior of the building is proposed to avoid sharp corners and flat lines in favor of a pine grove imitation. Along with the façade solution, other elements contributing to the building's naturalistic design are the winter garden and its recreation area (Fig. 4; photo courtesy of V.M. Supranovich, 2021).

- draft design solution for the Center for Preventive Care and Health, St. Petersburg, developed by graduate student Aida Ospanova as part of her master's thesis in 2020 (academic supervisor: N.M. Drizhapolova).

The Center is proposed to be located in the gray belt of St. Petersburg in the premises of the former factory, which is supposed to be relocated [3]. Earlier in the 19th century, the place housed municipal water and mud baths [3]. The purpose of the Center corresponds to the original purpose of the site and its natural resources. The Center is intended as a provider of preventive care and, mainly, outpatient treatment (with a small number of beds for non-severe patients), diagnostic studies, short-term rehabilitation and treatment courses (some part-time and not requiring visitors to interrupt their jobs), and healthcare recreational entertainment systems to offer, among other things, lectures, practical classes and conferences on healthy life-style [3] (Fig. 5,photo courtesy of N.M. Drizhapolova, 2020).

The facility is located on the bank of the Neva River and is a 2-section structure [2]. Its western section has an orthogonal design to symbolize the "western medicine" (evidence-based medicine), while the eastern section uses bionic, sinuous lines to

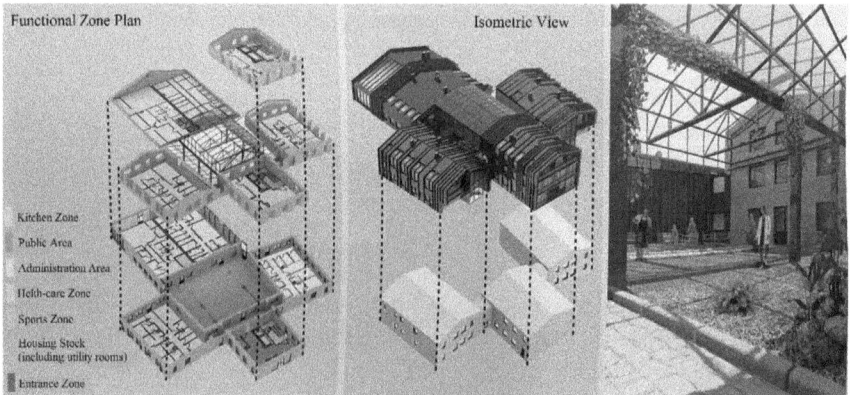

Fig. 4 Draft design solution for rehabilitation center for obesity treatment. Functional zoning integrates naturalistic design elements in general layout, facades and interior (student A. Safronova, academic supervisor V.M. Supranovich)

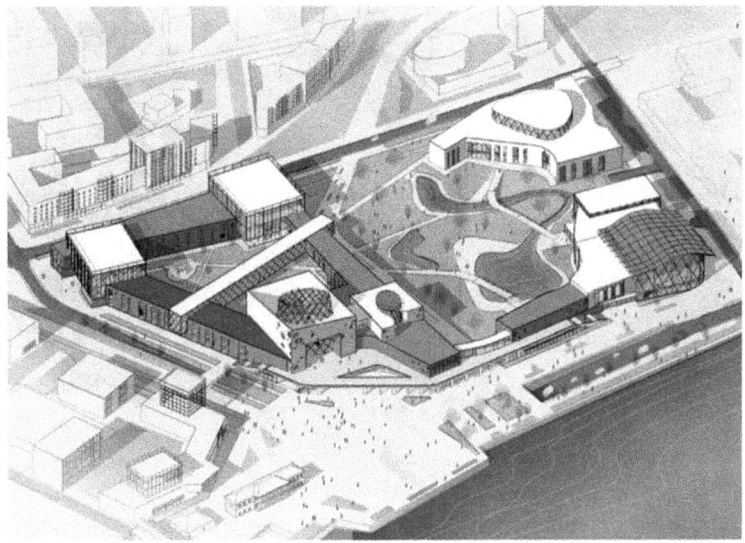

Fig. 5 Draft design solution for the Center for Preventive Care and Health, St. Petersburg. Bird's-eye view (developed by master's student A. Ospanova; academic supervisor N.M. Drizhapolova)

embody the "eastern medicine" (integrative medicine). The western section houses the premises to be found in a standard medical center (lobby, waiting room, doctors' offices, day and night hospitals, laboratory, research center, etc.). The eastern section houses SPA area, oriental medicine clinic, cafe, sports and wellness facilities. A separate building will house the assembly hall for lectures, workshops, conferences, as well as a hotel for visiting patients and doctors. The "nature block" constitutes a separate functional area and will house a large garden park with existing and recreated ponds and beach (that can function as an ice rink in winter), open areas for recreational gymnastics and walk therapy.

The two sections are connected by a spacious, multi-storied atrium with a clerestory that will house recreational and entertainment premises (cafes, family and friends meeting places, winter garden, etc.). The second level of the atrium will have an exit to an open terrace with a nice view of the Neva River and the city [2] (Fig. 6,photo courtesy of N.M. Drizhapolova).

5   Conclusion

The modern innovative medicine relies on an integrated approach that combines treatment and research, requiring a revision of the architectural and artistic solutions for healthcare buildings. Strict functionality and unification give way to humanization of healthcare landscapes, external and internal environments as elements that can

Fig. 6 Draft design solution for the Center for Preventive Care and Health, St. Petersburg. Perspective view (developed by master's student A. Ospanova; academic supervisor N.M. Drizhapolova)

improve patients' psychological status. This, in turn, requires new projects to involve novel architectural and space-planning layouts and artistic design techniques.

The proposed recommendations, tested as part of student theses, provide further evidence of the possibility for elements of functionality and producibility, which are integral characteristics of healthcare facilities, to be qualitatively combined with those of humanized environments and individual design solutions for clearer outdoor and indoor navigation.

References

1. Buka VY, Ivanova OG, Sheromova IA (2020) The theme of nature in the architecture of buildings. Archit Des Hist Theo Innov 4:187–192
2. Drizhapolova NM, Supranovich VM (2021) Pedestrian "House on the Embankment"(as exemplified by St. Petersburg Grey Belt). In: Contemporary problems of architecture and construction. CRC Press, pp 14–19
3. Drizhapolova NM, Matushkina A, Ospanova A et al (2020) Transforming the industrial territory of the southeastern part of Vyborg Side. In: X regional creative forum with international participation "Architectural Seasons in SPbGASU", pp 159-160
4. Ellard K (2016) Habitat: How architecture affects our behavior and well-being. Russ. ed.: Ellard K (2019) Sreda obitaniya: Kak arkhitektura vliyaet na nashe povedenie i samo-chuvstvie. Al'pina Didzhital Publ, Moscow
5. Gelfond AL (2013) The architectural design of public spaces. study guide for universities: N. NNGASU, Novgorod
6. Gozak A (1978) Alvar aalto. architecture and humanism. Progress, Moscow
7. Ivanchenko AV, Irimanova EV (2021) Analysis of architecture of children's medical centers. Matrix of Scientific Knowledge 3–2:198–201
8. Kurbatov YI (2017) Prerequisites for full-fledged configuration of an architectural form. Bull Civ Eng 4(63):23–25
9. Lebedev YS, Rabinovich VI, Polozhay ED (1990) Architectural bionics. Stroyizdat, Moscow
10. Levoshko S (2016) Alvar aalto. Direct-Media, Moscow

11. Makarova EA, Romanov OS (2017) Characteristics of architectural and space-planning orientation of medical rehabilitation centers in global and domestic practice. In: Current challenges of architecture: proceedings of the 70th Russian research to practice conference of graduate students and young scientists "Challenges of Modern Construction". Saint Petersburg State University of Architecture and Civil Engineering, SPb:220-225
12. Poydem EA, Mikhalychev AV (2017) Priority of patient interests in the design of healthcare facilities. In: Current Challenges of architecture: proceedings of the 70th Russian research to practice conference of graduate students and young scientists "challenges of modern construction. Saint Petersburg State University of Architecture and Civil Engineering, SPb:252-255
13. Yedike J (1970) History of modern architecture. Art, Moscow
14. Yerykov AA (2019) The emergence of healthcare architecture. Bull Sci Creativity 10(46):51–53
15. Zakieva LF (2016) Analysis of healthcare evolution. Int Res J 11–2(53):73–74

Building Information Modeling-Based Engineering Systems Design

I. N. Chikovskaya, I. I. Sukhanova, and K. O. Sukhanov

Abstract In the design of healthcare facilities' engineering systems (heating, ventilation, water supply, sanitation, outdoor networks, etc.), the building information modeling (BIM) is a highly effective tool that helps avoid design errors of various nature. Organizational deficiencies are revealed that affect combined activities. Measures are proposed for a more correct information modelling of engineering systems. Autodesk Revit software is analyzed for potential to be used as engineering design tool.

1 Introduction

The pandemic has forced all countries to focus on health facilities. In Russia, this has led to an increase in construction of social projects and tightening of construction timelines and monitoring by authorities.

In terms of design and construction, healthcare projects are much more complicated than public or residential buildings, recently requiring larger investments due to their expensive equipment and complex engineering systems [10].

The efforts to reduce construction timelines and the cost of healthcare projects lay much hope with building information modeling and, especially, careful design of engineering systems—heating, ventilation, air conditioning, water supply and removal, among other systems [5]. Multiple engineering systems make projects sophisticated and unique, so healthcare facilities set themselves apart from most civil construction projects by their complex layouts. The level of complexity of healthcare projects is comparable to that of industrial facilities.

In Russia, the customer for the majority of hospitals, out-patient clinics and related healthcare facilities is the state, the public–private partnerships and concessions accounting for only a small part of their number.

I. N. Chikovskaya · I. I. Sukhanova (✉) · K. O. Sukhanov
Saint Petersburg State University of Architecture and Civil Engineering, Saint Petersburg, Russia
e-mail: suhin@lan.spbgasu.ru

The healthcare projects being constructed in Russia are comparable in quality with international ones. Some of them are developed using building information modeling technologies (BIM technologies) [9].

2 Materials and Methods

Despite some difficulties associated with the application of BIM technologies by organizations [2, 8], the advantages they offer capital construction projects, and especially the engineering systems, throughout their life cycle are beyond doubt.

The global design practice uses different sections of the BIM analysis:

- energy modeling [7, 13],
- numerical modeling of air environment status [1, 3, 4, 6, 14, 15].

This study highlights the benefits of using BIM technologies in the design and installation of healthcare engineering systems.

The Russian and international experience of using BIM-based engineering solutions has shown the importance of BIM technology at the design stage.

Classified as social facilities, healthcare projects are installed with engineering systems as complex as those of industrial facilities. Unlike other types of social facilities, medical facilities operate a number of dedicated systems—medical gas supply, telehealthcare services, pneumatic installations (Fig. 1).

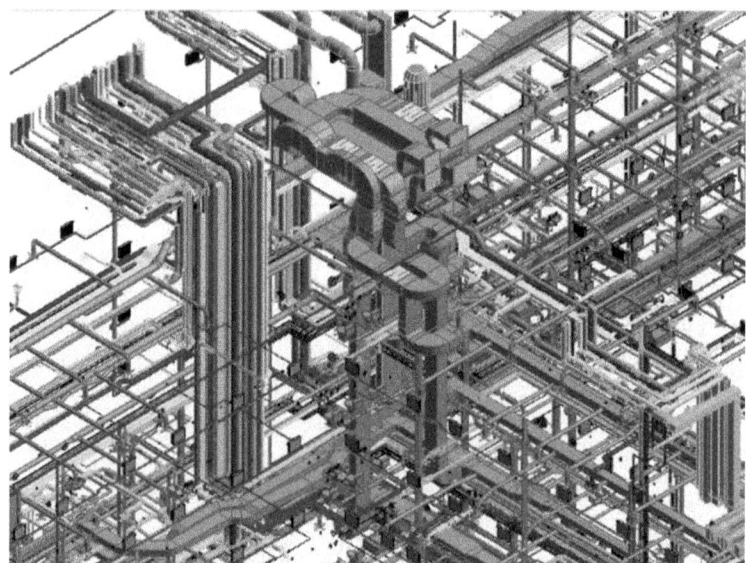

Fig. 1 The fragment of the information model of the outpatient clinic's engineering systems

While creating the digital information models based on ready-made healthcare design solutions and exploring the applications of information modeling technologies, a number of important points have been identified that are not being considered when implementing BIM.

3   Results

The utility tunnels run through corridors, basements and crawl spaces of buildings. The abundance and size of engineering systems can seriously challenge the design and installation teams, given the limited height of premises and the strict standards for laying of conduits.

Initially, the use of information modeling technologies was reduced to reproducing the traditional design solutions. Gradually, the design engineering process acquired its true form, 3D, and models to support design documentation development. The use of information modeling technologies per se does not lead to optimized project activities, specifically when it comes to combining of expertise. To date, work is in process to identify and improve ways to facilitate exchange between specialists with different backgrounds who are working on the same project.

A number of organizational deficiencies have been identified in the course of this study, that can lead to:

- lower height of conduits, which is completely unacceptable. In the operating rooms, there is a relation between the height of ventilation systems and the level of contamination of the operating field [11],
- inconsistencies in structural concept,
- misalignment of system elements on different floors (Fig. 2).

The paper proposes a method for positioning and aligning of shafts and utility tunnels in buildings that uses extensive solid-state elements (vertical and horizontal). Figure 3 shows location of the ventilation shaft embedded in the architectural solution (red), but not executed in its model (the shafts were introduced discretionally) and of the ventilation system's air ducts. It can be seen that the dimensions of the ducts exceed the volume allocated for the shafts of this given system. This error could have been avoided at an early stage of the project.

As a result of works, it was decided to move shaft modeling to Architectural Solutions section. The shafts were modeled within the given floors and the volume limited by the enclosing structures. This approach has made it possible to ensure, at an early stage of the design, the sufficiency of engineering systems path, on the one hand, and on the other it provided the design team with the visual image of the bearing part of the building, while also allowing to avoid penetration of structural elements into the shafts.

It was also decided to divide the engineering systems by models. It is recommended that a separate file should be created for each type of engineering systems (Fig. 4). Not only has this simplified the administration procedures and conflict

Fig. 2 Example of a fire
extinguishing system with
misaligned elements

Fig. 3 Example of the
ventilation shaft embedded
in the architectural solutions

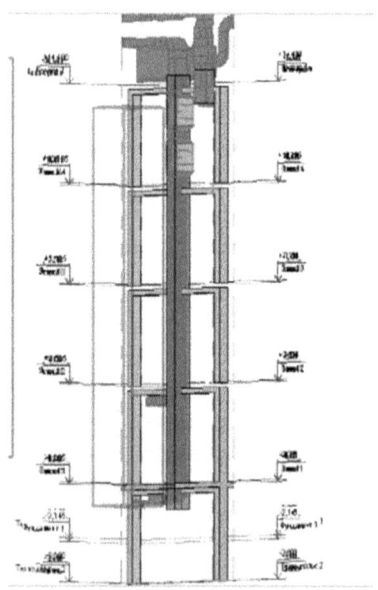

detection, it also allowed to make changes to models by system type.

Even though the existing building information modeling software has proved an effective tool in engineering systems design, design engineers, unfortunately, cannot always rely on the completeness and accuracy of the results obtained.

Fig. 4 The information model of the outpatient clinic's ventilation systems

Different software packages offer different ways of initial data processing and different forms of calculation data presentation. Every software seems to have its advantages and disadvantages [12, 14, 15] and has room for improvement.

Along with resistivity method, Autodesk Revit software was used in the design of heating systems (Fig. 5) and ventilation (Fig. 6).

The analysis of calculation results has enabled a conclusion that the software operates generally satisfactory, providing the main outputs, correctly selecting the cross-section sizes for pipelines and air ducts based on water and air velocity parameters, and calculating pressure losses along section. However, when calculating local pressure losses, deviations occur. While with ventilation systems the deviations did not exceed 15%, in heating systems they were as high as 30%. It is possible to create a relation between pressure losses and valve throughput in the valve family.

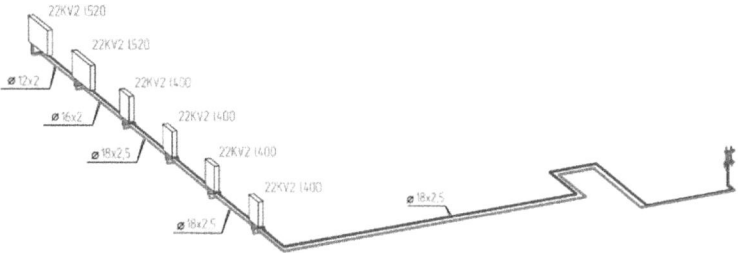

Fig. 5 The fragment of the heating system model

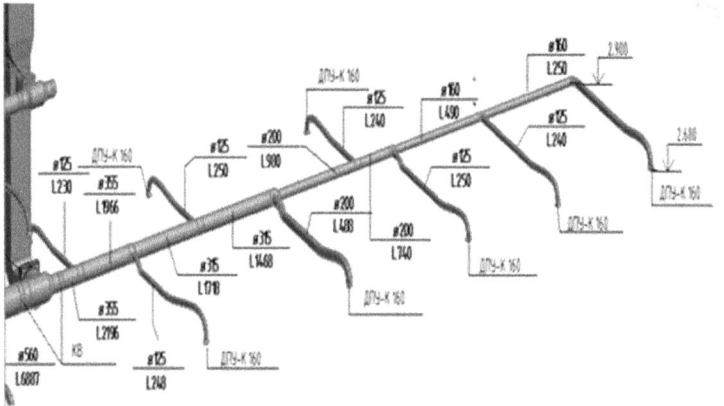

Fig. 6 The fragment of the forced ventilation system model

Given that all the families of connecting elements, fittings and equipment are config-
ured correctly, then the pressure losses will be calculated correctly as well, and the
hydraulic calculation in Revit will prove adequate for practical use by designer teams.

There are also dedicated software packages for detailed calculation of heating
and ventilation systems (liNear, MagiCAD, to name a few). In other words, it would
only expedient to integrate the information model data with the calculation modules
and to develop two-direction integration.

4 Discussion

Presenting the results of its research, the RF Ministry of Construction, Housing and
Utilities (MinStroy) claims that BIM improves efficiency performance by achieving
a 30% reduction in construction and operation costs; nearly 40% reduction in
design documentation errors; nearly 50% reduction in project implementation period;
sixfold reduction in the time needed for model verification; fourfold reduction in the
percentage of budget planning errors; nearly 90% reduction in the time needed for
project reviewing and approval; 10% reduction in construction timeline; and 20 to
50% reduction in design timeline [10].

The conducted research allows to make proposals towards minimizing the
possibility of errors in the information modeling of healthcare facilities:

- For more accurate positioning and alignment of ventilation shafts and utility
  tunnels, the solid-state elements modelling should be made in Architectural
  Solutions section.
- It is recommended that a separate file should be created for each type of
  engineering systems.

- For more accurate hydraulic calculation results, the families of the connecting elements, fittings and system equipment should be adjusted in Autodesk Revit. Or dedicated software should be used.

5 Conclusion

By creating digital information models, the engineering systems can be analyzed, managed and monitored at all stages of their life cycle of medical facilities from design stage to installation to operation to dismantling.

Acknowledgements The study has been conducted as part of the project "BIM-ICE—BIM Integration in Higher and Continuing Education", funded by 2014–2020 CBC Programme "South-East Finland—Russia".

References

1. Andersson H, Cehlin M, Moshfegh B (2018) Experimental and numerical investigations of a new ventilation supply device based on confluent jets. Build Environ 137:18–33
2. Astafieva N, Kibireva J, Vasileva IL (2017) Advantages of using and difficulties in implementing of building information modeling. Constr Unique Build Struct 59(8):41–62
3. Averyanov V, Vasiliev V, Ulyasheva V (2018) Selection of turbulence models in case of numerical simulation of heat-, air-and mass exchange processes. In: E3S Web of conferences, vol 44. EDP Sciences, p 00005
4. Balocco C, Petrone G (2015) Microclimate and indoor air quality in an operating theatre under real use conditions—an experimental and numerical investigation. In: Current air quality issues. IntechOpen. https://doi.org/10.5772/59671
5. Borisoglebskaya AP (2008) Medical and preventive treatment facilities. In: General requirements to the design of heating, ventilation and air conditioning systems. AVOK-PRESS, Moscow
6. Chen H, Janbakhsh S, Larsson U et al (2015) Numerical investigation of ventilation performance of different air supply devices in an office environment. Build Environ 90:37–50
7. Grimitlin AM, Deniskina DM (2018) Energy modeling—a tool for improving the energy efficiency of buildings, BIM Modeling in the problems of construction and architecture: Proceedings of research-to-practice conference, Saint Petersburg State University of Architecture and Civil Engineering, pp 93–97
8. Leśniak A, Górka M, Skrzypczak I (2021) Barriers to BIM implementation in architecture, construction, and engineering projects—the polish study. Energies 14(8):2090. https://doi.org/10.3390/en14082090
9. Lyalin DO, Mashtaler SN, Dmitrenko EA (2017) Application of the autodesk revit software complex in project activities. Proc Donbas Natl Acad Civ Eng Archit 3:23–27
10. Medvedeva O, Sautkina T (2019) Specifics of utilities design using information modeling. Archit Eng 4(3):13–21
11. Memarzadeh F, Jiang Z (2004) Effect of operation room geometry and ventilation system parameter variations on the protection of the surgical site. In: Proceeding of IAQ
12. Miller VV, Zhilina TS (2011) Characteristics of the novel software for hydraulic and thermophysical analyses of heating and heat supply systems. Bulletin of Moscow State University of Construction, vol 7

13. Sukhanov KO, Planes MV (2019) The problems of software interaction during building information modelling, BIM modeling in the problems of construction and architecture. In: Proceedings of research-to-practice conference, Saint Petersburg State University of Architecture and Civil Engineering, pp 184–188
14. Sukhanova II, Gnedykh VS, Demshinka DA (2019) BIM-based hydraulic and aerodynamic analyses of heating and ventilation systems. Engineering Bulletin of Don 9
15. Sukhanova I (2019) Numerical modeling of the microclimate and air quality of an Orthodox church in Saint-Petersburg. In: E3S Web of conferences, vol 91. EDP Sciences, p 02002

Public–Private Partnership in Innovative Healthcare Construction

G. F. Tokunova

Abstract The tasks of modernizing healthcare system necessitate the use of private capital. Global practice shows that public–private partnership (PPP) has proven an effective tool in promoting innovation-driven healthcare services and wiser use of high-tech equipment. The paper discusses the international experience of PPP projects. The challenges and prospects of the PPP institute in Russia are highlighted using the example of the national project "Healthcare". The author proposes a methodology for monitoring PPP projects throughout the entire life-cycle of a municipal outpatient clinic project.

1 Introduction

The two most effective public–private interaction models that appeared in the UK in the early 1990s were Private Finance Initiative and Public–Private Partnership (PPP). These models enabled the construction, within a short timeline, of large projects such as Euro Tunnel and London Tube, as well as smaller facilities—schools, hospitals, prisons, etc. [1, 9, 17]. The idea of public–private interaction is not new, but at this stage its innovative nature lies in the relations between parties, that rely on principles of equal partnership, and in the areas being covered by such partnerships, that were previously jurisdiction of the state [14, 15].

Public–private partnership in healthcare is defined as formal or informal arrangement between public and private organizations, designed to achieve public health goals or create health-related products or services. Parties to a PPP share all potential risks, exchange resources, and focus on priority research and development [3, 4, 8, 11–13, 16].

G. F. Tokunova (✉)
Saint Petersburg State University of Architecture and Civil Engineering, Saint Petersburg, Russia
e-mail: tokunova.g.f@lan.spbgasu.ru

2   Materials and Methods

The study involved the analysis of statistical and theoretical data on the challenges of public–private partnership as an institute existing in Russia and abroad. The obtained results made it possible to develop a methodology for monitoring PPP projects in healthcare.

To analyze the current situation, the following scientific methods were used: analogy as a method allowing to identify dynamics emerging in relation to the key PPP performance indicators in healthcare; comparative analysis; and analysis of regulatory frameworks.

3   Results

Today, attracting private investment in healthcare represents a key policy priority which is being implemented within the framework of the national project "Healthcare", designed to expand the network of hospitals and outpatient clinics. In this regard, the mechanism of public–private partnership can be expected to produce a significant improvement of the quality of medical services, promote new high-tech healthcare providers on the market, and contribute to healthcare modernization efforts [2].

With limited financial resources, the state can benefit from PPPs in term of cost efficiency, competition among private and public healthcare providers, and new jobs. The private sector, in turn, enjoys access to government support and guarantees of long-term funding. In PPPs, the state retains its regulatory and monitoring function while delegating all economic issues of the project to the private sector [5–7, 10].

And yet, the institute of public–private partnership in healthcare has not received sufficient development in Russia. This is due to a number of circumstances, the main ones being instability of the external operating environment in healthcare sector; lack of sufficient experience in PPP project implementation; inefficiency and high risk for the state; lengthy period of documentation preparation and review (up to one year at the stage of project specification development and up to six months or more at tender stage); unfavorable project conditions for the private sector; and insufficient regulatory framework governing public–private partnerships in health services sector. As of mid-2019, there were approximately 160 PPP projects in different stages of their life cycle in the country, sharing a totaling funding of slightly more than RUB 100 billion. Of those 160, 70 had been put into operation. In March 2020, this figure rose to 100, and the investment expenditure totaled RUB 156 billion, of which RUB 107.9 billion (70%) were private investments. Among the most successful projects are the center for nuclear medicine in Buryatia (RUB 1 billion), the department of radiology in Irkutsk (RUB 5.2 billion), the multifunctional hospital in Samara (RUB 3.5 billion). Every project required an investment of RUB 1 billion on average and had a payback period of 19.5 years. At the same time, according to the official data, in 2017 the healthcare sector suffered a shortage of RUB 1 trillion in fixed assets investment to eliminate its current wear.

In this regard, in 2019, the Ministry of Health of the Russian Federation considered a number of initiative on updating the regulatory framework for public–private partnerships and concession arrangements for long-term infrastructure projects.

To date, the vehicles for implementing healthcare infrastructure projects with extra-budgetary funds include Public–Private Partnership Agreement; Municipal-Private Partnership Agreement (Law N 224-Ф3); Concession Agreement (Law N 115-Ф3); Life Cycle Contract (Law N 44-Ф3); Public Property Lease and Capital Committment Agreement (Civil Code of the Russian Federation, Law N 135-Ф3); Social Entrepreneurial Corporation involving public and private capital for joint development and management of public infrastructure facilities (Civil Code of the Russian Federation, Budget Code of the Russian Federation); long-term contract for the supply of goods, works and services with contractor's investment obligations (Law N 223-Ф3).

Pursuant to Federal Law N 224-Ф3 "On Public–Private Partnership, Municipal-Private Partnership in the Russian Federation and Amendments to Relevant Legislative Acts of the Russian Federation" dated July 13, 2015, the monitoring of PPP projects will take place throughout their life cycles. PPP projects involve four stages:

Stage 1 is design engineering. It starts on the date of signing of agreement and ends on the date when design documentation receives positive expert opinion and relevant construction permit is issued.

Stage 2 is construction and commissioning. It starts on the date of receipt of the construction permit and ends on the date of issuance of operational acceptance certificate.

Stage 3 is operation. It starts on the facility commissioning date and ends on the date the facility is transferred to the state upon completion or termination of the agreement.

Stage 4 is transfer of the facility to the state upon completion of the agreement. It starts on the date the facility has been transferred and ends on the date of signing the certificate of transfer and acceptance or other document confirming the transfer of the facility to the state.

Progress monitoring is designed to timely identify the problems and potential risks and to provide recommendations as to risk minimization. The monitoring system comprises and regulates information analysis, reporting, data collection responsibility and decision-making, and will be established jointly by all stakeholders before the start of the project. Based on the results of monitoring, corrective actions will be provided to facilitate further progress of the project. Table 1 proposes possible methodology for evaluating the fulfillment by the private partner of his obligations.

Monitoring makes it possible to identify deviations and/or inconsistencies with contractual obligations or key performance indicators (KPIs), which allows the state, represented by regulatory authorities, to take remedial actions in the form of a corrective action notice, notice of irregularity, notice to remedy breach or penalties (Table 1).

Table 1 General structure of private partner performance evaluation

Monitoring and control stages	Compliance with contractual obligations
1. Monitoring	Monitoring of partner's compliance with contractual obligations
2. Corrective action	Corrective action notice is sent in case of first-time breach. Repeated breach is followed by notice of irregularity
3. Warning threshold	A warning or notice is issued within a period of 3 months, during which the breach is identified. Not more than two warnings may be issued during each such period
4. Duration of warning	24 months following the date of issue
5. Notice of irregularity	If 3 warnings have been issued over 9 consecutive months
6. Duration of notice of irregularity	24 months following the date of issue

In the course of project implementation, the partner must fulfill all the requirements stipulated by the agreement. A methodology is proposed for quantifying the fulfillment.

According to this methodology, the assessment involves four stages.

Stage one. Each indicator is quantified using a scale from 0 to 1: $0 =$ the indicator is not met; $1 =$ the indicator is fully met.

Stage two. Multiplication of the assigned score by the weight of indicator and, for the purposes of complex (integral) assessment, summation of all the results obtained (Eq. 1):

$$K.i = S.1 + S.2 + ... + S.n \qquad (1)$$

where $K.i$—integral indicator at ith stage of the project; Sn—final score for nth indicator.

$S.n$ is calculated by using Formula 2:

$$S.n = Vn * Bn \qquad (2)$$

where Vn—weighing factor of nth indicator; Bn—nth indicator score.

The weighting factor is a numerical coefficient and parameter reflecting the significance, relative importance, weight of a given indicator in comparison with other factors influencing the process under analysis. The value of each weight is determined by expert evaluation method. The weight value is set from 0 to 1, and the sum of all weights should be equal to 1. The weighing factor values can be revised, but not more than once a year.

This method allows evaluating K at the following compliance levels:

K.a—level of compliance with design specifications;
K.b—level of compliance with construction specifications;
K.e—level of compliance with service and maintenance requirements;
K.m—level of compliance with requirements to the provision of medical services;

K.p—level of compliance of the facility to the transferred with its technical specifications.

K.e and K.m are evaluated for one monitoring period. To evaluate them, integral indicator K.i should be calculated.

K.i is the integral indicator showing compliance with service, maintenance, repair and medical service provision requirements (Eq. 3). Its calculation uses indicators K.e, adjusted for weighing factor 0.45, and K.m, adjusted for weighing factor 0.55.

$$K.i = 0, 45 * K.e + 0, 55 * K.m \qquad (3)$$

Stage three. This stage involves comprehensive (integral) assessment of the fulfillment of requirements stipulated in the agreement and has as its outcome the quantified level of partner's compliance with contractual obligations.

In accordance with the proposed method, the maximum value of partner's compliance level is 1 and the minimum is less than 0.59 (Table 2).

Depending on the partner's score, the degree of compliance can be high, insufficient or low, which allows determining compliance rating level and the amount of penalties for deviations and/or inconsistencies to be deducted from payment. Only the value of 0.9–1.0, as the maximum possible value of the integral indicator, does not entail any sanctions, provided that the detected violation in the first-time breach (Table 3).

Stage four. At this stage, the partner eliminates the deviations/inconsistencies identified. If elimination occurs within the remedial period specified in the agreement, penalties are not imposed (Table 4).

Monitoring should be carried out once every three months by a commission composed of authorized state bodies and partner's representatives; and by the partner

Table 2 Partner's compliance table

Numerical value of partner's compliance rating (K), points	Partner's compliance level/degree	Rating level
$0.9 \le K \le 1,0$	High	1
$0.6 \le K \le 0,89$	Insufficient	2
$K < 0.59$	Low	3

Table 3 Penalties for deviations and/or discrepancies to be deducted from payment, %

Rating level	Partner's compliance level/degree	Penalties for deviations and/or inconsistencies to be deducted from payment, %		
		First time	Second time	Third time
1	High	0	1	2
2	Insufficient	2	4	8
3	Low	3	6	12

Table 4 Remedial periods

Deviation/inconsistency	Urgency level	Urgency level name	Established remedial period
Life-and health-threatening emergencies	Category 1	High	30 min
Situations threatening the operation of equipment	Category 2	Medium	5 working days
Inconsistencies relating facility management system	Category 3	Ordinary	14 working days

himself, in accordance with procedures established for internal monitoring and quality control.

It is the partner's responsibility to monitor his consistency with performance indicators (individual indicators and series of indicators), based on the established internal monitoring system. This internal monitoring system can provide for both daily and monthly surveys of key indicators, as well as quarterly and annual surveys of KPIs.

4 Conclusion

Thus, the monitoring of PPP projects, at all stages of their life cycles, for partner's compliance with the obligations to achieve performance targets as set out in the agreement, allows for timely response to possible deviations during project implementation and timely adjustments. The proposed method contributes to enhancing the efficiency of public–private partnerships in Russia.

References

1. Alexandersson G, Hultén S (2007) Prospects and pitfalls of public-private partnerships in the transportation sector–theoretical issues and empirical experience
2. Berdnikova EF, Raiskaya MV (2014) Gosudarstvenno-chastnoe partnerstvo: osnovnye tendentsii i perspektivy razvitiya v RF [Public-private partnership: the main trends and prospects in Russia]. Vestnik Kazanskogo tekhnologicheskogo universiteta [Bulletin of Kazan Technological University] 17(11):275–279
3. Chesbrough HW (2003) Open innovation: The new imperative for creating and profiting from technology. Harvard Business Press
4. Denee TR, Sneekes A, Stolk P et al (2012) Measuring the value of public–private partnerships in the pharmaceutical sciences. Nat Rev Drug Discov 11(5):419. https://doi.org/10.1038/nrd3078-c1
5. Dobrusina ME, Zavyalova GN, Tulupova ON et al (2011) Public-private partnership as an innovative format for enhancing Russia's healthcare sector. Bull Tomsk Univer 1:142–147
6. Dyachuk EA, Salimyanova IG, Nikolaeva OK (2016) Promoting public-private partnerships in healthcare. Innov Act 1(36):38–45

7. Gilfanova LG, Rodnyansky DV (2015) On implementing public-private partnerships in healthcare in the Republic of Tatarstan. Curr Challenges Humanities Nat Sci 11–3:53–57
8. Goldman M (2012) Public-private partnerships need honest brokering. Nat Med 18(3):341–341
9. Gunnigan L, Rajput R (2010) Comparison of Indian PPP construction industry and European PPP construction industry: process, thresholds and implementation
10. Kaneva MA (2016) Public-private partnership in healthcare: lines of development in Novosibirsk Region. Reg Econ: Theo Pract 1:169–181
11. Kaplan W, Wirtz V, Mantel A et al (2013) Priority medicines for Europe and the world update 2013 report. Methodology 2(7):99–102
12. Moran M (2005) A breakthrough in R&D for neglected diseases: new ways to get the drugs we need. PLoS Med 2(9):e302
13. Pardoe D, Hunter J, Cooke R et al (2010) Assessing the value of R & D partnerships. Drug Discov World 11:9–17
14. Tokunova GF (2011) The international experience of public-private partnership. Econ Educ 4:206a–2211
15. Varnavsky VG (2009) Public-private partnership. IMEMO RAN, Moscow
16. Weir SJ, DeGennaro LJ, Austin CP (2012) Repurposing approved and abandoned drugs for the treatment and prevention of cancer through public–private partnership. Can Res 72(5):1055–1058
17. Yastrebov OA (2010) Concession mechanism of public-private partnership for investment and construction projects. Science, Moscow

Numerical Modelling of Heat and Mass Transfer Processes in Medical Operating Rooms

V. M. Ulyasheva, N. S. Ponomarev, V. F. Vasil'ev, and I. I. Sukhanova

Abstract The results of numerical studies of the air environment state in medical operating rooms are summarized. Trends in the choice of schemes for organizing air exchange and air distribution in identical rooms are revealed. The features of the initial data formation for the numerical experiment, in particular, characterizing the heat and humidity regime, are considered. The accepted principles of constructing computational grids and the choice of turbulence models using the well-known modern software systems ANSYS and STAR-CCM+ are analyzed. The problems that arise in the analysis of particular solutions from the point of comparison view with the requirements of regulatory documents are studied. Recommendations for using the results of numerical modeling in the development of engineering solutions are proposed.

1 Introduction

The experience of the Russian Federation and European countries shows that heat release during operations and other medical manipulations reaches rather high values [3]. In complex operations the number of doctors in the operating room is up 10, who are powerful heat sources. In addition to medical personnel an operating lamp, electronic technological and lighting devices give off heat up 2.0 kW. It's also important to gain heat through external solar radiation.

Now this rooms are trying to place in the center of the building to exclude at least the solar radiation effect. Nevertheless, specific apparent heat excesses are more than 40 W/m^3.

As known the circulating air flows in the above-mentioned rooms are a set of powerful turbulent supply jets interacting with less active currents near the exhaust openings, convective currents in the vicinity of heated bodies. Due to such a complex picture of the air spatial movement it's possible to assess the microclimate systems effectiveness only with the use of mathematical modelling. The numerical modeling

V. M. Ulyasheva (✉) · N. S. Ponomarev · V. F. Vasil'ev · I. I. Sukhanova
St. Petersburg State University of Architecture and Civil Engineering, St. Petersburg, Russia
e-mail: ulyashevavm@mail.ru

basics of heat and mass transfer processe were laid in the Wilcox D., Ferziger J.H., Peric, Nielsen P.V., Patankar S. and other authors works [5, 6, 10, 12–14].

2   Materials and Methods

Numerical modeling of heat and mass transfer and aerodynamics of ventilation processes is based on a solving a system of partial differential equations of Navier–Stokes, continuity, energy and impurities:

$$\frac{\partial \overline{\rho}}{\partial t} + \frac{\partial}{\partial x_j}\left(\overline{\rho u}_i + \overline{\rho' u'}_i\right) = S_m$$

$$\rho\left(\frac{\partial \overline{u}_i}{\partial t} + \overline{u}_j \frac{\partial \overline{u}_i}{\partial x_j}\right) = -\frac{\partial \overline{P}}{\partial x_j} + \mu\left(\frac{\partial \overline{u}_i}{\partial x_j} + \frac{\partial \overline{u}_i}{\partial x_i}\right) + \frac{\partial \tau'_{ij}}{\partial x_j} + S_i$$

$$\frac{\partial (\overline{\rho a})}{\partial t} + \frac{\partial (\overline{\rho a u_i})}{\partial x_j} = -\frac{\partial \left(\overline{\rho u'_j a'}\right)}{\partial x_j} + \overline{J_a}$$

where $t = $ time; $\rho = $ the density; $\mu = $ the coefficient of dynamic viscosity; $\overline{u_j} = $ components of the averaged velocity vector along the coordinate axes; $\tau_{ij} = $ turbulent stresses (additional Reynolds stresses); u'_i, u'_j, T', $C' = $ local pulsations of velocity, temperature and flow impurities; $\overline{a} = $ averaged values of the specific density of the scalar quantity; S_m, $S_i = $ intensity of sources of mass and momentum; $\overline{J_a} = $ intensity of impurity sources.

The k-ε turbulence model was used for calculations [1, 7].

In general, it's necessary to calculate a lot of options for organizing air exchange and air distribution, taking into account the peculiarities of the all objects location that affect the distribution of air parameters in the room for different periods of the year.

The main software systems in the numerical modelling of heat and mass transfer processes in rooms are ANSYS Fluent and STAR CCM+. As noted in the article [15] each software package has its positive and negative sides. In this study the software package STAR-CCM+ is adopted.

Taking into account the requirements of regulatory documents and the dimensions of medical operating rooms, designers are forced to use a limited number of both air distribution schemes and air distributors. At present, in connection with the introduction of a new regulatory document, the requirements for the frequency of air exchange have changed. According to the design rules, an inflow of a least 15-fold should be provided, for operating conditions—tenfold. Since the height of these premises doesn't exceed 3 m, the most rational option is the option with air supply above the operating table through special air-distributing units of the VBD and VBD-P type with high efficiency filters of the HEPA class [9, 11].

Fig. 1 Operating room model: 1—medical worker; 2—patient; 3—equipment; 4—VBD-P; 5—exhaust holes; 6—door

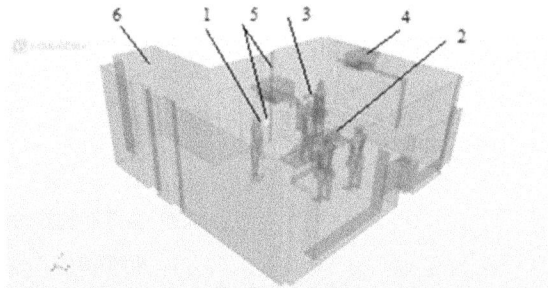

The numerical experiment formulation includes the spatial model construction of the research object (Fig. 1).

In this case, the computational grid is represented by hexahedrons in the amount 1,7 million cells. Thickening of the computational grid was carried out in places, where heat, water vapor and carbon dioxide are released, as well as in places, where a more detailed description of air movement is required (blades of air distribution blocks and sub-ceiling space).

The air inflow is supplied from top to bottom, and the location on the supply and exhaust openings should be such as to exclude the possibility of the formation of unventilated zones in the room (Fig. 2).

In operating rooms and anesthesia wards air extraction are organized from the upper and lower areas of the room. The external supply air is cleaned in filters of the central supply chambers or air conditioners. A three-stage cleaning system is used for operating rooms:

- a coarse filter G4 or M5;
- fine filter F7;
- high efficiency filter higher than H11.

To ensure the functioning of "clean premises" special air distribution units (VB) with filters high efficiency are used. The number and size of this units vary.

Fig. 2 Air exchange organization diagram in the operating unit

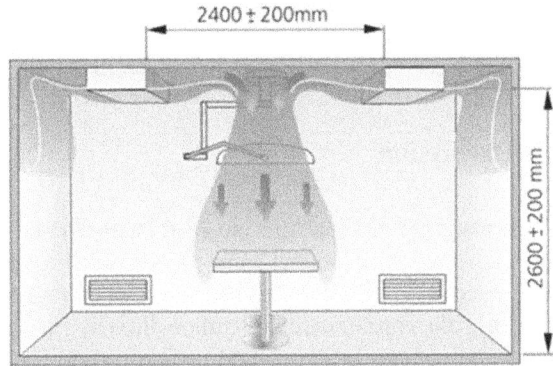

Fig. 3 Results of numerical modeling in operating rooms: installing 4 air distributors VBD 750 × 750, full fan-shaped spreading jets: velocity distribution

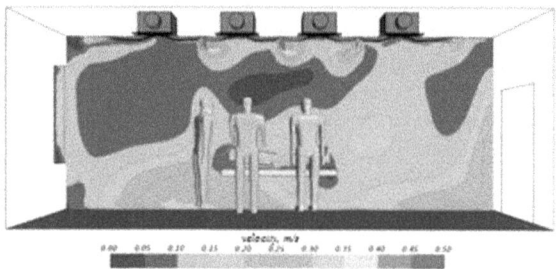

Depending on the type of air distribution panel, it'spossible to implement two ways of supplying air:

1. non-unidirectional ("turbulent") air flow;
2. unidirectional ("laminar") air flow [4].

With non-unidirectional air flow is fed in fan-shaped jets through the air-distributing units, usually installed in the ceiling above the operation area. On the one hand, the speed is reduced over the operating table. On the other hand, there is a mixing of the supply air with the room air, which deteriorates the air purity parameter.

The main sources of infections bacteria are scales or skin particles. They have a diameter of about 10 microns and are detached from exposed skin areas of both medical personnel and the patient himself. During a typical operation (lasting two to four hours), approximately 1.15×10^6 to 0.9×10^8 of these flakes are released. The results of a full-scale experiment to determine the dispersed composition of such dust are presented in the work Balocco [2].

3 Results

The results of numerical simulation in the Star-CCM+ program for various options for the placement of air distribution blocks are shown in Figs. 3, 4, 5, 6, 7, 8, 9, 10 and 11.

4 Discussion

Since there are currently no physical characteristics of dust pollutants, this does not allow theoretical studies to be carried out. Therefore, numerical modelling was performed only for such parameters as temperature, velocity and relative humidity, as well as the concentration of carbon dioxide.

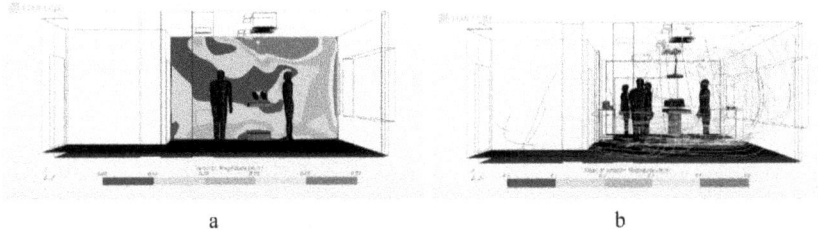

a   b

Fig. 4 Results of numerical modeling in operating rooms installing 2 air distributors VBD- 750 × 750, incomplete fan-shaped spreading jets, parallel outlet towards the window: velocity distribution; streamlines

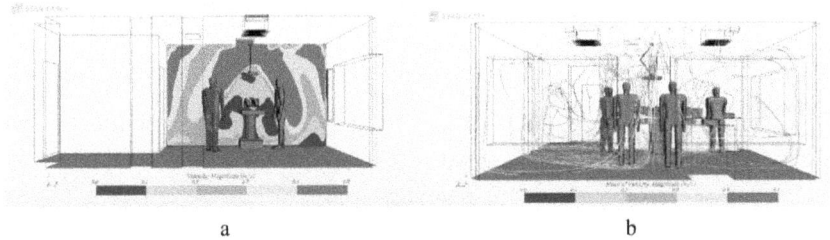

a   b

Fig. 5 Results of numerical modeling in operating rooms: installing 2 air distributors VBD-P 750 × 750, counter in-complete fan-shaped spreading jets: velocity distribution; streamlines

Fig. 6 Results of numerical modeling in operating rooms: installing 4 air distributors VBD 750 × 750, full fan-shaped spreading jets: relative humidity distribution

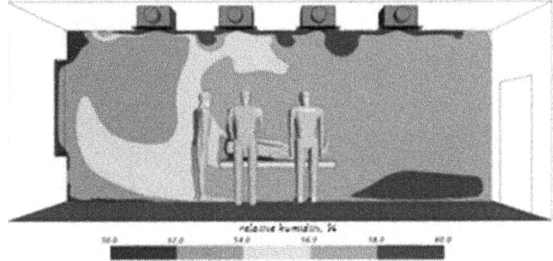

Fig. 7 Results of numerical modeling in operating rooms: installing 2 air distributors VBD- 750 × 750, incomplete fan-shaped spreading jets, parallel outlet towards the window: relative humidity distribution at 1.500 m

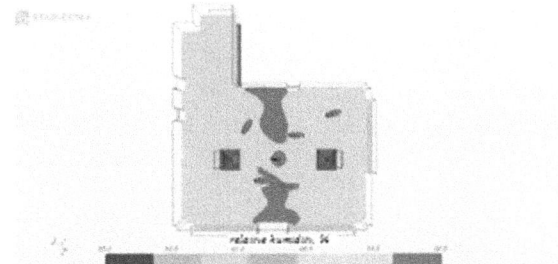

Fig. 8 Results of numerical modeling in operating rooms: installing 2 air distributors VBD-P 750 × 750, counter in-complete fan-shaped spreading jets: relative humidity distribution at 1.500 m

Fig. 9 Results of numerical modeling in operating rooms: installing 4 air distributors VBD 750 × 750, full fan-shaped spreading jets: carbon dioxide concentration distribution

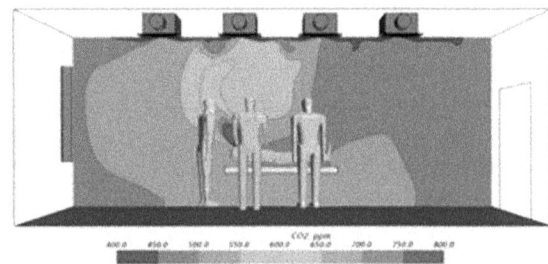

Fig. 10 Results of numerical modeling in operating rooms: installing 2 air distributors VBD- 750 × 750, incomplete fan-shaped spreading jets, parallel outlet towards the window: carbon dioxide concentration distribution

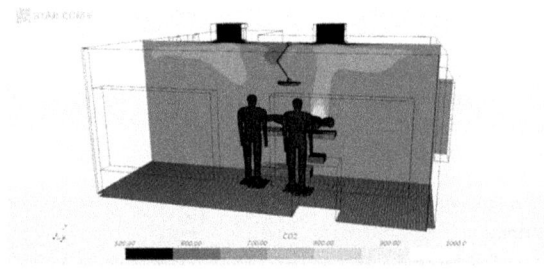

Fig. 11 Results of numerical modeling in operating rooms: installing 2 air distributors VBD-P 750 × 750, counter in-complete fan-shaped spreading jets: carbon dioxide concentration distribution

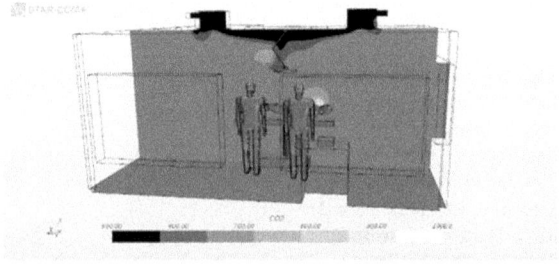

Because the temperature for any method of feeding corresponded to the normalized values, then is this article the results are not given.

The air velocity above the operating table is less than 0.15 m/s, and in general over the working area is 0.1…0.2 m/s. Such values of velocities are obtained due to the method of supplying the supply air, namely from top to bottom by horizontal overlapping jets (Fig. 3).

But in this case there will be higher costs due to the larger number of air diffusers [8].

Moreover, even when horizontal overlapping jets are used for feeding, there may be options with deviations in the zone of the microclimate parameters from the standard values. First of all, this concerns the speed of air movement (Fig. 4 and 5).

In all cases, the air velocity «on average» at the level of the serviced area corresponds to the standardized values of 0.15 m/s.

In the case of using 4 air distributors (Fig. 3), the best indicators were obtained in terms of the correspondence of the air speed, but there are problems with ensuring the normalized temperature (below 21 °C). In addition, this is costlier option in terms of economic indicators.

The relative humidity should be between 30–60%. When 4 air distributors are used (Fig. 6) the relative humidity value fluctuates within narrow limits from 50 to 60%, at which the accumulation of static electricity on metal surfaces is impossible. When two air distributors are used, regardless of their location, the relative humidity distribution at a level of 1.500 m above the floor corresponds to the standardized range and averages about 54% (Fig. 7 and 8).

The concentration of carbon dioxide in the working area of the room above the background concentration takes values of the order of 150…250 ppm, which corresponds to high air quality.

5 Conclusion

It's still difficult to choose the most rational one, despite the rather detailed studies—the computational grids contain up to 1.7 million cells; in addition to the speed of air movement, temperature distribution and relative humidity fields of air, carbon dioxide were obtained. In any case it's not possible to obtain the distribution fields of Colony Forming Units (CFU).

References

1. Averyanov V, Vasiliev V, Ulyasheva V (2018) Selection of turbulence models in case of numerical simulation of heat-, air-and mass exchange processes. In: E3S Web of conferences, vol 44. EDP Sciences, p 00005

2. Balocco C, Petrone G (2015) Microclimate and indoor air quality in an operating theatre under real use conditions—an experimental and numerical investigation. In: Current air quality issues. IntechOpen
3. Borisoglebskaya AP (2008) Medical and preventive institutions. In: General requirements for the design of heating, ventilation and air conditioning systems. AVOK-PRESS, Moscow
4. Cao G, Kvammen I, Hatten TAS et al (2021) Experimental measurements of surgical microenvironments in two operating rooms with laminar airflow and mixing ventilation systems. Energy Built Environ 2(2):149–156
5. Ferziger JH, Perić M, Street RL (2002) Computational methods for fluid dynamics, vol 3. Springer, Berlin, pp 196–200
6. Fletcher K (1991) Computational methods in fluid dynamics: part 2. Mir
7. Frick PG (2010) Turbulence: approaches and models. Regular and chaotic dynamics. Institute of Computer Science, Moscow-Izhevsk
8. Genco A, Viggiano A, Viscido L et al (2016) Numerical simulation of energy systems to control environment microclimate. Int J Heat Tech 34:3545–3552
9. Kuzmin AM (2020) Numerical simulation of microclimate in a small operating room. Symb Sci 11:28–31
10. Li Y, Nielsen PV (2011) CFD and ventilation research. Indoor Air 21(6):442–453
11. Lupashko VV (2021) Simulation of the operating room of a medical clinic in St. Petersburg. Mod Prob Environ Eng Urban Econ Collect Art FIEGH SPbGASU 2:116–127
12. Paskonov VM, Polezhaev VI, Chudov LA (1984) Numerical modeling of heat and mass transfer processes. Nauka, Moscow
13. Patankar S (1984) Numerical methods for solving problems of heat transfer and fluid dynamics, vol 152, no 2. EnergoAtomIzdat, Moscow
14. Wilcox DC (1998) Turbulence modeling for CFD, vol 2. DCW Industries, La Canada, CA, pp 103–217
15. Zou Y, Zhao X, Chen Q (2018). Comparison of STAR-CCM+ and ANSYS Fluent for simulating indoor airflows. In: Building simulation, vol 11, no 1. Springer Berlin Heidelberg, pp 165–174

On the Financing and Development of PRC's Health Care System

Tatiana Urzhumtseva, Anastasia Kovaleva, Tong Wei, and Valentina Larionova

Abstract Authors of papers have analyzed the modalities for the operation of PRC's system of financing the health care, including the pandemic period, and outlined spheres to pay attention to aiming at elimination of existing deficiencies. This requires increase of investments, complex and rational approach to investment projects, promoting new scientific and technical solutions regarding both research activity and infrastructure planning. The results of this study may be applied by other countries including Russia for improving their own counter-emergency strategies and increasing the effectiveness of health care system.

1 Introduction

Usually economics is all about numbers and indexes, but boring charts and digits are having an impact of the people's daily life. Those, who are following a new trend in the theory of economics, do believe that the effectiveness of a particular economy is to be estimated not by GDP only, but, on the opposite, the effectiveness is to be estimated by the level of people's happiness. There may be hardly any doubt about this. One of the major functions of the state is catering for health care system development and it's financing. The effectiveness of spending on health care is having a direct impact upon the quality of people's life and health, and, therefore,

T. Urzhumtseva (✉)
Eugeny I. Schwarz Molecular Medicine, Medical and Lab Genetics' Specialists Association, State University of Economics, St-Petersburg, Russian Federation
e-mail: apr@unecon.ru

A. Kovaleva
State University of Economics, St-Petersburg, Russian Federation

T. Wei
Central University of Finance and Economics, Beijing, People's Republic of China

V. Larionova
Eugeny I. Schwarz Molecular Medicine, Medical and Lab Genetics' Specialists Association, Mechnikov North-Western Medical University, St-Petersburg, Russian Federation

upon sustainable social and economic growth. These are exactly the extraordinary situations that prove the strengths and weaknesses of health care system as well as the readiness of a Government to protect its own population.

COVID-19 pandemic had taken the whole world by surprise, with no one being ready to respond. The world's economy had suffered a significant loss. In response to the pandemic challenge it was mandatory to provide the in-time coverage of the expenses on the confirmed and alleged patients' treatment and purchase of equipment and materials, construction of dedicated hospitals, as well as providing the basic health care services and keeping pandemic under control.

2 Materials and Methods

World scientific community's attention is attracted by China's successful campaign in countering the new COVID-19 pandemic. In this respect we've set a goal to analyze the modalities for the operation of PRC's system of financing the health care during pandemic, including it's strengths and weaknesses, as well as to outline ways to eliminate existing deficiencies. The results of this study may be applied by other countries including Russia for improving their own counter-emergency strategies. With the lack of medical infrastructure, staff deficit and explosive commercialization, Russian and others' healthcare systems are facing serious challenges [4, 8]. This study is based on the analysis of available statistical information as well as previous works on the issue in combination with a special case-study technique which allowed us to dwell upon operational patterns of Wuhan City and Guangdong Province's health care systems. This study's database includes statistical materials of PRC Ministry of Finance, PRC Ministry of Public Health, as well as data provided by Central University of Finance and Economics' Institute of Russian and East European Studies.

3 Results and Discussions

3.1 Health Care System Status and Development

3.1.1 Building-Up the Healthy China and the 14th Five-Year Plan

To provide the society's well-being and implementation of socialist modernization is based on building-up the healthy China. This is a key national strategy aimed at overall improvement of Chinese people's health as well as achievement of coordinated development of people's health, economy and society [6, 10, 12], also including an active participation in global health care management together with fulfillment of major international obligations up to 2030 in accordance with Sustainable Development Agenda.

In order to improve some drawbacks of Chinese health care system, PRC's Government had set a goal within the framework of 14th Five-Year Plan for National Economic and Social Development to increase the investments into health care and counter-pandemic infrastructure, as well as into operational systems and personnel training, etc. Guidelines to the Plan, set forth by the CPC Central Committee's 5th Plenary Meeting back in October 2020, include catering for building-up the healthy China and a focus on diseases prevention with an emphasis on a main role of state-owned hospitals. These later are to take care of monitoring and early warning, risks assessment and testing as well as timely counter-action in case of emergency. Hospital management is to become the core of public health care system, including faster building of infrastructure.

3.1.2   The Size and Tools of Chinese Health Care System Financing

The drawbacks of public health care should be cured by proper economic develop-ment, while growth of public health care quality should, in its turn, provide for an economic growth. It is revealed that healthcare expenditure is positively associated with labor productivity, personal spending, and GDP [9].

On top of this, growth of the public infrastructure reflects a level of society's civility. In many countries, in the context of the development of social infrastructure, there is also a need to improve the quality of services provided to the population, including with the involvement of non-state sector companies [5]. It was earlier that the Chinese Government had actively invested into housing construction and transport, now it's time to increase investments into public health' and counter-pandemic infrastructure. According to scientific data, only 30% of the investments made into railroads and automotive infrastructure do really add to GDP, while those made into public education, health care and culture do reach 60–70% in this respect [6]. Even though health care is more and more fully involved in the economic turnover of society, being facilitated by the active functioning of medical institutions on a market basis [3], but the role of state government is not going to decrease. The state should play more and more proactive role especially when it comes to fighting the emergency.

PRC Ministry of Finance's spending on local state-owned hospitals moderniza-tion are constantly growing. The effectiveness of the way hospitals are using these investments is being tightly monitored by the government in accordance with the strategy called «Spending of the funds should be effective, non-effective should be punished». In accordance with the PRC Ministry of Finance data, direct investments made by governmental bodies on all levels from 2013 till 2018 had increased from CNY 129.7 bln to CNY 270.5 bln, i.e. by 15,8% per annum on average. Govern-mental support of the Medical Insurance Fund is an indirect tool of state-owned hospitals support. According to the official data published by Ministry of Finance, n 2018 state-owned hospitals had received CNY 1233.9 bln from various Medical Insurance Funds, thus reaching 51.5% of their total incomes.

The major tools used in financing are government's bonds as well as adjustments of budget structure. An important role is played by special purpose municipal bonds, these being a key tool of a proactive fiscal and budgeting policy with regard to building the health care infrastructure. Currently the bonds of this type are having relatively low share. In 2019 total value of municipal bonds nationwide had reached CNY 4362.4, while special purpose bonds value nationwide was CNY 1774.2, including only 4% of those for social welfare, while major costs had fallen on housing, land and transport [6]. It had been outlined that China shouldn't blindly follow American and European examples in monetization of the fiscal deficit. In the long run this may lead to inflation as well to the gap between the rich and the poor. Instead, China, taking into consideration it's actual situation, should focus on structural reforms aimed at long-term economic and social growth [16].

3.1.3 Guangdong Province as a Successful Example of Effective China's Health Care Financing

Nowadays 30 high-class Guangdong hospitals are among 85 key national research centers, performing serious projects in science and technology. Guangdong Province is second to none with regard to the number of privately-owned hospitals, which prove being more effective than those state-owned from the top list. Thus, in order to speed up the building of a healthy Guangdong, extra CNY 6.5 bln had been assigned by the Province Government for construction of another 20 high-class hospitals. By 2025 50 of local high-class hospitals are expected to go up in overall national ranking, as well as in the fields of clinical and scientific research, in training highly qualified personnel, etc.

Within 5 years (2021–2025) financial institutions of the province will invest CNY 500 mln into local Support projects [CNY 100 mln per project]. Within the framework of these Support projects 5 local high-class hospitals are expected to extend their aid and support to local district hospitals, thus improving a coherence within the province. The monies for construction, assigned by provincial Department of Finance, will be wired to high-class hospitals directly, while these later will have a free hand in spending of the funds. This support is focused on aid in organizing of the special departments in the district hospitals, as well as on implementation of modern technologies and highly-qualified personnel training [via remote consultations, visits of experts, training courses etc.].

3.2 Financing of the Health Care System During the Pandemic

Combat against Corona pandemic proved to be successful, but, alongside with it, had highlighted China's shortcomings in the field of health care, diseases prevention and control, including shortage of special purpose financial resources. The pandemic highlighted the shortage of medical institutions per capita, lack of medical staff, insufficient amount of prevention and control tools in the hospitals and polyclinics, as well shortage of counter-pandemic items [e.g., protective gears] [6, 12]. Nowadays, with the exception of advanced cities like Beijing, Shanghai, Guangzhou, Shenzhen и Hangzhou, much of the large cities are facing shortage of 3rd degree hospitals and medical equipment, medium-sized cities and smaller towns are facing shortcomings in medical resources distribution alongside with gap in treatment level between them. This leads to big cities and big hospitals being often overcrowded.

Upon Corona pandemic outbreak China's Ministry of Finance together with the authorities in charge had launched an intense politics of lowering the loans' interest rates as well lowering the fiscal burden with the purpose of pandemic outbreak control and economic stabilization. In particular, huge investments were made directly into health care and countering the pandemic, including construction of special infectious units and converting hospitals into infectious units. Timely provision of subsidies for pandemic prevention and countering were used to cover benefits for the medical staff and medical items supplies. A victory in combat against pandemic comes hand in hand with its financing, from Hubei Province costs on pandemic prevention and countering, till subsidies for treatment of the patients nationwide; from subsidies for line medical staff till financing of Huoshengshan and Leishengshan hospital construction in Wuhan City; from basic counter-pandemic costs till spending on vaccines R&D and Central Medications Reserve. On Jan. 22, 2020 the Ministry of Finance together with the National Medical Insurance Authority had issued a joint «Emergency notice on medical insurance coverage with regard to pneumonia pandemic, caused by new Coronavirus infection», followed by a dedicated notice, clearly stating that 60% of the treatment costs for the confirmed COVID patients will be covered by central government subsidies [1]. By Feb. 23, 2020 an amount of subsidies granted at all levels had already reached CNY 99.5 bln. By Mar. 2, 2020 the amount of funds allocated by the Ministry of Finance for pandemic control at all levels had reached CNY 108.75 bln. According to China's State Council data, delivered at the press conference on Apr. 5, 2020, the amount of medical items, exported from Mar.1 till Apr. 4 nationwide, had reached 3.86 bln. masks, 37.52 mln protective gears, 2,41 mln infrared thermometers, 16 thou lungs ventilators, 2.84 mln Coronavirus detection sets, 8.41 mln protective eye-glasses. Draft 2020 budget stipulates special transfer payments at CNY 605 bln in value to support various areas of the country with a focus on providing people with the livelihood, building up serious pandemic prevention and control systems, creating the emergency reserves.

3.2.1   Wuhan City

Half of China's finance had been sent to Wuhan. Within one month, from February till March, tremendous changes had occurred. Budget incomes went significantly down, while the costs grew up. According to «China Financial News», local Wuhan's authorities had granted CNY 7.116 as on Mar. 12 to prevent and combat the pandemic.

Another CNY 500 mln were invested by the Central government into Hubei Province infrastructure. All funds were spent on hospitals construction, patients' treatment and isolation. Support was granted to 88 private companies engaged in enlarging of their units and converting these into hospitals. On top of this, hospitals Huoshengshan and Leishengshan have been built, while 33 hospitals were converted into COVID patients' centers. The shelters provided 35,673 beds, 775 isolation centers with 51,925-beds capacity have been organized. 315 medical teams [35591 staffs] were granted meals and accommodation.

3.3   Medical Infrastructure and Logistics

Construction of medical infrastructure designed for use in emergencies, is a key element of counter-pandemic strategy. Since the severe acute respiratory syndrome (SARS) outbreak in 2003, the provincial and regional governments of China have begun to pay more attention to public health issues and carried out many public health projects. Even though public health project management is often character-ized by low level of effectiveness and efficiency [14], we should admit that China made huge step forward in terms of implementing medical infrastructure projects. While building the hospitals in Wuhan City the following techniques were applied: special fast construction tech, digital hospital management; dedicated logistics to avoid in-house infections; water purification facilities; waste decontamination and management, etc. [15]. It is worthy of notice that modern facilities like those of Xiaotangshan hospital in Peking and Huoshengshan hospital in Wuhan should corre-spond to local standards of particular cities and municipalities and be integrated into local systems of diseases prevention and control. The systems should include plan-ning of shelter hospitals, transformation of college campuses, sport arenas, cultural venues and even shopping malls into temporary hospitals if needed in emergency [2]. Trouble-free transportation and logistics are the core elements in pandemic preven-tion and countering. The role of finances is to provide for logistic costs lowering and timely supplies of necessary materials, as well as for effective communities protection via organizing an appropriate microcirculation and people isolation without leaving homes. In addition, fully integrated information technology systems should support the delivery of information to ensure patients, carers and health professionals can access information they need, when they need it [7].

3.4   Changing the Approach: From Treatment to Prevention

It is crystal clear now that health care system as well as the funds invested into it, should be focused on diseases prevention and health protection. This should become not only part of anti-pandemic campaign, but a fundamental target.

Based on the experience in countering the SARS outbreak as of 2003, the following ratio between health care expenditures to treatment expenditures had been outlined: «decreasing of the first one leads to the growth of the second one». According to WHO data, provided that medical standards are equal, ratio between health care investments to treatment and resque expenditures amounts to 1:8,5:100 [2]. This means, that CNY 1 increase in expenditures on prevention will lead to CNY 8.5 saving in treatment costs and CNY 100 in resque operations costs. Therefore financial system and risks prevention and control system in public health should focus on combining of risks management and risks response on a full circle basis. Thus, it confirms the conclusion that the first lesson to learn from COVID-19 is to invest more in prevention [11].

4   Conclusion

The study of some aspects in developing and financing of Chinese health care system had outlined national priorities in this field as well as means and goals of financing. Corona pandemic had detected some vulnerabilities of the system and led to search of ways of troubleshooting. The authors would like to stress three following issues:

1. It is mandatory that shortage of funds in the health care system be bridged. This requires increase of investments into health care and pandemics prevention, going ahead with medical reforms, paying special attention on staff training, R&D, thus leading to improvement of state-provided medical services and well-being of the people. Investment climate in the field of medicine as well as active growth of key production chains in the field will be improved by combining the direct finance growth with applicable fiscal tools. In particular, Chinese scientists expect subsidies for practitioners to cover their rental and R&D costs, reduction of patent requirements on medications, reward for prescriptions public release, stimulation of pandemic studies and pandemic-related diseases with a purpose of reaching a new economic growth via medicine [13].

2. It is mandatory that structural shortcomings in production and technologies be overcome. Technological innovations are a powerful tool in combating the pandemic. The key elements leading to improving the R&D, are scientific and technological staffs training, creation of scientific and technological infrastructure, innovative equipment implementation, as well as science and technology promotion. Mrs. Li Yan, MP and CEO of Qilu Pharmaceuticals Group, had outlined her proposals for supporting the innovative pharmaceutical industry growth. According to Mrs. Li Yan, Chinese pharmaceutical industry is still left behind global companies, especially in the field of fundamental clinical research and China should have ideas and opportunities in order to make our contribution into people's wellbeing and health.

3. Investment projects require both complex and rational approach. On the one hand, provincial health care and prevention systems should include enlargement of the existing hospitals and construction of new ones, increase in number of beds

per capita, more ICUs, more modern equipment, operational systems, nets, etc. [6]. The growth of hospitals should correspond to the growth of the population in the area. On the other hand, one should consider construction of multipurpose venues, which may be used normally for sports, concerts and fairs, but quickly be transformed into hospitals in case of emergency.

References

1. Dong D (2020) The combination of fiscal policies is to contribute to the epidemic prevention and control. Special Plan 3:52–53
2. Du J, Jiang Z (2020) Building a financial system mechanism for the prevention and control of public health risks. Macroecon Manag 50(5):37–42
3. Gabueva LA, Mochalov DV (2007) Improvement of organizational and economic mechanisms of health care management in the conditions of state regulation and entrepreneurship development. Doctor Ru 4:60–66
4. Gaffney A, Himmelstein DU, Woolhandler S (2020) COVID-19 and US health financing: perils and possibilities. Int J Health Serv 50(4):396–407
5. Gerasimenko OA (2018) Improvement of infrastructure in the implementation of PPP projects in the social sphere. In: Ensuring the sustainable development of regions in the spatial structure of the Russian economy, pp 101–105
6. Li X, Long X (2020) Support the construction of public health infrastructure with special bonds. Financ Res 27–28
7. Luxon L (2015) Infrastructure–the key to healthcare improvement. Future Hosp J 2(1):4–7
8. Pesennikova EV, Perkhov VI (2020) Directions of development of medicine and healthcare in the post-pandemic world. Contemp Prob Healthc Med Statistics 4:535–551
9. Raghupathi V, Raghupathi W (2020) Healthcare expenditure and economic performance: insights from the United States data. Front. Public Health. https://www.frontiersin.org/articles/ https://doi.org/10.3389/fpubh.2020.00156/full. Accessed 12 May 2021
10. Tao L, Xiufeng W (2016) Connotation and realizing route of healthy China. J Health Econ Res 1:4–10
11. Tarricone R, Rognoni C (2020) What can health systems learn from COVID-19? Euro Heart J Suppl: J Euro Soc Cardiol 22(Suppl Pt):4–8
12. Wang L, Liu L (2020) Comparison between the current situation of medical and health facilities in China and worldwide under the new crown pneumonia epidemic. Sci Technol Herald 38(4):29–38
13. Wang Y, Ren H, Wang S (2020) China's financial and tax measures to deal with the epidemic. J Hunan Taxation College 4:23–27
14. Wu H, He P, Luo Y et al (2013) Thinking on system construction of basic public health services in Chongqing. Chin Health Serv Manage 30:760–762
15. Wu M (2020) Value of Huoshenshan hospital model for the construction speed of public hospitals. Constr Econ 41(S1):46–49
16. Yan M (2020) Practice, characteristics, and reflections of fiscal policies of various countries during the Covid-19 pandemic. Econ Res Ref 13:29–43

Experience of Adapting the Premises in Special Economic Zone for Use as Innovative Biotechnology Facility

E. Yu. Chernyakevich

Abstract The paper presents the Russian experience of adapting an operating building for use as a medical biotechnology research and production facility. Description is provided by the author of the office and utility building located in a special economic zone and of design solutions that helped to convert its premises into a full-cycle facility for diagnostic test systems production and research. The paper defines the primary and secondary factors to be considered when adapting a premise in a SEC for use as biotechnology facility. Recommendations are provided, based on the experience gained.

Keywords Innovative production · Biotechnology · Special economic zone · Ergonomics · Biotechnology company · Building

1 Introduction

The recent years have seen a major increase in the demand for innovative biotechnology products as a potentially promising investment area in Russia and globally. Experts and analysts forecast that biotechnology may soon become twenty-first century's fastest growing and highly profitable business. That said, the opportunities for starting innovative biotechnology projects should be analyzed not only in terms of their research potential and new construction prospects, but also from the perspective of using the existing urban infrastructure.

2 Materials and Methods

Biotechnologies will undoubtedly continue to be promoted all over the world to drive social innovation and development for many years ahead [12]. The Russian Academy

E. Yu. Chernyakevich (✉)
Saint Petersburg State University of Architecture and Civil Engineering, Saint Petersburg, Russia
e-mail: chernik.72@mail.ru

© The Author(s), under exclusive license to Springer Nature Switzerland AG 2023
D. Ivanov et al. (eds.), *Proceedings of ECSF 2021*, Lecture Notes in Civil Engineering 257, https://doi.org/10.1007/978-3-030-99877-6_33

of Sciences ranks among the top ten institutions in the world with tangible contribution to biotechnology research [1]. Not only does research in biotechnology improves the knowledge about environments, nature and diseases, it offers new ideas that can be turned into economically valuable assets. But, further success is hampered by a number of challenges. Foreign authors note that newly launched biotech companies has come a standstill in the European Union due to lack of capital and less favorable tax regimes for investors [7]. Achieving maximum progress with innovative biotechnologies implies profound involvement of and interaction with stakeholders. As many firms, from startups to large internationals, are facing the ever increasing difficulties associated with the changes in healthcare organization system, the require more support from governments for their innovation-driven activities [3]. Moreover, regulatory authorities are increasing their monitoring efforts and quality requirements for services and products [5]. Companies often have problems registering their products. Many of their products don't live to see the registration stage for lack of support from social institutions [8].

A widespread practice worldwide, special economic zones are still an emerging concept in Russia. Special Economic Zones represent a unique environment with opportunities for innovation-driven businesses. They are a space for the production and launching on the Russian and international markets of competitive, technology-intensive products.

The experience of creating special economic zones (SEZ) in different countries has shown that they are a development factor and effective tool for improving regional economies and stimulating new business activities and investments in cities [4]. At the same time, as an economic mechanism with beneficial effects for local development and national business environment, the special economic zones experience fierce competition with unequal initial opportunities. The specificity of special economic zones consists in their being geographically differentiated areas with shared administration. SEZs offer tax incentives, simplified customs procedures, milder economic and legal frameworks. However, the research conducted by the Russian scientists shows that the Russian Government's approach to initiatives seeking development of innovation-driven facilities is often an obstacle to productive business activity and joined efforts of SEZ residents [9].

The process of infrastructure preparation varies between special economic zones. Their managing companies suggest that this variability may be the reason why some SEZs are developing more intensively than others. Today's success model, as well as international experience, suggest that efforts to attract investors should be preceded by creating decent infrastructure first [2, 6]. Nevertheless, some SEZs first try to attract investors and build required infrastructure only later [10, 11]. In the Russian conditions, it seems difficult to build the infrastructure first and then look for investors, as there may be problems finding them. For this reason, the first thing to be sought is contract with investors with real intentions and financial capabilities to implement the project. Once the contract is in place, infrastructure starts to be developed on a territory designated for the venture [9].

Given the mission of SEZ as a cluster serving the needs of innovation-driven companies, we decided to accommodate our biotechnology facility on its territory.

Oce problem we encountered was finding a building suitable for accommodating our biotechnology facility.

3 Results

As a resident of Novoorlovskaya Special Economic Zone in St. Petersburg, our company, Biopalitra, a young innovative business engaged in full-cycle development and production of diagnostic test systems, was looking to be housed by an operating building.

The premises we received are in an ordinary nine-storey office and utility building and had to be converted to house a four-department facility with an R&D team, facility for producing trial batches of immunoenzymatic assay, and quality control team. Aesthetics, ergonomics and convenience were an important part of the conversion, too. One essential requirement to planning the layout of the premises intended for use as biotechnology facilities is quality zoning. The flows of the materials with high degree of epidemiological threat should be separated from all other flows using the appropriate design solutions and equipment. Therefore, the production facilities should be placed in such a way as to isolate the employees dealing with tests and testing materials. To minimize possible intersecting of the flows, the premises had to be redesigned.

Other challenges related to water removal, since the building's structural design didn't allow for gravity drainage, and the installation of our freezers on the 5th floor, which involved having to move some of the units to the top floor and the roof. The interior fit-out had to be changed as well and required the use of special materials to suit our purpose. These materials were to be sourced from domestic suppliers but there were items that weren't available from them, so we had to opt for more expensive solutions from abroad.

In the chapter below, we provide solutions we arrived at when dealing with the above challenges to adapt our premises for use as an innovative biotechnology facility.

3.1 Space and Layout Design

The task of converting the premises to a medical device production facility followed the relevant space planning and interior design. The premises are on the 5th floor of the technology transfer center with a business incubator, located in technology development special economic zone.

The main objective was to redesign the premises so that they could house all our operational processes and be consistent with our technical requirements. Conversion was performed in compliance with Construction Rules 1.13130.2009 "Fire Protection Systems". The design layout provided for installation of the following premises:

"infectious zone", "clean zone", cold storage rooms, change rooms, sanitary check-points, etc. The required width of horizontal sections of evacuation routes, had to be not less than 0.8 m taking into account the arrangement of equipment. The exits lead to the corridor and further into the smoke controlled staircases H2. Artificial and natural lighting is consistent with Construction Rules 52.13330.2016. To separating the "infectious zone" from the "clean zone", solid partitions were installed, made of moisture-resistant gypsum boards in 100 mm thick metal frames (Construction Rules 163.1325800.2014). These partitions are one-layer and reach the ceiling board. Other partitions are Rockwool Light Batts mineral wool boards. The finishing material for the moisture-resistant gypsum boards was plaster-based putty. The boards had their seams covered with self-adhesive mesh tape and coloring wallpaper. The solid partitions and internal walls received two layers of moisture-resistant acrylic washable paint of Caparol type (or counterparts). Door units are unglazed and made of moisture-proof fiberboard with PVC film. Some premises have plastic doors. All door units are hermetically sealed. The stationary partitions are 2.4 m high. Their frames are aluminum and have been painted by manufacturer with powder dyes; these partitions are filled with sandwich-type PVC boards. The side-hung doors are made of MDF laminwood. Door handles are latch equipped. The doors to rooms from the corridor are lockable. The color scheme chosen for the ceiling, walls and doors is white. Flooring: linoleum, joints taped. 300 mm high floor border along the perimeter; acrylic composition used for sizing. Transfer devices are made in the form of transfer boxes. Bathrooms and showers untiled.

3.2   Water Removal

As we mentioned earlier, drainage by gravity wasn't possible due to the building's structural features. The problem was solved by SOLOLIFT sewage pumps. Water removal system had to be arranged via the ceiling (Fig. 1). For safety reasons, given that the premises are on the 5th floor, Neptune shutoff system was installed.

Additionally, extra waterproofing was installed. Waterproofness was the key criterion for the choice of the materials. Changes were made to the power supply scheme; the industrial equipment used a separate connection.

Installing the freezers on the 5th floor was a difficult process and involved moving their units to the roof, following which the additional GPS module OwenCloud was installed to monitor performance. OwenCloud module monitors the temperature inside the chamber remotely and provides immediate emergency alerts.

The ergonomics of the office itself underwent repeated changes. The main emphasis was laid on the convenience of research, as well as comfortable performance of the work-related duties by all employees. For more comfortable interaction, open space solution was chosen as the most optimal for the workspace. The workspace was divided into zones, and among them negotiations area, with the help of partitions and arrangements of furniture. The chosen color solution is white as a symbol of cleanliness and freshness. A set of rules has also been adopted to be

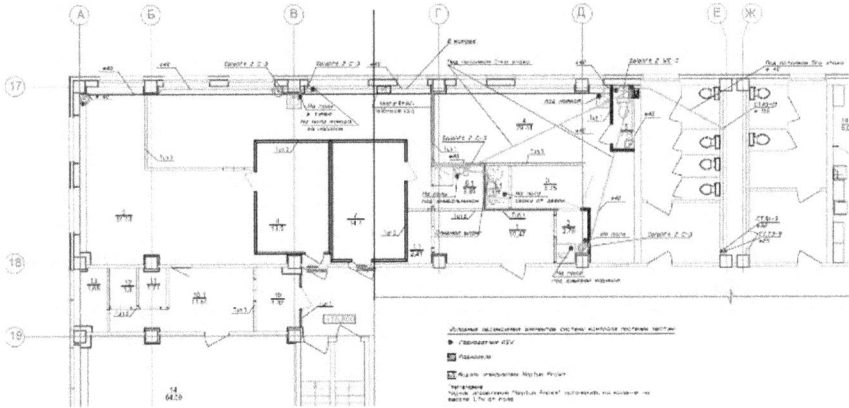

Fig. 1 SOLOFIT pump-assisted water removal

observed by employees in open office space. One unwritten rule concerns the use of cell phones: phone calls must be made and taken outside of the office. Employees benefit from open office space by being able to communicate more effectively as information can be communicated "undistorted" to all team members and problems can be handled quickly and more constructively.

4 Conclusion

Adapting a premise in the building in a special economic zone for use as innovative biotechnology facility can be defined as an ambivalent experience:

On the one hand, the location of the building in the SEZ has its advantages that come in the form general orientation towards investment projects, federal investments in SEZ infrastructure, tax incentives, simplified customs procedures, etc. On the other hand, additional arrangements need to be entered concerning, for example, waste disposal, extra power supply.

Secondly, the adaptation of existing premises for use as R&D complexes can be a complex process in terms of layout solutions, water removal, waterproofing, power supply, whereas a premise constructed from scratch and tailored to the needs of R&D would have none of these problems.

At the same time, our experience demonstrates that starting an innovative biotechnology business in a building in a special economic zone that wasn't originally intended for use as an R&D facility, is possible.

The main recommendation, therefore, is that biotechnology projects first carefully consider if the building or premise that haven't been designed for use as biotechnology facility, is expedient. Then, redesigning options should be studied carefully, and analysis be made of the availability of technical conditions meeting the redesign purposes.

References

1. Adams J (2013) The fourth age of research. Nature 497(7451):557–560
2. Aggarwal A (2010) Economic impacts of SEZs: theoretical approaches and analysis of newly notified SEZs in India
3. Aghmiuni SK, Siyal S, Wang Q et al (2020) Assessment of factors affecting innovation policy in biotechnology. J Innov Knowl 5(3):180–190
4. Cocconcelli L, Medda F (2018) Regional effects of port free economic zones on real estate speculation: a Korean case study. In: Finance and risk management for international logistics and the supply chain. Elsevier, pp 269–291
5. Cornelissen M, Małyska A, Nanda AK et al (2021) Biotechnology for tomorrow's world: scenarios to guide directions for future innovation. Trends Biotechnol 39(5):438–444
6. Gupta KR (ed) (2008) Special economic zones: issues, laws and procedures, vol 2. Atlantic Publishers & Dist
7. Krauss J, Breitenbach-Koller L, Kuttenkeuler D (2021) Intellectual property rights and their role in the start-up bioeconomy–a success story?. EFB Bioeconomy J 1:100002
8. McNamee LM, Cleary EG, Zhang S et al (2021) Late-stage product development and approvals by biotechnology companies after initial public offering, 1997–2016. Clin Ther 43(1):156–171
9. Sosnovskikh S (2017) Industrial clusters in Russia: the development of special economic zones and industrial parks. Russ J Econ 3(2):174–199
10. Tantri ML (2015) Fiscal implications of special economic zones (SEZs) expansion in India: a resource cost approach. J Int Econ Commer Policy 6(01):1550006
11. Wang J (2013) The economic impact of special economic zones: evidence from Chinese municipalities. J Dev Econ 101:133–147
12. Yeung AWK, Tzvetkov NT, Gupta VK et al (2019) Current research in biotechnology: exploring the biotech forefront. Curr Res Biotechnol 1:34–40

Rules for Creating COVID Hospitals (Adapted and Repurposed Facilities): Organizational Management Requirements

A. Y. Chudakov, D. N. Zhidkov, and K. P. Indyk

Abstract The paper discusses the current updates to organizing COVID hospitals (adaptation for use as makeshift hospital, repurposing of existing healthcare facilities) and requirements to their operational management (in the context of innovative medicine and construction) in the Russian Federation.

1 Introduction

The authors discuss the following areas of organizing healthcare facilities: general requirements for the construction (repurposing) of COVID hospitals, prompt construction of COVID hospitals (modular construction, RODER hospitals), COVID hospitals design layout and performance.

2 Materials and Methods

The methodological framework of the study relies on generalization and analysis. These methods made it possible to identify key patterns in organizing COVID hospitals (adaptation for use as makeshift hospital, repurposing of existing healthcare facilities) in the Russian Federation, as well as their performance patterns from the perspective of medical and organizational management.

A. Y. Chudakov · D. N. Zhidkov (✉)
Ministry of Internal Affairs of Russia, St. Petersburg University, St. Petersburg, Russian Federation
e-mail: dmitry_jidkov@mail.ru

K. P. Indyk
Saint Petersburg State University of Architecture and Civil Engineering, St. Petersburg, Russian Federation

3 Results

3.1 General Requirements for Organizing (Repurposing) COVID Hospitals

1. COVID-19 Units Will Be Housed in Separate Building(s) of a Healthcare Facility [7, 11, 12].
2. Where COVID-19 units cannot be housed in separate building(s), they can be organized in the existing healthcare facilities provided that they allow for split-flow patient management (separate entrances isolated from other premises) and are installed with isolatable supply and exhaust ventilation systems [1, 12].
3. COVID-19 Units Classify into Two Types [7, 12]:

    Type I is COVID-19 units for patients in grave to extremely critical condition. Type I will have beds for patients in grave condition [3, 14] not requiring artificial pulmonary ventilation, beds for patients in critical condition requiring artificial pulmonary ventilation; beds for patients in extremely critical condition requiring artificial pulmonary ventilation; and beds for patients in moderately grave condition. Type I units are licensed to carry out medical activities involving works/services in the field of Roentgenology, Clinical Laboratory Diagnostics or Laboratory Diagnostics, Functional Diagnostics, Ultrasound Diagnostics, Endoscopy, Anesthesiology and Resuscitation, Nursing (hereinafter referred to as "Type I COVID-19 units");

    Type II is COVID-19 units for rehabilitation of patients [3, 7] transferred from Type I COVID-19 units for further treatment, and patients in mild-to-moderate condition. Type II is intended for follow-up medical care and licensed to carry out medical activities involving works/services in the field of Therapy and Nursing (hereinafter referred to as "Type II COVID-19 units").

3.2 Prompt Construction of COVID Hospitals (Modular Construction, RODER Hospitals)

To ensure prompt response to the increasing epidemiological situation in the area of their location, hospitals are advised to erect modular buildings or so-called RODER hospitals (RÖDER Group is headquartered in Germany and specializes on tents and modular space solutions). RODER designs offer a great deal of flexibility that allows its modular solutions be tailored to any purpose including backup COVID hospitals.

With modular facilities right by their side, the general hospitals can move patients that must be isolated in a prompt and comfortable manner.

Modular facilities must have beds with oxygen, and medical staff workplaces and recreation areas. Heating should be provided by hospitals'. If impossible, electric sources are acceptable. Toilet facilities will be installed in every ward. Modular designs offer a modern solution that ensures organizing the medical care in prompt manner, as well as isolating the infected citizens in order to prevent the disease from further spread. Quickly erectable structures, modular hospitals use lightweight, durable aluminum alloy frame. Their relatively light weight makes it possible to do without foundation.

The RODER-type modular facilities unburden the hospitals with high load of scheduled care, serving as backup facilities with quickly expandable bed capacity for the quality treatment and isolation of patients with new coronavirus infection [12].

The medical modules and laboratories will be connected to central heating system and installed with air conditioning. Their interior design must use wear-resistant (non-flammable) materials designed for long-term operation. Such an approach to the arrangement of modular medical facilities is intended to simplify their disinfection and sanitary treatment.

COVID hospitals should be designed as multi-purpose modular T-tents as a most optimal solution allowing to create large hospital spaces within the shortest timeline possible.

If the epidemiological situation should aggravate, such T-tents of RODER type can be assigned the role of backup mobile hospitals.

A modular medical facility can be described as high-quality if it is:

- quick to erect;
- lightweight and hence allows to do without heavy machinery;
- does not require a foundation. Temporary modular hospitals are erected with only a minimal scope of excavation works or without them if the site is large and flat enough;
- energy efficient. Temporary modular hospitals use the structural materials that meet high thermal efficiency standards;
- durable. The durability of temporary modular hospitals is achieved by the aluminum alloy as a material for their load-bearing and critical parts that boasts high robustness against negative environmental exposure;
- self-sufficient. Temporary modular hospitals must be installed with sewage and pumping stations, power generators, diesel fueled blow heaters.
- fully out-fitted. Temporary modular hospitals must be installed with engineering systems including air conditioning, ventilation, heating, as well as fire and medical alarms.

3.3 COVID Hospitals Layout and Operation

Depending on clinical and epidemiological indications, and in order to prevent the spread of the new coronavirus infection, patients are placed in isolation units as a measure designed to stopping the further spread of the infection and to ensure quality therapy.

One of the core epidemiological requirements to the design layout and operation of a COVID hospital is ensuring the protection of other patients and medical personnel from nosocomial transmission [4, 10, 13]. Unlike other types of hospitals, COVID hospitals must be equipped with the facilities allowing to provide full scope of operations involved in managing the COVID patients, especially when COVID cases increase. Along with admission units, wards, boxes, intensive care units, COVID hospitals must have X-ray rooms, disinfection units, centralized sterilization facility, catering unit, physiotherapy rooms, ultrasound examination, endoscopy, staff rooms, showers, toilets, materiel and medicines storage rooms [5, 7, 12].

For the effective performance purposes, the COVID hospitals must operate split-flow patient management. This means that all incoming patients must be divided into groups depending on the severity of their symptoms. The split-flow model ensures that COVID patients are separated from patients with other pathologies. Boxes in admission units allows for every patient to be managed individually, decreasing the possibility of infection.

With newly admitted COVID patients staying in these boxes, the medical staff can safely examine other arriving patients.

The box units must be insulated properly. However, if the layout of a stationary or modular COVID hospital does not allow to have separate boxes, partitions must be installed. The height of such partitions should be 2 to 2.5 m. Such boxes are open-type [12].

In addition, the admission departments must have exits for patients and exits for medical personnel.

In our opinion, the layout of the admission unit of an adapted (repurposed) COVID hospital should be as follows:

1. A duly adapted (repurposed) COVID hospital, like a specialized hospital, should have a separate admission ward with street-side entrance. This admission ward must have an illuminated sign for easy recognition at night time [7].
2. When organizing the admission of patients with fever, the "three zones two passages" is recommended for use as the most optimal solution. This principle implies dividing the admissions ward into three zones—infectious zone, potentially infectious zone, clean zone. These zones are properly separated from one another to ensure safe transfer of infected objects from the potentially infectious area to an isolated ward. Due measures should be in place to exclude removal of untreated items from the infectious zone.
3. An important element to combat nosocomial transmission in COVID hospitals is the presence in admission units of individual (Meltzer) boxes. Meltzer boxes allow patients to be examined in a quality manner and with minimal risks of infection for medical personnel. Anterooms are mandatory and must never be entered by patients because they lead to infectious diseases department. Anterooms are places where the medical staff can put on their protective equipment before entering the patient's ward and conduct thorough disinfection when leaving.

It is essential to see to it that when opening the door of the isolation box, the door of the anteroom is kept tightly closed to prevent the spread of COVID-19 viruses through the air. Each individual (Meltzer) box normally has one bed. After its patient is discharged, the box must be thoroughly disinfected.

Each box is provided with items required for patient care and cleaning. Dirty laundry and garbage, duly disinfected prior to removal, are taken out of the box in special bags for further processing (washing, boiling) or incineration.

Importantly, the layout of the COVID hospitals should enable zoning some of some of the premises into:

– examination room;
– laboratory;
– wards;
– intensive care unit.

It is advisable to zone infectious diseases units in the COVID hospitals in such a way that medical personnel are able to provide full scope of operations from monitoring to intensive care. It is essential that patients with suspected coronavirus infection are separated from those with confirmed COVID [6].

When organizing the work of COVID hospitals' infectious diseases department, an important role belongs to nurses.

For nosocomial transmission prevention purposes, the nurses handling COVID patients must at all times observe the following rule: a patient in acute phase of the disease should not be admitted to the ward with recovering patients.

It is important for the nurse to observe the numbering of beds and items that belong to them—vessels and dishes are intended for individual use. Patients are prohibited to move their beds. The distance between beds should be at least 1.5 m. Patients' dishes will be boiled in 2% sodium bicarbonate solution. Spatulas, beakers, pipettes, etc. must be properly sterilized [10]. Urea and stool will be disinfected in vessels or pots with bleach or chloramine prior to discharge into the sewer.

Nurses and doctors should wash their hands thoroughly before performing each routine manipulation, as well as during transition from one patient to another. Sprays with disinfectant solution for hand treatment should be placed at the doors of wards. All reusable medical instruments will be sterilized in steam autoclaves.

Nurses play an important role in preventing nosocomial transmission by monitoring the sanitary and hygienic condition of the wards and other premises of COVID hospitals. Nurses contribute to the prevention efforts by ensuring that wards are aired regularly and undergo quartz treatment. They see to it that patients' bedding and underwear are changed if smeared by vomit, feces, urine or other biological fluids. After the patient is discharged, their ward must receive final disinfection.

3.4 The Structure of a Standard COVID Hospital

COVID hospitals offer three kinds of services:

- diagnostic and treatment (admission unit with boxes, treatment section with boxes and wards, intensive care and resuscitation unit, etc.)
- administration and utilities; and
- guidance and methodology.

In COVID hospitals, treatment sections should have boxes and two medical checkpoints, one for the staff and one for newly admitted patients.

COVID hospitals are subject to the 3 V principle—video surveillance, ventilation, and visualization. Patients or those infected with coronavirus infection (COVID-19) are placed in COVID hospitals for the purpose of isolation and receiving therapy in individual boxes with individual entrance for patients and individual entrance for service personnel to prevent further spread of the infection [8].

Larger (400 + beds) infectious diseases hospitals should have a specialized intensive care unit installed with appropriate equipment and staffed with relevant personnel, while smaller, repurposed departments of hospitals will have the intensive care units capable of providing emergency therapy, emergency diagnostics and intensive monitoring of patients in a grave (critical) condition. The facilities adapted for use as COVID hospitals and repurposed hospitals will have at least 3% of their bed capacity allocated for intensive care. The repurposed infectious departments of district hospitals are recommended to have 1 or 2 emergency treatment boxes or wards, while those in municipal hospitals (under 400 beds) should have 2 emergency treatment wards/boxes and 1 intensive care ward in each of their departments [3, 5, 7].

Million-plus cities in emergency state are recommended to deploy in each of their repurposed hospitals a 200-bed section for patients with coronavirus infection (COVID-19)—at a rate of 20 cases per 100,000 residents.

Personnel distribution:

- 1 epidemiologist per 200 beds;
- 1 infectious disease doctor, 2 nurses and 2 nurse's aides per 10 patients in mild-to-moderate condition of coronavirus infection (COVID-19). There should be two shifts and at least 3 teams (1 for day shift and 2 for night shift);
- with mild to moderate severity of coronavirus infection (COVID-19), for every 10 patients—1 infectious disease doctor, 2 nurses, 2 nurses. The work should be organized in two shifts and at least 3 teams (1 team—day, 2—night).

Video surveillance will be installed in an infectious hospital for the purposes of monitoring the compliance with epidemiological regime by medical workers and patients (moving between wards, remote monitoring); and monitoring the observance of disinfection requirements by medical workers (PPE, hand treatment with antiseptics) and patients (moving between wards) [1, 12]. It is important that specific resuscitation measures are provided in accordance with prescribed continuity (prolonged artificial pulmonary ventilation, altichamber, hemodialysis, etc.). Ideally, each ward should have one bed so as to provide free access to patient from either of sides. The beds should be of standard rigidity (so-called "functional beds"). In their absence, rigid beds with upliftable head pieces. Wards must be equipped with nurse call buttons, centralized oxygen supply or oxygen cylinders, taking into account safety regulations.

Additional local lighting will be provided in wards, such as mobile lamps or dental lamps [7]. The essential equipment includes electrocardiograph, electric pump (OH-2), artificial pulmonary ventilation devices (RPL-2, DP-2, or DP-5; RO-2, RO-5, RO-6, or RO-6-03 for larger hospitals). The devices commonly used for reliving severe respiratory failure are extracorporeal membrane oxygenation (ECMO, EMO) devices. Conventional breathing equipment and electrocardiographs will be placed on mobile tables. The list of essential items also includes portable, preferably ultrasonic, inhaler (UIP-1), laryngoscope, portable X-ray machine, video bronchoscope with endoscopic stand, and ECG device.

Items such as standard express diagnostics and sampling kits, acid-base balance microanalyzers (ABBM), ACT (activated clotting time) devices should be within easy reach. The wards will be equipped with portable quartz treatment or bactericidal lamps, disinfectants, infusion warmers (electric). The wards must have the supply of the most essential medications, mainly for parenteral administration [7] and access to the Internet [7, 12] for remote counselling with specialists (patient management of specialists, video consultations) [2].

4 Discussion

Modern science has accumulated a sufficient amount of methodological guidance for creating COVID hospitals and organizing therapy for COVID patients in the absence of required infrastructure or well-structured interaction between therapy providers.

At the same time, the published research into organizing thee medical care for COVID patients cannot be said to reflect the whole picture and scale of the problem [4, 9].

It is evident that the issues of creating and operating the medical facilities intended for prompt admission and treatment of COVID patients should be examined in closer detail.

5 Conclusion

The study has enabled the following conclusions:

- to ensure the effective provision of treatment for COVID-19, the COVID units should be housed in separate building(s) of a healthcare facility;
- COVID-19 treatment facilities must operate the split-flow patient management (separate entrances isolated from other premises);
- the design layout of innovative healthcare facilities, including COVID hospitals, must provide for isolatable supply and exhaust ventilation systems;

- the design layout of innovative healthcare facilities, including COVID hospitals, must provide for the division of space into two zones: type I zone, intended for COVID patients in grave to extremely critical (with further subdivision into two zone, one for cases requiring artificial pulmonary ventilation and the other for cases not requiring artificial pulmonary ventilation), and type II zone intended for rehabilitation of COVID patients transferred type I zone for further treatment, and patients in mild-to-moderate condition;
- to successfully combat the coronavirus infection, COVID hospitals should be organized in prompt manner, the preferred design being RODER;
- COVID hospitals should necessarily provide for isolation of patients and split-flow patient management;
- COVID patients must be managed in strict accordance with regulatory documents, with special emphasis on prevention of nosocomial (community-acquired) transmission;
- the layout of the boxes for isolating COVID patients should provide for anteroom, ward, sanitary unit with a bathroom, and gateway for staff;
- organizationally, COVID hospitals are intended to provide three types of services: diagnostic and treatment (admission unit with boxes, treatment section with boxes and wards, intensive care and resuscitation unit, etc.); administration and utilities; and guidance and methodology;
- for more effective monitoring of the personnel and patients, it is expedient to install COVID hospitals with video surveillance;
- wards should be arranged in such a way as to enable the staff to perform their duties most effectively and the patient to have access to emergency call buttons.

References

1. Babayan AR, Fisenko AP, Sadeki NMY et al (2020) Coronaviruses: biology, epidemiology, prevention. Russ Pediatr J 1:57–61
2. Bai Y, Yao L, Wei T et al (2020) Presumed asymptomatic carrier transmission of COVID-19. JAMA 323(14):1406–1407
3. Belotserkovskaia YG, Romanovskikh AG, Smirnov IP (2020) COVID-19: a respiratory infection caused by new coronavirus: new data on epidemiology, clinical course, and patients management. Consilium Medicum 22(3):12–20
4. Briko NI, Kagramanyan IN, Nikiforov VV et al (2020) COVID-19 Pandemic Measures to combat the spread in the Russian Federation. Epidemiol Vaccination 2:4–12
5. Fisenko PV, Chichkova NV (2020) Modern pandemic COVID-19 and pharmaceuticals. Exp Clin Pharmacol 4:43–44
6. Khismetova ZA, Samarova US, Tokanova SE et al (2020) Organizing the activities of infectious diseases hospital during coronaviruses infection. NAO Semey Medical University, Methodological recommendations, p 12
7. Liang T (2020) Handbook of COVID-19 prevention and treatment. The First Affiliated Hospital, Zhejiang University School of Medicine. Compiled According to Clinical Experience
8. McIntosh K, Hirsch MS, Bloom A (2020) Coronavirus disease 2019 (COVID-19). UpToDate Hirsch MS Bloom 5(1):873

9. Pasechnik IN (2020) Circulation of medical devices during their trial phase and operation in special legal regime. Med Law 3:39–44

10. Pasechnik IN (2020) Nutritional support for coronavirus patients in critical conditions. Anesthesiology Resuscitation 3:70–75

11. Pshenichnaya NY, Veselova EI, Semenova DA et al (2020) COVID-19—Humanity is facing a new global threat. Epidemiol Infect Dis 1:6–13

12. Sokova E (2020) Healthcare performance in high-alert regime/E. Sokova//Chief Physician 3:58-61

13. Tsaranov KN, Zhiltsov VA, Klimova EM et al (2020) Perception of personal safety threat by medical specialists during COVID-19 pandemic. Health Care Manag 4:15–19

14. Voitenkov VB, Marchenko NV, Skripchenko NV et al (2020) Instrumental methods for diagnosing pneumonia in coronavirus infection. Consilium Medicum Pediatrics 1:20–25

Environmental Impact Assessment of Construction Waste-Based Composites

Alexander Scherbakov, Elizaveta Lukashuk, Mariya Subbotina, and Oksana Karnaukhova

Abstract Theoretical and practical studies highlight the advantages of recycling municipal solid waste and construction wastes. In an attempt to lessen the negative environmental impact from solid waste and to enhance sustainability of energy- and resource-intensive construction, significant efforts have been channeled into re-use of solid wastes for environmentally friendly building materials. This study seeks to identify the ways towards more efficient use of municipal solid waste and construction waste as recycled materials for construction industry. The paper discusses possible applications of solid municipal waste and construction waste as substitute materials in the production of sustainable geopolymer composites. Even though inclusion of these fractions in geopolymer composites is found to have a negative effect on composites' performance characteristics such as strength, processibility and durability, satisfactory performance can be achieved by adding the correct percentage.

1 Introduction

The alarming increase in the volumes of municipal solid waste is a global environmental concern. In an attempt to eliminate the negative environmental impact of solid waste and to enhance sustainability of energy- and resource-intensive construction, significant efforts have been channeled into re-use of solid wastes for environmentally friendly building materials [10].

Recycling and recyclable materials offer a huge potential for reducing the production costs and are seen as a means of mitigating the effects of climate change by ceasing the use of primary raw materials. For the recyclables to become a preferred choice among manufacturers, their environmental performance and cost-efficiency must be higher than that of primary raw materials.

A. Scherbakov (✉) · E. Lukashuk · M. Subbotina · O. Karnaukhova
Saint Petersburg State University of Architecture and Civil Engineering, Saint Petersburg, Russian Federation
e-mail: shurbakov.aleksandr@yandex.ru

© The Author(s), under exclusive license to Springer Nature Switzerland AG 2023
D. Ivanov et al. (eds.), *Proceedings of ECSF 2021*, Lecture Notes in Civil
Engineering 257, https://doi.org/10.1007/978-3-030-99877-6_35

Construction industry is one of the world's largest consumers of plastic, accounting for 19.8% of the total demand for converters. The main types of plastics in use are polyethylene, polypropylene and polyvinyl chloride, each of which can be used as ingredient of thermoplastic composites. In addition, the amount of wastes generated by the construction industry often exceeds that by other sectors. For example, in 2019 construction wastes represented 38.1% in the total amount of wastes produced in the EU. On a global scale, construction debris and demolition wastes average 3 billion tons annually [5].

The cost of products made from recycled materials depends on several factors and among them transportation costs, landfill fees, recycling process, volumes and quality of recyclables, throughput fees, operating costs, and waste taxation. Environmental protection measures operate eco-taxes, eco-procurement and standardization of recycled materials [12]. The traditional applications of recycled thermoplastic composites include the production of wood-plastic composites, semi-finished products, molding and various types of cladding. The potential scope of recycled composite applications is expanding gradually to include such technically complex applications, such as molded door panels and molded holes [7].

Our analysis of dedicated literature has shown that construction waste can produce a tangible positive impact from economic and environmental perspectives. Thus, a comparison of the environmental and economic efficiency of pressed cellulose-plastic composite and its wood-plastic counterpart has found that the use of cellulose sludge as a filling material is more environmentally efficient than the use of the recycled wood flour [7].

Good progress has been achieved in solid waste recycling for geopolymer composites. Geopolymers –alkali-activated materials—are products of the chemical reaction between alumosilicate precurcors and alkaline activators, and is generally considered to be an alternative to ordinary Portland cement (OPC). The last three decades have witnessed rapid development of geopolymers for academic purposes due to their excellent performance in various fields. In general, geopolymers demonstrate excellent mechanical properties and other inherent characteristics such as superior strength, immobilization of toxic pollutants, versatility and even intelligence. Further, geopolymers are characterized by low greenhouse gas emissions, lower energy consumption, and higher recyclability waste reuse, which are critical for the future sustainability of construction industry. Therefore, the use of municipal solid wastes as a component of geopolymer composites will certainly contribute to achieving more environmentally friendly and safer building materials. As a rule, municipal solid waste includes fractions of household waste, industrial waste, construction waste and agricultural waste.

This study seeks to identify the ways towards more efficient use of municipal solid waste and construction waste as recycled materials for construction industry.

2   Materials and Methods

The paper discusses the anthology of progress towards the use of solid household and construction wastes in the production of geopolymer composites. The resultant waste-based geopolymer composites are analyzed for benefits and downsides. The study uses as its methods literature analysis, benchmarking and data comparison.
The study pursued to:

- analyze the existing practices for recycling solid waste and construction waste into secondary environmentally-safe products;
- identify potential applications of recycled materials based on their core ingredient; and.
- highlight the benefits of recycling.

3   Results

Municipal solid waste (MSW) classifies as household and commercial waste generated within jurisdictions of municipal authorities. In most cases, MSW consists mainly of organic materials, waste paper, waste glass, plastic waste, cans, textiles, etc. In today's conditions of rapid urbanization, the growth rate of MSW outruns urbanization itself. It is reported that by 2025, MSW is expected to reach the level of 2.2 billion tons per year, compared to 0.68 billion in 2002. Taken together, these circumstances have intensified the search for ways to increase re-use of wastes, one being the manufacture of geopolymer composites [2].

Incineration remains a widely used practice in MSW management. Incineration can reduce the volume and mass of waste by nearly 90% and 70%, respectively. In addition, incineration makes wastes a source of energy. Incineration generates two types of ash, namely, municipal incinerated bottom ash (MIBA) and municipal incineration fly ash (MIFA). MIBA is a residue with large-size particles occurring on the bottom of the incineration plant, whereas MIFA is composed of very fine particles trapped by air pollution control systems. Since MIBA and MIFA represent two different ashes, their use in geopolymer composites will be discussed separately.

MIBA accounts for about 80% of waste combustion residues. Compared to MIFA, it contains much less toxic organic substances and therefore has a greater potential for recycling instead of landfilling. Despite the significant efforts to promote MIBA as a raw material for cement production and aggregate for road construction, there are significant disadvantages that limit wider use of MIBA, one being heavy metal leaching.

Initially, MIBA was used as a partial substitute for precursors in synthesized geopolymer composites. Some authors have demonstrated that MIBA is a suitable feedstock for geopolymers mixed with methacaolin, with the precursor content of up to 70% [11]. Subsequent studies explored the possibility of using MIBA as sole precursor of geopolymer. By analyzing the microstructure and determining the characteristics of the resultant composition, the scientists discovered highly satisfying levels of geopolymerization and newly formed crystalline phase consisting of silica, aluminum and sodium.

A number of studies have involved analysis of the effect of heavy metal binding in MIBA-based geopolymer composites. It is generally believed that geopolymerization is able to immobilize effectively most of the hazardous elements in MIBA, the general conclusion being that the resulting geopolymer composites represent an environmentally safe and non-hazardous material.

However, it should be noted that resultant MIBA-based geopolymer composites usually have low mechanical performance and a highly porous structure. This is mainly due to the fact that the metallic aluminum, contained in MIBA, can react with alkaline solution and then release hydrogen gas. Consequently, MIBA can be used as a partial or complete precursor for synthesized aerated geopolymer composites.

One study has produced aerated MIBA-based geopolymer pastes with dry density between 600 and 1000 kg/m^3, indicating that parameters such as alkali concentration, liquid-to-solid ratio and mixing duration were important factors in managing the physical and mechanical performance of resultant aerated MIBA geopolymer. Similarly, another study into aerated geopolymer pastes, synthesized by integrating MIBA and spent glass powder, has shown low density values (between 494 and 1295 kg/m^3) and low thermal conductivity (between 0.14 and 0.38 W·m*K. In addition, compared to traditional aerated concrete, the resultant geopolymer concrete had fewer spherical air voids and a wider distribution of air voids by size [9].

Further, researchers have used MIBA as a gas-forming additive for aeration of geopolymer composites. The comparison of the effect of MIBA and commercial aluminate powder on the light aerated geopolymers has found that MIBA had a reaction rate and gas-forming ability comparable with that of commercial aluminate powder. Moreover, the resultant aerated MIBA geopolymers had a density of only 860 kg/m^3 and a thermal conductivity of 0.33 W·m*K, which are comparable to the reference aerated geopolymers based on commercial aluminate powder.

Series of studies have been conducted into pretreatment—alkali treatment, vitrification, and wet grinding—to eliminate the foaming and expansion effect of the metal aluminate contained in MIBA. One series of experiments used alkaline treatment: MIBA was mixed with sodium hydroxide solution to form a suspension which was matured within 4 h prior to converting into MIBA-based geopolymer composites. Meanwhile, several more additives were tested to further improve the performance of geopolymer composite. The test results have shown that the obtained geopolymer composites have satisfactory compressive strength and durability due to a high degree of geopolymerization and dense microstructure [4].

Regarding the use of MIBA as a precursor or gas-forming additive, the studies involved testing MIBA's capacity as a replacement filler for geopolymer composites. Thus, some studies proposed to use MIBA as a substitute for up to 50% fine aggregate (by volume) in geopolymer solution. Although MIBA was found to have a negative effect on the strength of geopolymer solution's porous brittle structure, there was no expansion or cracking due to MIBA's metal aluminate. Ultimately, it is possible to achieve a compressive strength of 35 MPa to 56, which implies that MIBA has a broad potential for reuse in geopolymer composites. Besides, the resultant products' leaching behavior is found consistent with statutory regulations, which speaks in favor of geopolymer composites.

MIFA is a fine powder extracted from flue gases by air pollution control devices. Although the weight of MIFA is only 2–5 wt.% of the MSW before incineration, its global generation is huge, growing together with urbanization and population. In China, for example, the amount of MIFA is estimated to reach 1.0×10^7 tons per year by 2022. MIFA contains large amounts of heavy metals such as chromium, cadmium, lead, zinc, etc., and is therefore considered a hazardous waste. Along with heavy metals, another concern is MIFA's soluble salts. Therefore, scientists are looking into ways to reduce its harmful trace and find useful applications of MIFA [8].

Since geopolymer composites can serve as immobilizing agents for hazardous wastes' stabilization/solidification (S/S), there have been numerous studies into the effectiveness of geopolymer composites in MIFA's S/S. Particular attention was paid to the impact of such synthesis parameters as precursor type and content, alkaline activator type and dose, as well as curing process, on S/S. In general, geopolymer composites have proven to be a highly effective material for MIFA's S/S, causing a significant reduction in toxic elements leaching. For example, after some authors have included MIFA in a geopolymer matrix based on coal fly ash (CFA), the release of heavy metals from the geopolymer composite was found to be much lower compared to MIFA's initial state: chromium leaching reduced from 1.57 to 0.02 mg/l, copper from 3.80 to 0.04 mg/l and lead from 11.5 to 0.1 mg/l.

Moreover, one recent study has demonstrated excellent long-term S/S-effectiveness of MIFA-containing geopolymer composites even in aggressive environments. In particular, heavy metals leaching (chromium, copper, lead, zinc, mercury and cadmium) remained relatively low after immersion in aqueous alkali and exposure to acid rain. A number of studies have revealed the mechanism of heavy metals immobilization in MIFA-containing geopolymer composites. It is believed that this mechanism is vehicle by both physical and chemical processes, including physical encapsulation by a geopolymer matrix, Friedel salt ion exchange and geopolymer adsorption, which lead to heavy metals becoming immobilized in the geopolymer network.

Since the content of chlorides and sulfates in MIFA is usually high, the negative effect of these compounds on the kinetics of geopolymerization cannot be ignored. Individual studies used pretreatment with water to remove the inorganic bar from MIFA, and then investigated MIFA geopolymerization to evaluate the effectiveness of the pretreatment with water. It was found that the water flush pretreatment has significantly contributed to early strength and led to higher tensile strength (22.7 MPa after 28 days), compared to the counterpart that didn't use water flush pretreatment. The highest heavy metals immobilization efficiency was found in the geopolymer composites based on water-washed MIFA. Therefore, viable pretreatment is an important part of using MIFA as a raw material for geopolymer composites in civil engineering, as well as for a more efficient stabilization process.

The use of raw waste paper as an ingredient of building materials is not a common practice. Instead, large amounts of waste paper are being recycled into new paper products to save wood and forest resources and thus lessen the environmental impact. At the same time, recycling of recycled paper into usable fiber often leads to the formation of a secondary flow, commonly referred to as waste paper sludge. This sludge has high water content, 50–70%, and is usually dried before further processing to facilitate handling, incineration and other potential applications. In addition, waste paper sludge contains approximately equal amounts of organic substances (mainly residual cellulose fiber) and inorganic fillers (such as kaolin clay and calcium carbonate) [1]. Previous research has mainly focused on the use of waste paper sludge in OPC-based building materials, while the use of waste paper sludge in geopolymer composites represents a relatively advanced development.

In general, the studies on the use of waste paper sludge in geopolymer composites show two main approaches. Chemical analysis has shown that the waste paper sludge is compatible with the chemical composition of geopolymers and can serve as a potential additional additive to geopolymer composites. In the first approach, this material is used in its raw form.

The other approach to converting waste paper sludge into geopolymer composites uses waste paper sludge ash, which is formed as a result of thermal processes—incineration of waste paper sludge. During combustion, the latent energy of organic component can be recovered, and there form highly reactive metakaolin-type phases and calcined limestone. This ash was found suitable for use as a replacement precursor in geopolymer composites.

Disposal of the increasing volumes of rubber waste represents a challenging task. With three-dimensional network structure, rubber degrades very slow. One major source of rubber waste is car tires. Rubber waste is estimated to exceed 200 million annually by 2030. Traditionally, waste tires are warehoused or landfilled as temporary solutions. Stored tires can become a favorable environment for reproduction of insects and mosquitoes. They as a source of negative environmental impact, as toxins can be easily washed out from tires to contaminate soil and groundwater. Recycling of these used tires is currently a global environmental challenge that should dealt with urgently. The idea to use recycled tires in geopolymer composites appeared only in recent years [3].

The preferred method of rubber tires processing is grinding their crushed pieces to granules of the desired size—rubber crumb. In geopolymer composites, rubber crumb can be used as replacement, partial or even complete, of large and small aggregates. There is a consensus among existing studies that rubber crumb can significantly change the properties of geopolymer composites: compressive strength tends to reduce consistently with an increase in the amount of added replacement rubber crumb. In addition, at a certain percentage, the loss in strength at the expense of mass replacement appears greater that at the expense of volume replacement. There are several factors contributing to the loss of strength, one being the hydrophobic nature of rubber, which explains the weak bond between rubber and the geopolymer matrix.

The microstructure tests confirmed imperfect adhesion between the rubber fillers and the geopolymer matrix, as evidenced by deep cracks and voids at the interface.

Another reason for the loss of strength is the low elastic modulus of rubber, which can lead to premature cracking near the junction of rubber and geopolymer matrix. However, mechanical strength can be retained if rubber replacement coefficient is within the appropriate range, which makes rubber a suitable material for construction purposes. The use of geopolymer composites with high rubber replacement coefficients is limited to secondary or non-critical structures.

The previous studies have also covered the impact resilience of geopolymer composites with three different levels of rubber crumb filler, exposed to falling weight. The tests have shown that impact energy absorption was higher in geopolymer composites with a higher content of rubber crumb. This can be explained by rubber's high elastic performance, especially under substantial deformations, and its capacity to absorb energy. In other words, rubber is by nature capable of absorbing sudden shocks. This high absorption cannot be achieved with the help of natural fillers due to their fragility. It was also found that inclusion of rubber crumbs increases the viscoelasticity and damping properties of geopolymer-based building mixes.

In addition, increased insulating properties (acoustic impedance and thermal conductivity) have been reported for geopolymer composites containing rubber chips. The authors concluded that the thermal conductivity of geopolymer concrete reduced significantly due to the inclusion of rubber chips, decreasing from 1.284 and 0.237 W·m*K. This is due to the fact that rubber has low thermal conductivity, ranging between 0.1 and 0.25 W·m*K, as compared to the thermal conductivity of a normal filler, approximately 1.5 W·m*K.

Another source of grave ecological concern is plastic waste. The rate of plastic production exceeds that of its disposal. It is well known that plastic is a non-biodegradable material that takes a long time to disintegrate. Disposal of plastic wastes is a heavy burden on the environment. Moreover, since plastic production uses harmful chemicals, the disposal of plastic waste will lead to their release. One of the best solutions to reduce these negative effects is to use recycled plastic as ingredients for new materials such as mortar or concrete.

A number of studies have been conducted to evaluate the properties of geopolymer composites containing plastic waste as a filler. In one study, plastic waste was melted into lumps and then crushed into particles with a diameter of about 2.1 mm to act as a fine aggregate for a geopolymer composite. Another study replaced the granules of spent PET bottles with a particle size of less than 4 mm with fine fillers in geopolymer solution at different levels (20–100%). In one of the papers it was proposed to include in geopolymer concrete the polystyrene foam with particle size of 2.36–4.75 mm, obtained from discarded packaging foam. As a rule, the density of geopolymer composites decreases with an increase in the replacement coefficient of spent plastic filler, mainly due to plastic material's low density. An increase in plastic aggregate replacement coefficient has also led to a decrease in mechanical properties, including compressive strength and bending strength. However, achieving the appropriate ratio of plastic aggregate can give geopolymer concrete the acceptable strength and density comparable with that of lightweight structural concrete. Also, a decrease in abrasive resistance and an increase in porosity and water absorption were observed as the amount of waste plastic aggregate increased.

Since plastic has low thermal conductivity, the inclusion of waste plastic additionally provides geopolymer composites with lower thermal conductivity and, hence, better thermal insulation.

Along with the MSW presented above, other types of MSW have, too, been studied for their potential for reuse in geopolymer composites. One common recyclable is glass. The world annually produces approximately 65 million tons of glass waste, a fraction that accounts for 5% in the total amount of solid waste. However, recycling of glass waste remains low and ineffective. In the United States, only 28% of the 11.54 million tons of glass waste undergo recycling. Mainland China produces 40 million tons of glass waste annually, but only 13% of it is recycled. The chemical composition and mineralogy of glass suggests that it contains a lot of amorphous silicon and calcium. Glass is known to have high reactive capability. Its suitability for use in the production of geopolymer composites has been confirmed by extensive research. In general, glass waste can be reused as aggregates, precursors and alkaline activators in geopolymers.

4   Discussion

Construction waste, as an unavoidable by-product of construction, repair and demolition, includes a wide range of materials and among them concrete, metals, brick, wood, ceramics, asphalt, soil, gypsum and polymers. These wastes constitute the largest source of solid waste in the majority of countries. As a result, the issues of waste management are causing a serious concern from the perspectives of economy, environment and public safety. In recent decades, numerous studies have been devoted to way of increasing recycling and reducing wastes disposal.

Concrete is the most widely used building material due to its relatively low cost, availability of ingredients, high mechanical performance and durability.

Waste concrete recycling efforts target mainly applications such as coarse and fine aggregates in geopolymer composites.

The second commonest building material after concrete is clay brick. Clay brick waste forms not only as a result of demolition, but also from defective bricks during production, transportation and construction. Clay bricks are made by mixing crushed clay with water and giving the resultant mass the desired shape before drying and firing.

Ceramic materials and products are often used as finish materials such as floor tiles, garden tiles, brick ware and sanitary ceramics. The production of ceramics is similar to that of clay bricks: it begins with mixing of the raw materials, molding, firing, polishing and glazing. Ceramic materials are usually fired at a higher temperature than bricks so as to allow silicon dioxide to recrystallize and form a vitrified material of greater density, strength, hardness, resistance to chemicals and frost, and greater dimensional stability. The use ceramic waste as an ingredient of geopolymers is of great academic interest [13].

5 Conclusion

As can be seen from the above data, waste recycling represents a highly relevant activity designed to reduce environmental load and to promote material re-use in different. Consequently, there is an obvious need for effective solid waste management plans and strategies. Waste sorting is a key step towards increased recycling. The fact is that municipal and construction waste often contains all sorts of factions, while mixed and contaminated waste is not suitable for recycling. Sorting is intended to separate wastes into fractions according to basic components and seems to be the only way to ensure that the target fractions end up at recycling plants. This strategy implies the use of effective separation and sorting methods, as well as appropriate equipment, regardless of the processes occurring at recycling plants or beyond. Awareness raising and involvement of the public and relevant stakeholders are important components in waste management and recycling.

In this regard, no less important is the state support. Measures can include tax refunds to the contractors that take care that their wastes are duly recycled; creation of recycling markets; incentives and interest-free loans to small businesses engaging in waste recycling-related projects.

References

1. Adesanya E, Ohenoja K, Luukkonen T et al (2018) One-part geopolymer cement from slag and pretreated paper sludge. J Clean Prod 185:168–175
2. Aly ST, Kanaan DM, El-Dieb AS et al (2018) Properties of ceramic waste powder-based geopolymer concrete. In: International Congress on polymers in concrete. Springer, Cham, pp 429–435
3. Azmi AA, Abdullah MMAB, Ghazali CMR et al (2019) The effect of different crumb rubber loading on the properties of fly ash-based geopolymer concrete. In: IOP conference series: materials science and engineering, vol 551, no 1. IOP Publishing, p 012079
4. Bugayan SA (2019) Recycling of solid household waste as an integral element of the rational use of natural resources. J Econ Regul 10(1):90–99
5. Fořt J, Vejmelková E, Koňáková D et al (2018) Application of waste brick powder in alkali activated aluminosilicates: functional and environmental aspects. J Clean Prod 194:714–725
6. Hong J, Li X, Zhaojie C (2010) Life cycle assessment of four municipal solid waste management scenarios in China. Waste Manage 30(11):2362–2369
7. Medvedev VS, Tokarev AS, Panin PA et al (2019) Plastic—ways to improve the planet. Problems Sci 5(41):14–16
8. Mukhamedova NB, Abdukarimova SM (2019) Analysis of modern technologies for sorting solid household waste. Achievements Sci Educ 13(54):15–16
9. Nuaklong P, Sata V, Chindaprasirt P (2018) Properties of metakaolin-high calcium fly ash geopolymer concrete containing recycled aggregate from crushed concrete specimens. Constr Build Mater 161:365–373
10. Rozina VE, Dagbaeva YB (2019) Management of the construction waste recycling system. Universum: Tech Sci 6(63):32–34
11. Sarmiento LM, Clavier KA, Paris JM et al (2019) Critical examination of recycled municipal solid waste incineration ash as a mineral source for Portland cement manufacture–a case study. Resour Conserv Recycl 148:1–10

12. Solomin IA (2019) Organization of the municipal organic waste management system. Environ Manage 2:60–65. https://doi.org/10.34677/1997-6011/2019-2-60-65
13. Wu A, Peng Y, Huang B et al (2020) Genome composition and divergence of the novel coronavirus (2019-nCoV). Cell Host Microbe 27(3):325–328

Characteristics of Cost-effective Prefabrication in Healthcare Construction

Alexander Scherbakov, Elizaveta Lukashuk, and Oksana Karnaukhova

Abstract The tense epidemiological situation has caused the need for rapid construction of healthcare facilities and deployment of additional hospital beds. The paper discusses new cost-effective steelwork modules as a design solution for healthcare facilities with improved performance. These structures have proved effective for use in medical emergencies. Further use of prefabricated structural steel frames in healthcare construction will allow us to respond promptly to emergencies, should they occur in the future. Prefabricated steel modular units owe their recognition as the best choice in healthcare construction to their exceptional strength and rigidity. The possibility of using modular breather panels, mounted on angle posts and designed to improve the indoor air quality and reduce the spread of diseases, is considered. The technical solutions discussed in this article make it possible to organize affordable medical care over a short time, ensuring the survival of the population during the pandemic and preserving the quality of life for those who have undergone treatment and rehabilitation in hospitals.

1 Introduction

Modular construction is an alternative construction method, which, unlike conventional techniques, has the major part of the construction works performed off-site (i.e. in the controlled environments of factories). The only works to be performed onsite are installation, finishing, and utility connection [12]. Modular buildings significantly increase the efficiency of the construction sector.

Modular construction uses blocks made of steel, wood, concrete or hybrid materials. However, modular blocks with a steel frames are superior to other materials due to their structural and environmental advantages. With most of the operations being performed outside of the construction site, the modular construction method offers high rate of construction, high quality, increased safety, accuracy, profitability,

A. Scherbakov (✉) · E. Lukashuk · O. Karnaukhova
Saint Petersburg State University of Architecture and Civil Engineering, Saint Petersburg, Russian Federation
e-mail: shurbakov.aleksandr@yandex.ru

© The Author(s), under exclusive license to Springer Nature Switzerland AG 2023
D. Ivanov et al. (eds.), *Proceedings of ECSF 2021*, Lecture Notes in Civil Engineering 257, https://doi.org/10.1007/978-3-030-99877-6_36

environmental performance, and reduced number of jobs. These inherent benefits have made modular technologies a preferred choice around the world for application in housing, commercial premises, schools and healthcare construction.

As known, there has been an acute shortage of beds for patients diagnosed with COVID-19 during the pandemic globally. Since the existing hospitals and repurposed facilities appeared insufficient, the shortage was to be made up for with new hospitals suitable for infectious patients. Those new hospitals had to be built within a very short timeline [2].

Among the countries that have been demonstrated the effectiveness of modular design in emergency situations like COVID-19, is China. In early February, a temporary 1000-bed hospital was built in Wuhan in just 10 days. It made it possible to increase the volume of emergency medical care and reduce patients mortality [3].

Similar experience is found in other countries, where the modular concept is being successfully applied in the construction of healthcare facilities (Table 1).

The past year kept construction specialists all over the world busy conducting in-depth studies of the possibilities for enhanced use of modular structures in global emergencies.

The majority of publications focus on the development of structurally stable lightweight volumetric modular units for emergency healthcare. Such modular units achieve high performance due to the improved characteristics of their components— beams, columns, joints and connections—and the new technologies. The expediency of the proposed improvements has been validated by the results of optimization studies and physical tests [4].

This study is designed to analyze the key characteristics the cost-effective prefabricated medical buildings.

Table 1 The use of modular construction technologies in the construction of medical buildings in different countries

Country	Purpose	Construction timeline
UK	Intensive care	3 weeks
United States of America	Social distancing	4 months
United States of America	Mobile testing facility	2–3 weeks
Australia	Emergency Medical Care and Consultation	–
Armenia	Extension of Nork Hospital in Yerevan	10 days
China	Huoshenshan Hospital	2 weeks
China	Leishenshan Hospital	2 weeks
Georgia	Temporary hospital for infected patients	4 weeks
United States of America	Quarantine	2 weeks

2 Materials and Methods

The study focuses on the methods and approaches to the use of prefabricated modular structures in healthcare construction.

3 Results

Modular construction sets itself apart from the traditional construction methods by using as its basic component volumetric modules. These modules are prefabricated structures that are built at factories and deployed on the construction site for further assembly and utilities connection. The process combines various types of technologies for rapid construction and installation. Higher durability of modular buildings, as compared to conventional designs, is achieved due to offsite construction.

Modular volumetric blocks comprise walls, floors, ceiling panels and bracing beams (if necessary). Corner posts are usually hot-rolled steel or hollow profiles. Notably, prefabrication is either element-based or panel-based.

In practice, modular units can be subdivided into load-bearing wall modules and corner post-supported modules, depending of the load. In load-bearing wall modules, the load is transferred to the foundation through the walls, while in corner post-supported modules it is through corner and intermediate posts. In a load-bearing steel module, the wall pieces are spaced at 300 or 600 mm. The modular construction industry uses modules of various shapes—steplike modules, faceted modules, conical modules, the commonest design being rectangular [10].

It should be noted that wall-supported modules are compatible with any shapes, unlike angle-supported modules. Corner post modules are best fit for buildings where large open spaces are important, in which case the modules can be joined to form a diversity of building configurations. With highly stable concrete or steel frames, modular blocks can be assembled vertically up to 25 storeys.

It is commonly believed that steel is a preferred choice in modular construction because of its excellent performance as a material for offsite engineering. Modular steel blocks allow excellent precision and boast durability, increased fire resistance, exceptional strength, low weight and high stability. The benefits of modular buildings with steel frames are highlighted in many previous studies [5].

Another advantage of steel modular structures is their reusability. The authors have estimated that reused steel can save approximately 50% of the material mass and 80% of energy, while materials such as wood and concrete demonstrate lesser potential for reuse.

A standard steel modular block weighs approximately 15–20 tons, while a standard modular concrete block approximately 20–35 tons. Steel modular blocks are 20–35% lighter than their concrete counterparts.

Steel modular buildings are quick to assemble. They are connected using bolted and riveted joints, whereas concrete modular blocks use on-site pouring methods, that extend the on-site working time. Modular blocks can be disassembled and transported to another site for re-assembly. Given the current and future environmental conditions, the use in modular structures of environmentally friendly materials such as light steel becomes vital.

The outbreak of the new coronavirus has had a huge impact on the lives of all people around the world. With many communities, the capacity of healthcare infrastructures, especially those that deal with individual testing of patients, appears insufficient to respond to the growing needs.

Modular structures have attracted increasing attention during the current COVID-19 situation due to their high usability in construction, as the pace of the pandemic has led to the urgent need for rapid deployment of emergency care. Along with test centers, the range of vital facilities includes intensive care and first aid units, command centers, administrative offices, disinfection facilities, toilets, distribution centers for basic services, temporary training facilities and medical supplies storages.

The authors emphasize the benefits modular construction in hospital outbuildings. Prefabricated structures can significantly reduce malfunctioning in medical facilities. They also provide shielding from noise and dust, which is very important when creating or expanding the facilities for patients with weakened immune system, reducing patients' exposure to excessive noise and polluted air [4].

As quarantine centers, modular buildings allow for increased number of beds and capacity of intensive care units. All things considered, prefabricated modular buildings for on-site assembly and installation represent the best solution to the complex problems being faced by the healthcare system.

One more important benefit of modular construction relates to occupational safety. Since modular structures are constructed office, i.e. in controlled environments, the risk of work-related injuries is less.

The delivery of modular units to the construction sites is a matter of few days and uses well-managed supply chains enabling timely response to the expected growth in demand. This is possible because the production of modular structures is off-site and can takes place concurrently with the on-site assembly of multiple elements. The completed modules can be adapted for use as residential buildings or healthcare facilities.

Modular blocks and their components can be placed for storage before sending to their final destination.

Modular buildings offer an optimal solution to the main problems experienced by health sector, since traditional construction using brick, wood and concrete is never rapid. Modular design is expected to gain wider use in several countries to provide quarantine facilities, isolation rooms, testing laboratories, recreation facilities for medical personnel, etc. Thus, modular solutions offer healthcare systems a unique advantage for use in crisis situations.

4 Discussion

The structural performance of the modular systems for use in different areas of construction, including healthcare, have been explored in a wide range of studies. The importance of consideration of modular buildings' structural characteristics may vary depending on the location.

One method proposed in literature is for cold-formed steel beams, a unique approach that can be used in the design of structures of increased efficiency. Its main focus is on reduced material consumption. Optimization tests worked with the cross-section moment load: different shapes of cross-sections with reduced material consumption were tested for suitability for given moment loads [8].

The recent pandemic situation has highlighted the need for ecologically safe components in the design of buildings. The idea is not new but has gained even greater relevance today. The experience of fighting the virus has revealed that people with unhealthy living and working conditions are the first to be affected [1]. The statistics highlights the need for ecologically safe buildings with quality indoor air. What also adds to the need for safer buildings is the fact that people tend to spend most of their time indoors.

The concept of ecologically safe building should be a standard not only for hospitals but also offices and residential buildings. One of the main requirements is enhanced ventilation necessary to "dilute" the airborne pollutants and reduce the rate of diseases transmission. It has been found that an environment with low humidity contributes to the survival of viruses. The optimal humidity range is 40% to 60%. The future modular construction should focus on integrated use of ventilation systems (clean air; air displacement) and filtration technologies. One option could be "breathing" walls. Such walls can be used as a non-load-bearing and as part of corner post-supported module.

Modular "breathing" panels are a convenient system consisting of insulating material and a casing. They enable distribution of the supplied air without any additional costs. Moreover, they are a kind of air filtration package for the entire life of the building, that can be easily installed in steel-frame modular. It is evident that after the coronavirus pandemic the construction industry will be even more focused on energy-efficient and environmentally friendly performance, and for this reason the use of modular technology represents a tool for promoting the environmental friendliness of buildings and structures.

Research focus is mainly on the corner post-supported module systems with improved characteristics for emergency situations. The corner module is usually a combination of more than one modular unit, designed to create larger unpartitioned work areas. Lengthy modules may require intermediate posts [5].

The strength of corner post-supported module depends solely on the corner posts, since they carry and transmit the entire load of the module. High-rise construction normally uses 100×100 mm or 150×150 mm sections and low-rise projects 80×80 mm.

The use of hollow profiles as corner posts is associated with their high resistance to longitudinal bending. Hollow steel columns can be sometimes filled with light concrete to maintain the size of columns on each floor and to avoid greater thickness or larger size of columns on lower floor levels. This enables the use of the same inter-module connections across the entire modular design.

Optimal beams are proposed for use as ceiling and floor beams. These optimal beams, such as flanges and sigma sections, can withstand equal loads at 24% lower weight. The resultant steel-frame modular unit is lightweight. Such lightweight installations can solve the load-related issues (tower cranes' lifting capacity; transportation) of modular designs.

Such corner post-supported modular blocks involve a simple and fast inter-module connection by cutting and bending. This cut-bend connection method does not require any additional materials because the cross-piece of the support acts as a connecting plate. The holes formed in the bearing elements can be used to accommodate service pipelines. Thanks to this simple inter-module connection method, the time needed to manufacture the modules at the factory is reduced. Thus, in any emergency situation, modular units can be delivered within a short timeline to extend the life of the hospital and to meet other needs [2].

For ensuring the transverse stability of the proposed system, belts (X-shaped brackets) are more preferable than conventional brackets. The experience of fighting the infection has shown that premises should have good indoor air. Therefore, modular breathing panels are proposed for use as sidewalls of the corner post-supported modular elements. These sidewalls are non-load-bearing component, since the gravitational load is transmitted through the corner post. The filter material inside the modular breathing walls "dilutes" the airborne pollutants and reduces the rate of disease transmission. Thus, the proposed corner post-supported modular system can provide a safer indoor climate and improved air quality for residents.

The proposed modular system for emergencies, such as spread of infectious diseases, takes into account not only health-related improvements, but also the improvements in structural, fire-fighting and light aspects. Thus, the proposed modular system represents a complete package with improved performance.

It should be noted that the process of designing a prefabricated building may not always fully consider the needs of the actual production and assembly, which leads to design-construction conflicts and collisions. With the emergence of the building information modeling (BIM) concept, a technical platform is now in place for the easy exchange of information and feedback at all stages of design engineering process.

With the help of BIM technology, various design, production and installation specifications can be considered in full detail, while the virtual concept of a project can be built using BIM model, that covers design coordination, installation modeling, schedule modeling, etc., enabling to avoid possible problems.

BIM technology allows to build an information interaction platform and to ensure information exchange for better coordination of the engineering design. This capacity of BIM models is used by modern modular hospitals: information exchange involves a lot of specialists who can use is to jointly complete the model. Various types of design models can be processed on such information interaction platform in a timely and efficient manner; for joint design purposes, interaction and combination can involve several models [9].

At an early stage of BIM software-assisted healthcare project, a three-dimensional model can be built to visualize the relationship between the project and its site, as well as to align the site and its spatial positioning based on the analysis results. Consequently, BIM technology proves an effective tool in prefabricated construction of healthcare facilities.

Standardized production and regulated assembly will be carried out using a standardized method. The process of design engineering and construction should be provided with user-friendly coordination tools. When designing a structure, aspects such as load and specifications of electromechanical pipeline and related equipment must be taken due account of. At the initial design stage, optimal communication should be established between parties to construction projects regarding timeline, handling and transportation, human resources, equipment, supplies, onsite construction techniques, etc.

Digital models allow developers to control construction process using a large number of high-tech tools such as big data, artificial intelligence, drones and 5G. Modular design allows to increase the speed of construction by more than 50% while reducing costs. Labor productivity increases as well, since multiple construction works can be performed simultaneously.

5 Conclusions

The recent crisis of the health care system, induced by COVID-19, has led to a dramatic increase in the need for health infrastructure such as hospital wards, testing centers, isolation wards, etc. All these facilities should be delivered fast in order to provide timely treatment to patients and control the spread of the disease. To meet this requirement, modular construction methods are widely used all over the world.

The study investigates how the existing steel-frame modular blocks can be improved in terms of health care, construction, firefighting, ease of manufacture and reliable use in emergency situations. This study has enabled the following conclusions:

1. Modular construction represents the only possible solution for meeting the infrastructure development needs, as compared to the traditional construction methods, and is being widely used in healthcare construction around the world.
2. Installed in Steel-Frame Modular Units, the Modular Breathing Panels Can Potentially Improve the Indoor Air Quality and Reduce Disease Spread.
3. Modular building system has proven an expedient solution with improved performance for use in the current crisis situation and in the post-crisis world. Further research is focused on full-scale experimental and numerical studies of the proposed modular installation.

References

1. Belenky IG (2020) COVID-19 challenge: what has been done and what needs to be done? Traumatol Orthop Russ 26(2):15–19
2. Dhanapal J, Ghaednia H, Das S et al (2019) Structural performance of state-of-the-art VectorBloc modular connector under axial loads. Eng Struct 183:496–509
3. Eryk B (2020) Coronavirus outbreak: China to complete 1000-bed hospital in under a week. https://www.stuff.co.nz/world/asia/119139230/coronavirus-outbreak-china-to-complete-1000bed-hospital-in-under-a-week. Accessed 18 May 2021
4. Stephen F (2020) Modular construction firm manufacturing ward for Surrey hospital. https://www.insidermedia.com/news/national/modular-construction-firm-manufacturing-ward-for-surrey-hospital. Accessed 18 May 2021
5. Hough MJ, Lawson RM (2019). Design and construction of high-rise modular buildings based on recent projects. In: Proceedings of the institution of civil engineers-civil engineering, vol 172, No 6. Thomas Telford Ltd, pp 37–44
6. Howickltd (2020) Fast build construction helping meet overwhelming demand for hospital beds. https://www.howickltd.com/stories/fast-build-construction-helping-meet-overwhelming-demand-for-hospital-beds?sslid=MzMysjA0NrEwM7AwBgA&sseid=MzIwMzU0MzGyNAQA&jobid=db75b5fd-19ee-4e69-bd48-553106632c09. Accessed 05 Jun 2021
7. Kazakov Y, Birjukov A (2017) Fast assembly of quality suspended ventilated facades. Archit Eng 2(1):32–40
8. Lacey AW, Chen W, Hao H et al (2019) New interlocking inter-module connection for modular steel buildings: experimental and numerical studies. Eng Struct 198:109465
9. Mojtabaei SM, Ye J, Hajirasouliha I (2019) Development of optimum cold-formed steel beams for serviceability and ultimate limit states using Big Bang-Big Crunch optimisation. Eng Struct 195:172–181
10. Prabowo PA (2019) Multi-storey modular cold-formed steel building in Hong Kong: challenges & opportunities. In: IOP conference series: materials science and engineering, vol 650, no 1. IOP Publishing, p 012033
11. Scherbakov A, Monastyreva D, Smirnov, V (2019) Passive fluxgate control of structural transformations in structural steels during thermal cycling. In: E3S Web of Conferences 135:03022
12. Vechersky GS (2021) The tasks of public administration system in the fight against the pandemic in 2020. Student 1
13. Vechorko VI, Silaev BV, Tanshina OV et al (2020) Preparation and performance of a multidisciplinary hospital during the pandemic. Semashko National Research Institute of Public Health, Bulletin of N.A, p 4

Additive Technologies for Manufacture of Formwork

A. F. Yudina, D. A. Zhivotov, and Yu. I. Tilinin

Abstract Improved solutions for monolithic construction formwork systems are designed to decrease specific weight of formworks, improve the quality of concrete surface and accommodate complex walls. Having analyzed the conventional designs of formwork systems, that consist of prefabricated panels with waterproof plywood bottom attached to metal frame, the authors came to the conclusion that formwork systems need a cardinally new solution in order to achieve the above specifications. A series of experiments were performed on samples of additively manufactured synthetic fiber-reinforced plastic composites. The use of additive technologies opens up the prospect of creating formwork elements for complex walls, while the use of plastic can decrease specific weight of formworks and improve the quality of the front surface of monolithic concrete structures. *Purpose of study.* Create a carbon fiber composite formwork for complex monolithic walls with high-quality front surface. *Methods.* The carbon fiber composite formwork for complex walls is manufactured using additive technology. The resultant 3D samples have been tested for strength and deformation in laboratory conditions. *Results.* The experimental testing of the 3D-printed carbon fiber samples has shown their sufficient strength and deformation resistance.

Keywords Formwork · Complex walls · 3D model · Additive manufacturing technologies · Prototype · Tensile test

1 Introduction

As a versatile technology, cast-in-place construction is widely used in urban development. However, the formwork systems that it uses are mostly flat and therefore unsuitable for concreting of the facades with raised or protruding elements [1, 8, 9]. The facades of the buildings being constructed in the historical part of Saint Petersburg have pilasters, cornices, dripstones and other architectural and structural

A. F. Yudina · D. A. Zhivotov (✉) · Yu. I. Tilinin
Saint Petersburg State University of Architecture and Civil Engineering, Saint Petersburg, Russia
e-mail: d.zhivotov@mail.ru

© The Author(s), under exclusive license to Springer Nature Switzerland AG 2023
D. Ivanov et al. (eds.), *Proceedings of ECSF 2021*, Lecture Notes in Civil Engineering 257, https://doi.org/10.1007/978-3-030-99877-6_37

Fig. 1 Architectural and
structural elements of walls:
1—up-stand wall; 2—main
cornice; 3—window rabbets;
4—pillar; 5—string cornice;
6—intermediate cornice;
7—dripstone; 8—basement;
9—horizontal waterproofing;
10—pilaster;
11—semi-column;
12—recess

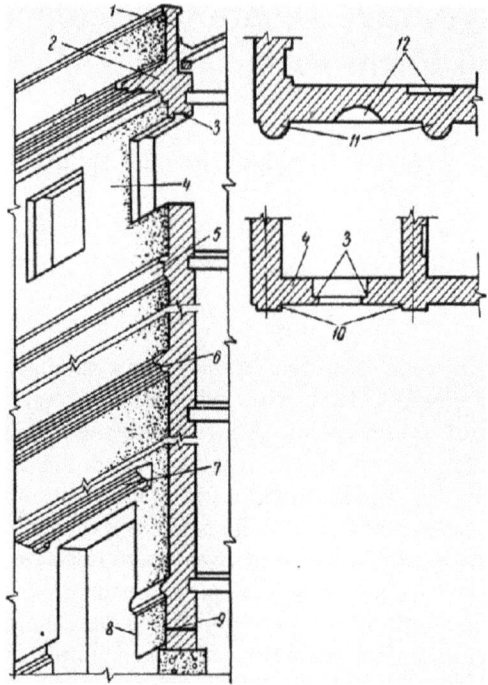

elements designed to integrate the walls into the old-style architectural environment (Fig. 1).

Considering the design of formworks for accommodating complex facades with raised/protruding members, the authors assume that additive technology might prove an effective solution.

Carbon fiber and fiberglass are suitable as materials for the manufacture of complex formworks.

Carbon fiber, which is lighter than fiberglass and has sufficient tensile strength, was chosen as the formwork material and samples for further testing of its basic physical properties. Despite its high strength, fiberglass is heavier and thus increases the weight of the formwork [5, 2, 3].

2 Materials and Methods

The additive manufacture of the composite plastic as a material for formwork was followed by laboratory testing of the samples of domestically produced carbon fiber suitable for temperature range from −50 to + 50 °C (Table 1) [4, 6, 7].

Table 1 Test sample material and sequential number

Sequential number	Carbon fiber (material) grade
1	TOTAL GF-30
2	
3	TITAN GF-12
4	
5	TOTAL GF-30 (N)
6	

3 Results and Discussion

The manufacture of the plastic samples used additive technology and a special 3D printer. The material of the samples and their sequential numbers are presented in Table 1.

The samples were tested in the laboratory of Saint Petersburg State University of Architecture and Civil Engineering using *Instron* 5982 tensile testing machine with tensile strain of up to 100 kN.

Prior to testing, the samples were measured along their edges and in the middle. Measurements recorded, each sample was placed inside the grip so that its longitudinal axis coincided with the load application axis. Strain sensor (gauge length: 10 mm) were installed in the middle part of the sample to determine axial deformation (Figs. 2 and 3).

Short-term tensile loads were applied at normal temperature range (18–22°C).

The experiment was intended to investigate the mechanical properties of plastics—behavior under load, deformation stages and displacement patterns.

The results are presented in table (Table 2), deformation versus load graph (Fig. 4) and displacement versus load graph (Fig. 5). The number of samples of each material was limited to two for reason of results identity.

Fig. 2 Samples and measuring instruments

Fig. 3 Sample placed inside *Instron* 5982 tensile testing machine

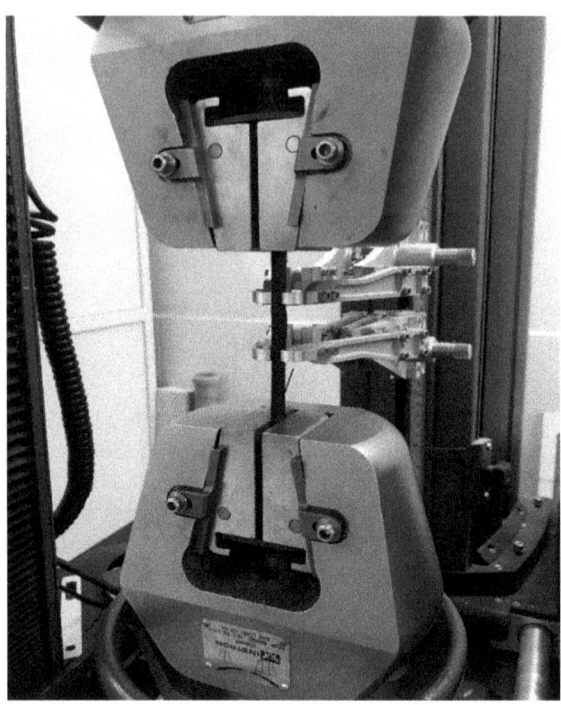

Table 2 Experimental results

Pos	Max.load [kN]	Tensile limit [MPa]	Modulus [MPa]	Relative elongation [%]	Note	Section, a*b, mm²
1	1.22	32.41	282.19	25.01	GF30(1)	4 × 10
2	1.21	32.60	321.30	29.44	GF30(2)	4 × 10
3	2.08	51.49	4256.21	1.72	TitanGF12(1)	4 × 10
4	2.06	52.49	4702.36	1.54	TitanGF12(2)	4 × 10
5	1.60	41.93	313.70	28.97	GF30(N)(1)	4 × 10
6	1.56	39.14	311.58	30.59	GF30(N)(2)	4 × 10

The nature of rupture in the samples is shown in Figs. 6, 7 and 8.

The results of the experiment correlate with the nature of deformation, indicating the reliability of the data obtained. The above graphs and figures show the load-induced changes characteristic of brittle materials.

Exposed to load, the samples of materials *GF30* and *GF30(N)* were deforming gradually, the values of their deformation increasing step-wise from plastic stage to brittle fracture. The tensile failure graph also shows that the rupture pattern of these plastics is identical to that of wood.

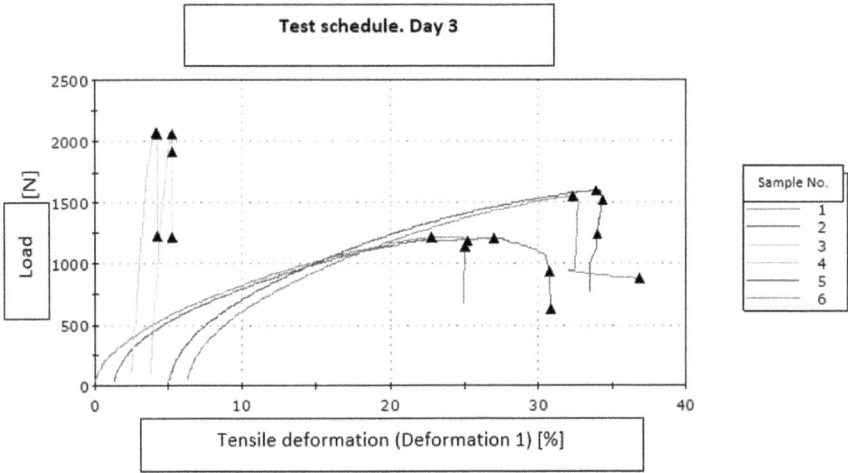

Fig. 4 Load versus deformation

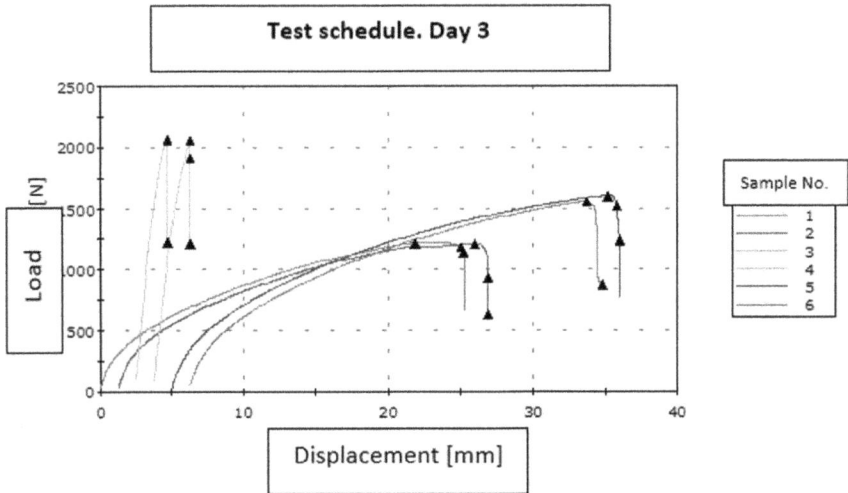

Fig. 5 Load versus displacement

The tensile failure of *Titan GF12* samples was accompanied by rapid increase in load and brittle fracture with minimal deformations.

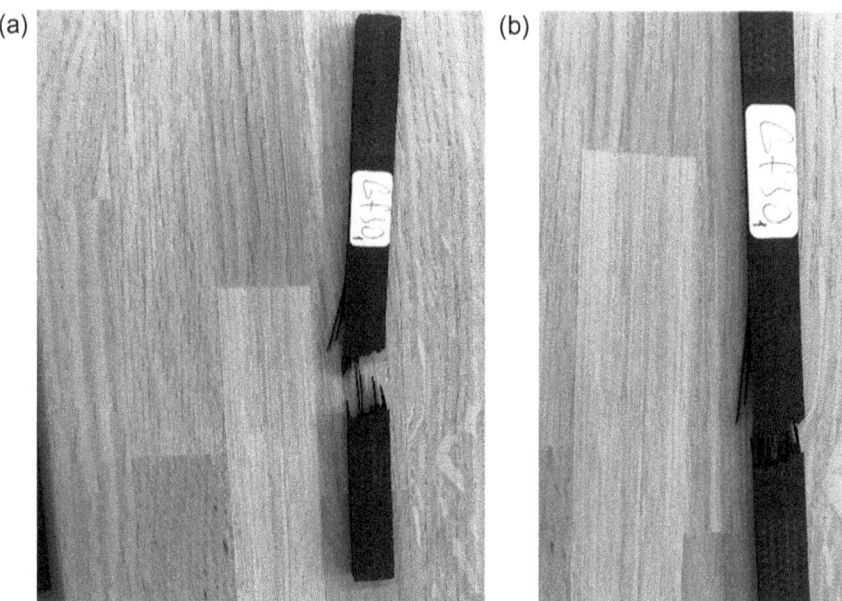

Fig. 6 Ruptured sample *GF30*: **a**, **b**—fiber split

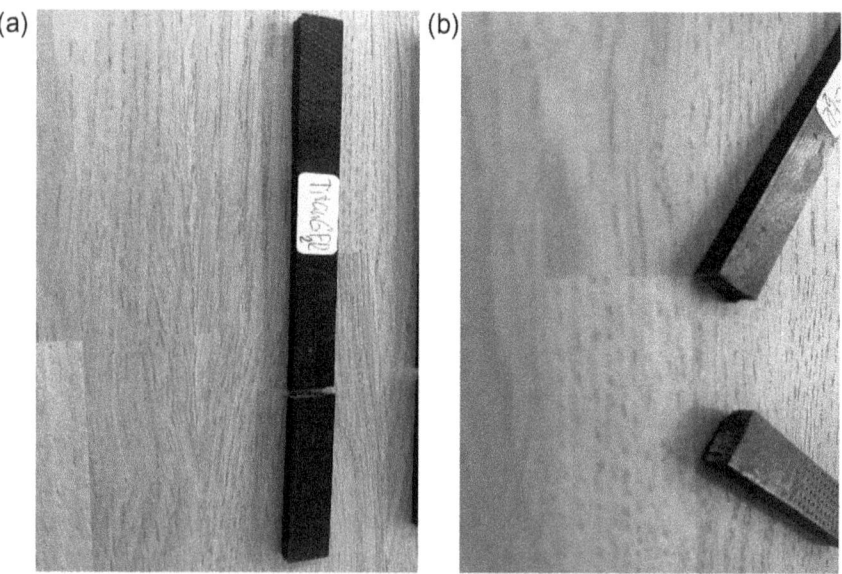

Fig. 7 Ruptured sample *Titan GF12*: **a**, **b**—fiber split

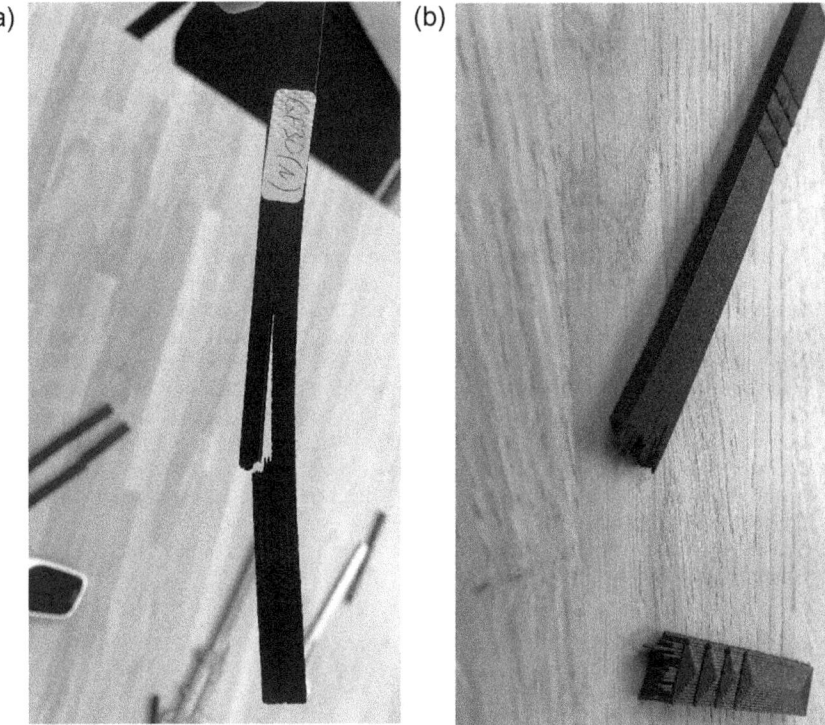

Fig. 8 Ruptured sample *GF30(N)*: **a, b**—fiber split

4 Conclusion

The experimental testing of *TOTAL GF-30, TITAN GF-12, TOTAL GF-30 (N)* plastic samples has shown that they have different mechanical performance.

The obtained results allow us to identify *TOTAL GF-30 (N)* as the most optimal grade for the manufacture of formwork elements capable of resisting the external forces and operating in elastic stage for concreting the sections of facades with raising/protruding elements. The use of carbon fiber formwork systems has a promising potential. Such systems offer a wide range of applications both in reconstruction/restoration of facades with decorative protruding elements and in new construction.

References

1. Golovina SG (2019) Architectural and structural characteristics of the evolution of St. Petersburg housing development the 18th—early 20th century. Bull Civil Eng 6(77):36–43

2. Kurnosov AO, Vavilova MI, Melnikov DA (2018) Technologies for manufacture of glass fillers and the effect of finishing agent on physical and mechanical performance of fiberglass. Aviation Mater Technol 1(50):64–70. https://doi.org/10.18577/2071-9140-2018-0-1-64-70
3. Peshcherenko E (2014) Modern technologies for manufacture of composite products by ESI Group. Aerosp Courier 4:2–6
4. Rybnov EI, Yegorov AN, Gorovaya NS (2018) Development of contour construction technology. Bull Civil Eng 2(67):135–140
5. Stroy-server. http://stroy-server.ru/notes/svedeniya-o-stenakh. Accessed 12 Jun 2021
6. Yegorov AN, Gorovaya NS (2019) Systematization and evaluation of 3D printing technologies in construction. In: Fundamental, investigatory and applied research conducted by the Russian Academy of Architecture and Construction Sciences in 2018 to support research-based architecture, urban planning and construction industry of the Russian Federation. Russian Academy of Architecture and Construction Sciences, Moscow, pp 177–184
7. Yegorov AN, Gorovaya NS (2018) 3D printing technology in construction. In: Fundamental, investigatory and applied research conducted by the Russian Academy of Architecture and Construction Sciences in 2017 to support research-based architecture, urban planning and construction industry of the Russian Federation. Proceedings of the Russian Academy of Architecture and Construction Sciences, Moscow, pp 192–194
8. Yudina AF, Tilinin YI, Yevtyukov SA (2019) Development of housing construction technologies in St Petersburg. Bull Civil Eng 1(72):110–119
9. Yudina AF, Tilinin YI (2019) Identifying the criteria for comparative evaluation of house building. Archit Eng 4(1):47–52

Factors Affecting the Efficiency of Façadism. Ensuring the Stability of Free-Standing Walls

A. F. Yudina and D. I. Kulakova

Abstract The paper discusses the historic building reconstruction technology that involves dismantling the building while retaining its facade for the entire period of reconstruction. The role of various factors influencing the stability of free-standing walls during the dismantling process is analyzed. The overview is provided of façade-retaining solutions to ensure the stability of free-standing walls.

1 Introduction

The careful and comprehensive approach to analyzing factors that may affect the performance of works is prerequisite to identifying the most optimal and cost-effective method for reconstructing buildings and structures. The landmark buildings with cultural heritage status deserve special attention. In the historic center of St. Petersburg, cultural heritage sites include 1378 apartment buildings, of which 426 are in poor condition and 48 are in emergency maintenance state [4].

On the one hand, the reconstruction of cultural heritage buildings should follow the legal regulations existing for cultural heritage sites [2] according to Federal Law of the Russian Federation of June 25, 2002 No. 73-FZ. "About objects of cultural heritage (historical and cultural monuments) of the people of the Russian Federation", and on the other hand, it is supposed to make use of modern standards and specifications for their further operation. In this regard, technical solutions are constantly being sought to maximize the efficiency of reconstruction. Among the modern reconstruction methods is façadism, that allows to retain structure's facades [6].

This method allows adapting a structure for new use while taking into account the modern requirements to its layout, insolation, energy efficiency and comfort level and preserving its architectural value and the historic image of the urban environment.

Technically, the method involves dismantling structure's internal frame and retaining its front facade. Façade preservation makes this method highly complex in nature.

A. F. Yudina (✉) · D. I. Kulakova
Saint Petersburg State University of Architecture and Civil Engineering, Saint Petersburg, Russia
e-mail: yudinaantonina2017@mail.ru

2 Materials and Methods

Façadism usually involves arranging of the underground space.

This, in turn, makes it necessary to dismantle the old foundation under the dismantled elements and to lay a new foundation. The new foundation can use any configuration regardless of the bearing capacity of the old foundation. Accordingly, the need arises to separate the foundation under the retained elements from that under the dismantled ones, while maintaining the stability of free-standing walls.

3 Results and Discussion

There are, however, a number of factors that can affect the stability of the retained walls during the dismantling works and reconstruction (Fig. 1).

The first group of factors comprises architectural and structural features of the retained elements:

- the state of elements to be retained;
- the area of elements to be retained;
- the height of elements to be retained;
- presence/absence of basement.

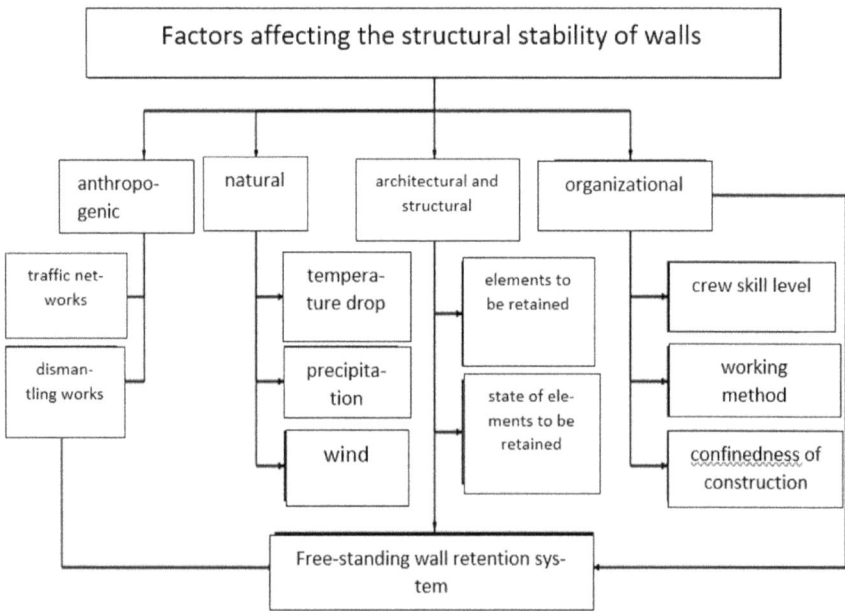

Fig. 1 Factors that affect the structural stability of walls

The physical condition of the elements to be retained is a factor crucial to the choice of dismantling technology, equipment and retention system. Ageing and degree of wear determine the duration and efficiency of works and expenditure.

The area of elements to be preserved varies. These elements can include the façade, or the façade and one side wall, or the façade and two side walls. The presence or absence of window openings is a factor influencing the specifications of the stabilizing structures, resistance to wind load, and gravity load. By calculating horizontal and vertical loads it is possible to determine the required number of supporting elements [1].

The height of free-standing elements is a factor influencing the arrangement of facade retention system. The higher the building, the bulkier the retaining elements should be.

The presence of basements and cavities will be considered when spacing the retention system's concrete blocks.

The second group includes anthropogenic factors. Dynamic loads and vibrations that occur during the dismantling works can cause the retained walls to subside.

Vibration induced by traffic and subway can cause the existing cracks in brickwork to expand [5].

The third group includes natural factors:

- wind load;
- atmospheric precipitation;
- temperature changes.

While precipitation negatively affects the unprotected brickwork, temperature changes cause soils to freeze and thaw, leading to uneven subsistence in the foundation. It is important to consider the duration of the use of the retaining system, as factors such as weather, underlying soil layer, season change can negatively affect the condition of the retained facade and the retaining system.

The fourth group includes organizational factors:

- inspection of structures;
- project documentation;
- specialists' skill level;
- confinedness of construction site.

The quality and completeness of the engineering and construction surveys, the competently developed technical solutions and dismantling technology, and the skill level of teams play a key role in the performance of works on the construction site.

In most cases, works in the historic downtown are performed in confined space. Factors such as surrounding buildings, utility networks, transit communications, roads, basements, and the construction site itself, limit the choice of equipment and technology, as well as arrangement of the wall-retaining system and placement of dismantled elements according to GOST 24,259–80—Rigging for temporary stabilization and alignment of structures. Classification and general specifications.

Due consideration of the above factors is essential to accurate calculation of the wall-retaining systems in the practice of façadism.

To preserve the geometric immutability of the free-standing walls during dismantling and reconstruction works, a facade retention system is used.

The facade retention system will be installed prior to the start of dismantling.

This system represents a metal structure designed for temporary retention of the facade.

Retention systems can be internal or external. The internal type involves spacers and bracing frames, while the external type uses counterforts and tower-type retention. Common retention solutions often use counterforts [3].

The frame of counterfort (Fig. 2) consists of vertical supports that are attached to the facade, and longitudinal and transverse tie spreaders.

To avoid backward fall, foundation wall blocks are used as ballast.

The retaining system (counterfort) is mounted from the outside (Fig. 3a–b). This technology is cost-effective, easy to install and fasten to the wall. Its only drawback

Fig. 2 The structure of counterfort

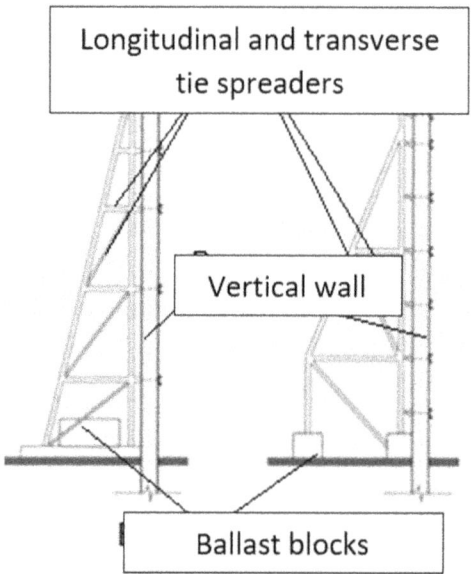

Fig. 3 Temporary façade-retaining counterfort: **a** wall without openings; **b** facade with openings

Fig. 4 Temporary retention of the facades from the inside: **a** using struts; **b** using braced frame

Fig. 5 Tower-type façade
retaining system

is that its elements require sufficient space along the facade, which, given the nearby public roads and sidewalks, is often difficult to provide. Also, there is a height limit to be observed—not higher than 2–3 floors.

To retain the facade from the inside of the building, strut systems and braced frames are used, which rest (are supported) on opposite walls (Fig. 4a–b).

While allowing for ample space around the building, this technology, however, limits access to works.

Tower-type facade retention system is a solution widely used in Europe (Fig. 5).

The tower-type retaining system uses couplers in its fastening mechanism, allowing to preserve the integrity of the façade finish of particular value. Its design also allows, if necessary, to place the retaining blocks with an offset.

In the practice of façadism, every project is approached individually.

4   Conclusion

As can be seen from the analysis, none of the factors influencing the stability of vertical walls (usually brick walls) has a decisive role. It is the combined influence of factors that is decisive to the stability of walls, each factor contributing its own correction factor. Therefore, it is the combination of factors, not one single factor, that should be considered when analyzing the possible influence.

Identification and systematization of the factors that are likely to arise during the reconstruction process would allow further analysis of the degree of influence of each of the factors at project development stage as a way to minimize risk of contingent works, delays and additional costs.

References

1. Alimov SG (2006) Assessment of the impact of transport vibration on the construction of buildings-monuments of architecture (on the example of Vladivostok). Doctoral dissertation, Far Eastern Federal University
2. Chainikova OO (2017) Preservation of the front facades as a way of recreating architectural monuments. Fundam Res 2:98–106
3. Charytonowicz J, Skowronski M (2013) Facade retention accomplishments in view of ergonomic design. In: International conference on universal access in human-computer interaction, Springer, Berlin, Heidelberg, pp 264–272
4. Dmitrieva OG (2018) The draft law on the reconstruction and overhaul of the housing stock in the center of St. Petersburg was discussed in Smolny. https://www.gov.spb.ru/press/government/138924/. Accessed on 15 Apr 2021
5. Kozhukhina ON, Popov VV (2019) The influence of dense urban development on repair and construction works. Tambov State University of Technology, p 181–183
6. Osman MME The study of stability and strength of free-standing brick walls during the reconstruction of buildings. Doctoral dissertation, RUDN University

The Architecture of Multifunctional Prefabricated Hospitals

Dmitry Yakovlev and Andrey Surovenkov

Abstract The paper discusses the principles of planning infectious diseases hospitals of a new type, the need for which arose in connection with the global epidemiological situation and the announcement by the WHO (World Health Organization) of the SARS-CoV-2 pandemic. Using the examples of the projects built in the Russian Federation, the prerequisites are analyzed for the decision-making as to functional zoning, architecture, space-planning and design solutions, as well as within the framework of the SDP (Site Development Plan). Options for the subsequent use of the developed solutions in new projects are proposed.

Keywords Infectious diseases hospital · COVID hospital · Multi-functional hospital · Pandemic · SARS-CoV-2 · COVID-19 · Functional zoning · Architectural and space-planning solutions · Prefabricated structures

1 Introduction

The outbreak of SARS-CoV-2 (Covid-19), first identified in Wuhan, China in December 2019 and subsequently declared a pandemic by WHO, required prompt solutions as to how to tackle the increasing demand (higher than the average annual) in medical services in a timely manner. When the pandemic reached Russia, it was already known that about 15–20% of the reported cases were severe and likely to require intensive therapy [4]. Beds for infected patients, especially those with artificial lung ventilation and oxygen supply, were insufficient.

Before the Covid-19 outbreak, there were approximately 55,000 beds for patients with infectious diseases in Russia, of which 12,000 were in intensive care units [7]. This number was significantly lower than the estimated 182,300 thousand (1250 beds per 1 million people) and 29,200 thousand (200 beds per 1 million people), respectively according to Methodological Guidelines of the criteria for calculating stocks of preventive and therapeutic drugs, equipment, personal protective equipment

D. Yakovlev · A. Surovenkov (✉)
Saint Petersburg State University of Architecture and Civil Engineering, Saint Petersburg, Russia
e-mail: 9107977@gmail.com

© The Author(s), under exclusive license to Springer Nature Switzerland AG 2023
D. Ivanov et al. (eds.), *Proceedings of ECSF 2021*, Lecture Notes in Civil
Engineering 257, https://doi.org/10.1007/978-3-030-99877-6_39

and disinfectants to be available in the Russian regions during the influenza pandemic by Federal Service for Supervision of Consumer Rights Protection and Human Well-Being. Despite the fact that there was capacity for up to 100,000 additional bed, the shortage of about 80,000 became a serious challenge for the domestic healthcare system.

In addition to repurposing of the existing hospitals, adapting of the military hospitals and deployment of temporary facilities in exhibition areas (a measure that received a mixed response from experts [3, 6], the need became evident for permanent infectious hospitals of a design that could be built within the shortest possible time [1]. Importantly, this new design was a means to facilitate the proactive approach to fighting against the SARS-CoV-2 pandemic [2]. Given the wave-like spread of the disease, the new design solution made it possible for hospitals to be built before the onset of the second wave. The projected hospitals were to be made ready for operation in the shortest possible time, while meeting all requirements to their premises and equipment.

2   Materials and Methods

Our analysis of the principles of planning multifunctional infectious diseases hospitals uses three case studies—the hospitals in Ufa, Chelyabinsk, and Ivanovo. Given that these principles incorporate a multitude of aspects governing organization of hospitals, it would be appropriate to focus on the main guiding principles:

- functional zoning of territory;
- structural concept;
- functional zoning of hospital;
- architectural and space-planning solutions.

3   Results and Discussion

3.1   Territory Functional Zoning

The functional zoning schemes for the above-mentioned hospitals are inclusive of the key principles of infectious diseases hospitals design and operation and are consistent with all relevant regulatory standards and recommendations according to SP 1.3.3118–13:

- Infectious diseases hospitals will be located in an isolated area and have their clean zones separated from infected zones.
- The layout of the hospital should ensure proper isolation of patients, enable diagnostic and therapeutic measures, exclude cross-infection, as well as allow maintaining necessary sanitary, hygienic and anti-epidemiological conditions.

– The corridors, entrances and exits from infectious diseases hospital's buildings and premises will be located so as to ensure that the clean flows of patients, hospital staff, instruments and materials do not cross paths with infected flows.
– Utility buildings will be placed with observance of sanitary zones.
– A disinfection station or facility will be arranged at the exit from hospital's infected zone for decontaminating of vehicles.
– The personnel should enter and exit using the clean zone; the passage from the infected zone to the clean one will be through sanitary checkpoints.
– Discharged (healthy) patients leaving their boxes will be exposed to sanitary treatment before entering the clean zone according to the Instructional and methodological guidelines on hygienic issues of design and operation of infectious diseases hospitals and departments by Ministry of Health of the USSR.
– Waste water from infectious departments will be disinfected.
– Garbage, solid waste and soiled linen will be disposed of in incinerators.
– Air supply to units will be provided by separate supply ventilation systems.

Considering the above regulations and conditions, the most optimal functional zoning scheme is the one with stand-alone modules (Fig. 1), which, despite its large area, offers a number of important advantages. These modules, oriented outward, make it possible to organize stand-alone units of boxes, each with its own entrance. They are one-storey buildings and therefore allow for transport accessibility to any of their parts with no need to install elevators and stairs, which reduces the cost of their construction and facilitates evacuation in case of emergency.

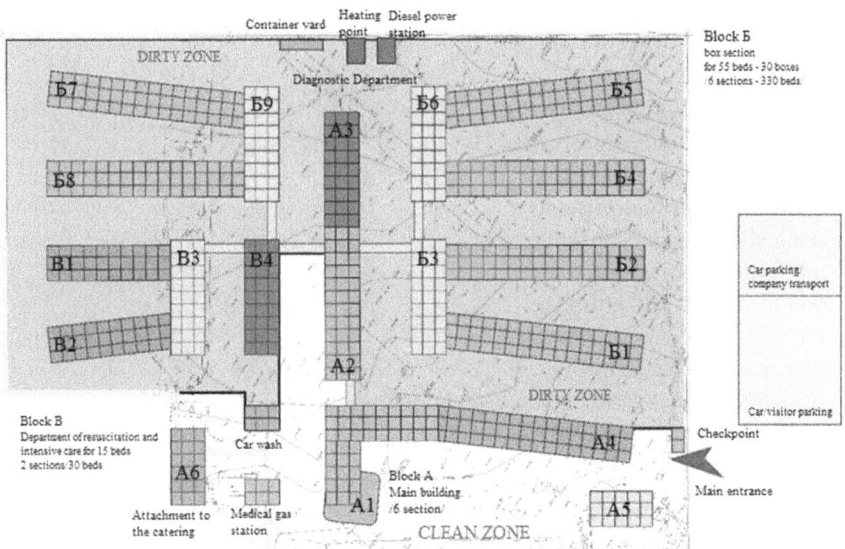

Fig. 1 Prefabricated infectious diseases hospital in Ivanovo: functional zoning scheme

3.2 Structural Concept

In conditions of extremely tight deadlines set for the design and construction of hospitals, light gauge steel framing (LGSF) was chosen as a structural framing solution and sandwich panels as enclosure structures of the roof, walls, partitions, and ceiling. This combination of materials and technologies allowed to reduce the design and construction cycle to 2–3 months for the projects with a total area of about 15,000 square meters.

However, the use of LGSF limits the structural performance of a hospital. This is one of the reasons why the majority of prefabricated hospitals are one-storey. The limited load-bearing capacity of LGSF makes it unsuitable for heavy medical equipment. At the same time, the use of LGSF significantly reduces the load on the base, which makes it possible to use prefabricated, lightweight foundations [5].

3.3 Hospital Functional Zoning

The location of hospital's functional modules varies with medico-engineering assignments of the projects to be built. The essential modules to be considered at the design stage include:

- units of boxes (quantity and bed capacity);
- intensive care units (quantity and bed capacity);
- diagnostic department (CT, X-ray, endoscopy, ultrasound, functional diagnostics);
- operating unit (quantity of operating rooms);
- laundry;
- disinfection and disposal;
- central sterilization department;
- dinnerware treatment station;
- pharmacy distribution point;
- clinical and diagnostic laboratory;
- ambulance substation;
- administration and utilities;
- department of morbid anatomy;
- patient discharge unit;
- dispatch service, garage, vehicle disinfection station;
- staff dormitory.

A project may be subject to significant changes depending on its mode of use, bed capacity, etc. However, unification of hospital buildings is a priority, as a unified design allows changing the configuration through minor structural changes. Configurability is achieved by dividing the hospital into functional modules, each of which is housed in a separate building. Thus, adding new departments or changing bed capacity takes adding/downsizing the buildings in the campus, which, given the

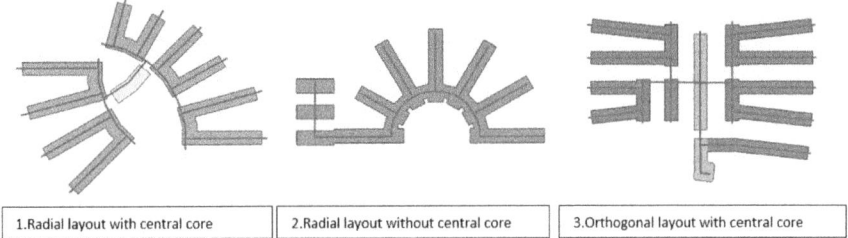

| 1.Radial layout with central core | 2.Radial layout without central core | 3.Orthogonal layout with central core |

Fig. 2 Multifunctional hospital layout options

tight deadlines in conditions of the pandemic, makes this solution scalable to other healthcare projects.

Hospitals may have different layouts, but the most optimal one—in terms of movement of staff—is the radial scheme. The orthogonal scheme uses a central core, which usually houses administrative service and from which all functional modules of the hospital are located at an equal distance. Some radial schemes can include the central core, too (Fig. 2).

Buildings will be arranged so that the clean and the infected flows do not cross paths. Discharge and admission of patients, as well as their movement between departments in the infected zone, should be through outdoor area (use of special vehicles is recommended). Admission of patients will be through reception and diagnostic boxes according to Order of the Ministry of Health and Social Development of Russia dated 31.01.2012 N 69n about the procedure for providing medical care to adult patients with infectious diseases. Transportation of waste, soiled linen, tools, tableware, and patients' belongings from the infected zone should take place outdoors using special trolleys and containers. Clean flows (linen, food, tools, materials, medicines, etc. to be delivered to patients) should use mainly the corridors of the clean zone.

This design stage has among its outcomes the scheme for organizing clean and infected flows (Fig. 3).

3.4 Architectural and Planning Solutions

The core function of a multifunctional hospital is placement of patients. Patents are placed in treatment sections, each housed in a separate building. Since treatment sections are structure-forming elements, their architecture and space-planning solutions will be considered in this paper in close detail.

In addition to boxes, treatment sections will have common premises. A substantial space saving can be achieved by interconnecting two treatment sections through the building housing the common premises—doctors' offices and utilities such as ventilation chambers, switchboards, wiring closets, heating units (Fig. 4).

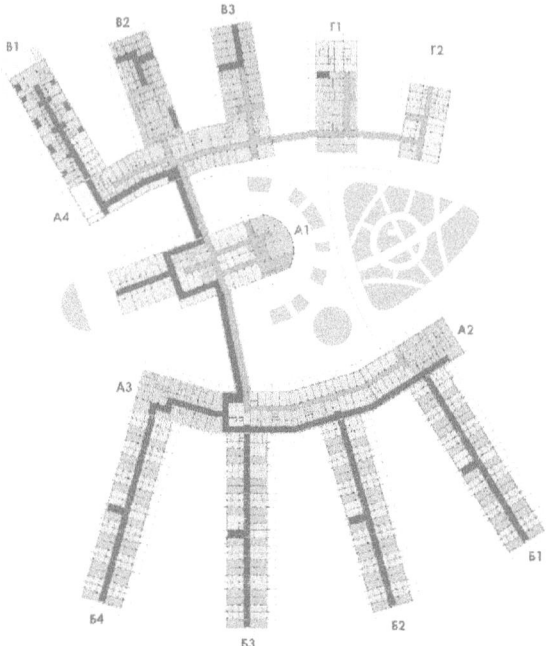

Fig. 3 Clean and infected flow scheme

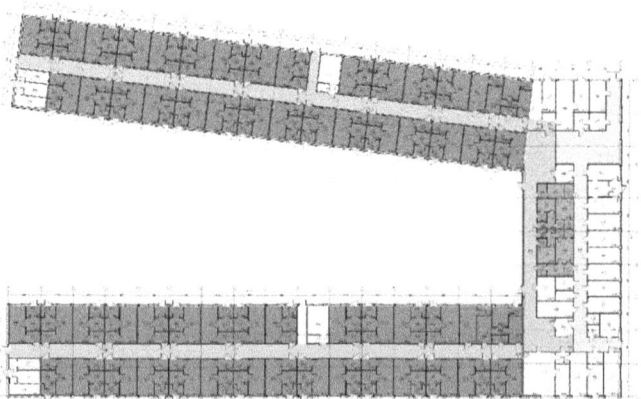

Fig. 4 The module consisting of two treatment sections and common premises

Thus, a fully autonomous module of premises is formed, that allows for increasing the bed capacity.

The space-planning solution where buildings are oriented at an angle (radial layout scheme) allows to ensure due spacing between buildings [2] while occupying less space.

Every treatment section classifies as contaminated but is divided into infected zone (wards/boxes) and conditionally infected zone (corridor system) with access exclusively for medical staff. The planning solution provides for boxes with anteroom, toilet and sluice.

Patients are introduced into the boxes via anterooms. Upon admission of every new patient, the toilet in their box will be disinfected and patient's clothes taken for further treatment and storage. The internal corridors are intended for use only by the staff. The treatment section has the nurse's area and the room for the preparation of infusion solutions.

The staff enters via the sanitary checkpoint located in the adjacent building and exits in the same manner. The transfer of patients between buildings uses special vehicles that are based in the clean zone and are disinfected every time they leave the infected zone.

The service building houses common premises—offices of chief doctors and senior nurses, staff rooms, clinical nurse manager's office, and junior medical staff rooms; storerooms and clean zone storages; medical and auxiliary rooms of the infectious zone. The treatment sections include dressing and manipulation rooms, designed to assist patients of box departments. Patients enter these sections from outdoor via special anterooms. The treatment sections are equipped with necessary portable medical and diagnostic equipment to assist the patients in the boxes. At the border between clean and infected zones, there are airtight sluices for transmission of articles and sanitary checkpoints for personnel, divided into men's and women's zones.

4 Conclusions

The presented functional layout of new prefabricated hospitals achieves optimized design and construction primarily due to modular organization of buildings, which uses as its key element the stand-alone functional module, not premise. As a result, the stage of design and construction can be tackled within shorter timelines.

Single-storey structures, these modules do not require complex fire safety solutions, especially in terms of escape routes. The single-storey design can benefit from light gauge steel framing [5], which, in turn, allows for a unified design calculation and accelerates the construction process, while fully observing the requirements to the arrangement of clean and infected zones. The successful practice of constructing and operating the prefabricated hospitals describes them as optimal solution for increasing bed capacity in within the shortest timeline and in any locality.

References

1. Avksentiev NA, Agranovich ML, Akindinova NV et al (2020) Society and the pandemic: experience and lessons of fighting COVID-19 in Russia. RANEPA, Moscow
2. Bateneva T (2021) What lesson did the pandemic teach healthcare system. https://rg.ru/2020/06/02/kakoj-urok-prepodala-pandemiia-zdravoohraneniiu.html. Accessed 18 May 2021
3. Galeeva V, Vaganova E (2020) Escape from Lenexpo: why are patients of the temporary hospital on Vasilyevsky so scared?. https://www.fontanka.ru/2020/05/09/69249583. Accessed 18 May 2021
4. Government of the Russian Federation (2021) On public healthcare measures against the novel coronavirus infection. http://government.ru/news/39218. Accessed 15 May 2021
5. Kondrakov S (2019) Prefabricated metal structures: pros and cons. https://www.radidomapro.ru/ryedktzij/proyzvodsvo-materialov/stroymateriali/stroiteligstvo-na-osnove-bystrovozvod imych-metallo-66550.php. Accessed 17 May 2021
6. Olevsky T, Muglov M, Rozhansky T (2020) Another billion for Crocus. Doctors in temporary hospital in Moscow Region complain about the lack of medicines and labor contracts. https://www.currenttime.tv/a/agalarov-cro-cus-2/30629445.html. Accessed 15 May 2021
7. President of the Russian Federation (2019) Meeting with members of the Government. http://www.kremlin.ru/events/president/news/62010. Accessed 15 May 2021

Architectural and Artistic Strategies of Regionalism Towards Integrated Design of Biomedical Facilities in Built-Up Environments

Yulia Yankovskaya

Abstract The paper discusses the architectural and artistic strategies for the design of new civil projects, including universal buildings for innovative medical research and development, in the historic built-up environments. The analysis uses the case of refunctionalization of a number of neighborhoods adjacent to the floodplain of the Iset' River in the city of Yekaterinburg. The paper proposes ways for introducing regional trends into local architecture, as well as principles ensuring the inclusion of the regional specifics in contextual design. The proposed theoretical assumptions have been tested in the experimental design of a series of public and residential buildings, one being an innovative medicine research facility in the central part of Yekaterinburg.

1 Introduction

1.1 General

The tasks of designing modern biomedical facilities and innovative medical research complexes should be approached not only from the perspectives of function and technological design, but also from those of architecture, urban planning, artistic and aesthetic solutions, and region-specific characteristics. In architectural practice, we are increasingly faced with trends of globalization and loss of national and regional identity. This is due to the emergence and widespread use of autonomous architectural solutions that disregard the natural and climatic conditions, local traditions and social factors of their home area. Such an approach to shaping the surrounding environment seems unacceptable, given today's technological advance, information exchange, changing attitudes to traditional cultures, and reasonable criticism from consumers. Distinctive character and authentic appearance are especially important

Y. Yankovskaya (✉)
Saint Petersburg State University of Architecture and Civil Engineering, Saint Petersburg, Russia
e-mail: spbgrado@spbgasu.ru

© The Author(s), under exclusive license to Springer Nature Switzerland AG 2023
D. Ivanov et al. (eds.), *Proceedings of ECSF 2021*, Lecture Notes in Civil
Engineering 257, https://doi.org/10.1007/978-3-030-99877-6_40

for innovative medical research facilities because they act as facilitators of creativity and research potential in employees.

There is a reason why national architectural identity increasingly receives attention all over the world. The era of globalization threatens to erase the boundaries not only between national architectures, but also cultures. The search for ways to preserve the special in the architecture of various countries and regions is becoming increasingly relevant today.

1.2 Historical Perspective

Region-specific characteristics have always been a major focus in the domestic architecture, but, unfortunately, have not found broad and comprehensive application in the modern practice despite their having been become a distinctive line of research. This paper deal specifically with the creative task of shaping an innovative medical facility project concept from the perspective of regional traditions in modern architecture and the search for authentic images and contexts [1–3].

The regional approach has formed in architecture naturally in the course of evolution. The first attempts to identify regional identity date back to the late nineteenth—early twentieth century. The subsequent rise in the ideas of regionalism occurred in the 1950–1970s in the Western countries. The studies of those years considered regionalism in the context of the modern architecture, consciously opposing modernity to traditional architecture. Regionalism started to be perceived not as isolationism, but as an opportunity for expression of identity [4, 5].

1.3 Terminology

In this paper, we operate a number of key terms: biomedical facility (complex), innovative medical research facility, integrated approach to design, regionalism.

In terms of architecture and urban planning, we adhere to the view that design concept of biomedical and innovative medical research facilities should be comprehensive enough to integrate them into premises such as healthcare, research, hotel and insurance business, tourism and recreation. This statement is based on the scientific study conducted by the Institute of Statistical Research and Economics—National Research University Higher School of Economics and presented in "Biomedical clusters in the world: Success factors and success stories" [6]. In this article, we consider the integrated design of two city blocks reconstructed in Yekaterinburg, which have on their territory the premises of various functions and purpose—a universal building for innovative medical research, apartment hotel, theatrical entertainment.

As a starting point in our search for a contextual and authentic architectural and artistic solution for the innovative biomedical facility, term "regionalism" allows for various interpretations. "Regionalism" refers to quite a number of concepts including design in historic environment; use of national decor in modern construction; ancient architecture imitation; historic style design; among others. Having analyzed its current definitions and interpretations, we define regionalism in architecture as an integrated approach that identifies specifics of an architectural structure designed to ensure its artistic continuity within certain socio-cultural and geographical context. It is our understanding that regionalism is designed to provide comprehensive solutions to architectural forms that rely on historical heritage, area's landscape and climatic features, as well as local social and national characteristics and identity [7].

1.4 Cultural Perspective

The cultural context is an important factor to be considered in the integrated design of biomedical clusters oriented towards research and medical tourism as key areas of activity. In recent years, services related to treatment, rehabilitation and wellness have become some of the most sought-after services in tourism sector.

This paper highlights the role of regionalism as a key architectural and artistic concept in the integrated design of the facilities under analysis.

Regionalism evolves parallel to globalization, being a response to it. With regard to architecture, globalization leads to the widespread use of certain construction technologies, unification and typification of building elements, ousting the traditional culture and identity and disrupting the continuity of time. On the other hand, borrowing of elements sometimes leads to enriched architectural flavor, but this happens only when the global influence is balanced by the local initiative. There are also some manifestations of globalization in the architecture that can be called neutral—common compositional language, common constructive system, common software, etc.

Considering the regional and global trends in architecture, we come to a fundamental conclusion that in order to create something authentic and non-trivial, one need to stay open to the world. As we know from history, isolation and self-restraint are unproductive in the long term, curbing creativity and manifestations of true identity. It is by focusing on advanced trends coupled with active local initiative that sustainable development can be achieved in architecture [8–10].

2 Materials and Methods

2.1 Analysis of Global Experience

Our analysis of the domestic and international experience of promoting regional trends [11–14] has identified a number of approaches to interpreting regionalism in architecture:

- pictorial—introduction of monumental art and artistic heritage of earlier civilizations;
- "illustrative traditionalism"—interpretations of traditional forms in the architecture of modern buildings;
- natural-climatic—interpretation of climate issues through the use of non-conventional techniques to generate unusual architectural details;
- philosophical—pursuit of national architectural tradition;
- traditional—return to traditional materials to experiment with non-conventional form making;
- assimilative—revision of the philosophy of modern architecture through the prism of local natural and climatic conditions.

2.2 Principles of Regionalism in Architecture

The analysis of regionalization and the influence of globalism in architecture has identified a number of principles shared by the regionalism trends in all regions of the world. The key principle consists in as comprehensive integration of local features and environmental setting as possible. At the same time, a regionally oriented structure should be able to meet the changing tasks and means of architecture, while reflecting the entrenched cultural stereotypes (philosophy) of a given locality. Other important principles include environmental friendliness and preservation of the existing heritage. Unlike earlier, when regionalism was a natural product of closed borders, today it seems to be inextricably linked with globalization.

2.3 The Strategies of Regionalism in the Architecture of the Urals

Based on the generalized global experience and the principles of regionalism in architecture, we have formulated the "creative design concept of the regionalism in the architecture of the Urals", which uses the following strategies:

- consideration of local characteristics;
- priority preservation of existing heritage sites;
- adaptation of global trends such as energy conservation, environmental sustainability, green technologies, renewable resources, tolerance;
- humanization of forms and space-planning solutions; special focus on human dimension and detailing of architectural forms.

3   Results and Discussions

3.1   Experimental Design Testing

3.1.1   Design Context

The results of the experimental testing of our theoretical assumptions are presented in this paper using the example of an integrated design developed for two city blocks in the central part of Yekaterinburg to include a biomedical research facility along with public hotel infrastructure. The design concept was developed under the supervision of, and the contribution from, architect V. N. Kurshakova.

The site to be developed lies in the central part of Yekaterinburg along Maxim Gorky and Gogol Streets. It has its starting point in Malyshev Street and ends in Engels Street, bordering on the Iset' River to the west and Rosa Luxemburg Street to the east. On the territory there are several objects of typical estate development of the nineteenth—early twentieth centuries, but most of them have been changed almost beyond recognition. The site has a diverse history of its formation: its location was first outside the city limits, then on the outskirts and, by the turn of the millennium, appeared in the center of the city (Fig. 1).

In the course of its development, this part of the city has acquired a "scaled-down architectural flavor"—the soft and quiet atmosphere that is important to preserve as new buildings are introduced.

Despite its being located in the heart of the city, the site does not receive any meaningful, systematic development. It clearly has a historical value and some of its blocks definitely require comprehensive reconstruction (Fig. 2). According to the design concept proposed for these blocks, and after careful examination of the options, there are three facilities that can be built within their premises—a universal building for biomedical research; a multi-functional accommodation facility—apartment hotel; and a recreational and entertainment facility with a drama theater in the floodplain of the Iset' River.

The design concept (Fig. 3) relies on regional approach and is consistent with the principles of regionalism described above. Let us list the main points of the design concept:

- it is a fundamental condition that the existing fragments of cultural heritage are treated with utmost care. Subject to preservation is the historic grid of streets. The landmark buildings are either embedded in the body of development or given a new function and are "exhibited as stand-alone buildings";
- the existing pedestrian paths are transformed for better convenience of different groups of pedestrians and to avoid potential crimes; provisions are made for people with limited mobility;
- proximity to the river significantly increases the value of the place. New buildings are located so as to preserve the visual connection, while leaving space for parks and greenscapes;

Fig. 1 Photofixation of the site

- with some height restrictions, the new buildings set the hierarchy of large volumes, reflecting the intent to achieve a harmonious transition from the low-rise construction near the floodplain to the historic downtown to the existing high-rise buildings along the eastern border of the site;
- the presence of the landscaped recreational area near the Iset' River and the arboretum opposite to the site under reconstruction increases the value and uniqueness of this territory;
- landscaping of the floodplain area requires a comprehensive, well-reasoned approach; the river protection zone significantly limits the zone of development.

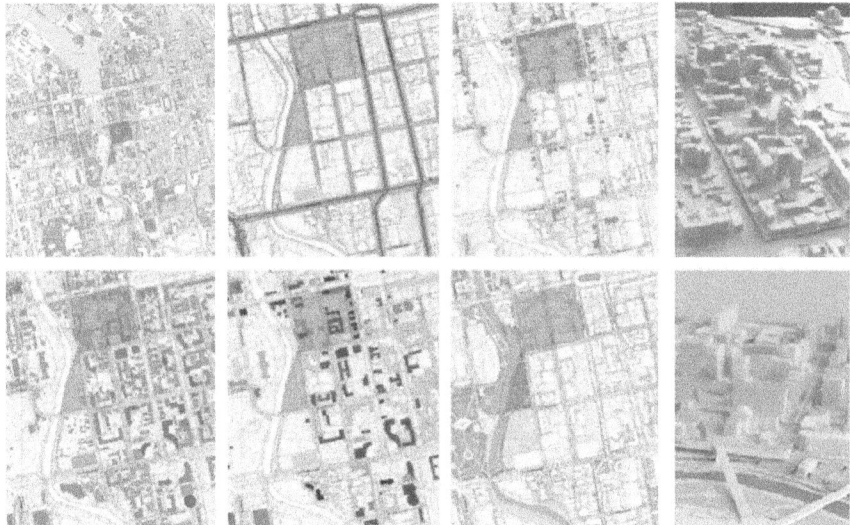

Fig. 2 Site-planning analysis. Mock-up models of the built environment

3.1.2 Creative Concept

Based on the proposed design concept, the most optimal strategies were identified for implementing the principles of regionalism for maximum optimization and more active economic development of the site with business facilities, housing and public recreational zones. Linking the proposed buildings is the diagonal pedestrian boulevard. The three functionally diverse buildings—biomedical research facility, multifunctional apartment hotel, and recreation and entertainment complex—embody three artistic strategies of regionalism: imitation of nature, neutrality, and contextuality (Fig. 4).

The "spirit of the place" and the Ural nature lay basis for the design of universal building for innovative biomedical research (Fig. 5). Its design was inspired the local natural forms (rocks, lakes, mosses), which found reflection in the swirling layout of the building, non-conventional landscaping on the adjacent territory, rock-like atriums and recreational and exhibition space.

The design of the apartment hotel used as its key principles neutrality, laconism, motives of the Ural constructivism, and machine forms (Fig. 6). This building integrates harmoniously in the dense urban environment sue to its neutral shapes and colors. Insolation is the decisive factor in planning its apartments. All of them have dual orientation and rooms located taking into account the movement of the sun during the day.

Contextuality and respect for the area's historical past are reflected in the design of the recreational and entertainment complex with a drama theater, located in the floodplain of the Iset' River (Fig. 7). Contextuality is key to this design: the historic

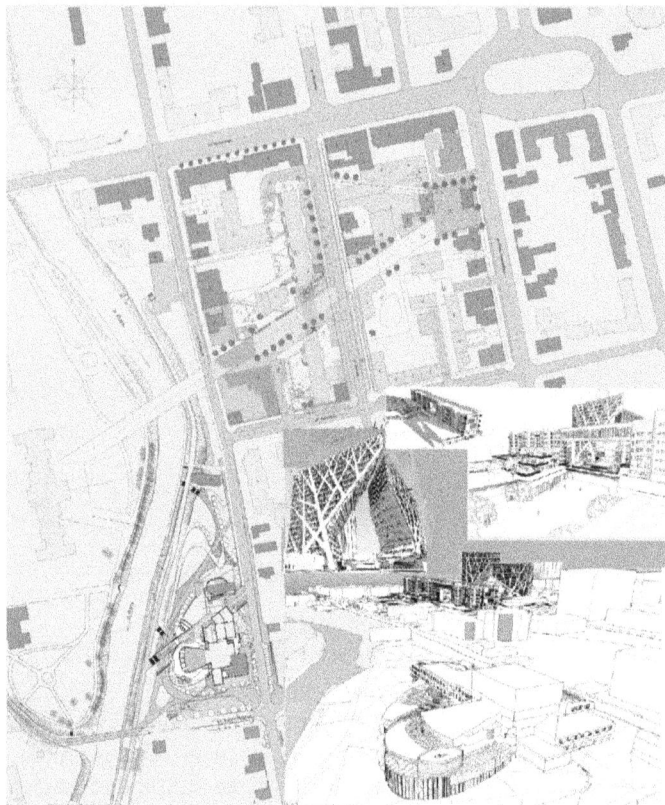

Fig. 3 The proposed design concept

buildings and the new construction are mutually integrated by embedding delicately some of the premises into the environment of the adjacent hill and using naturally occurring materials and prefigurations. The advanced trends find application in this design in the form of interactive projection installations illuminating the facade in dark hours, the sculpture park and the art objects placed next to the recreational and entertainment complex.

4   Conclusions

The proposed artistic strategies of regionalism—contextuality, neutrality, and nature imitation—have been tested in the design solution for three typologically dissimilar buildings and have proved productive tools for re-interpreting and unfolding local traditions in modern design solutions; these strategies also highlight the potential of

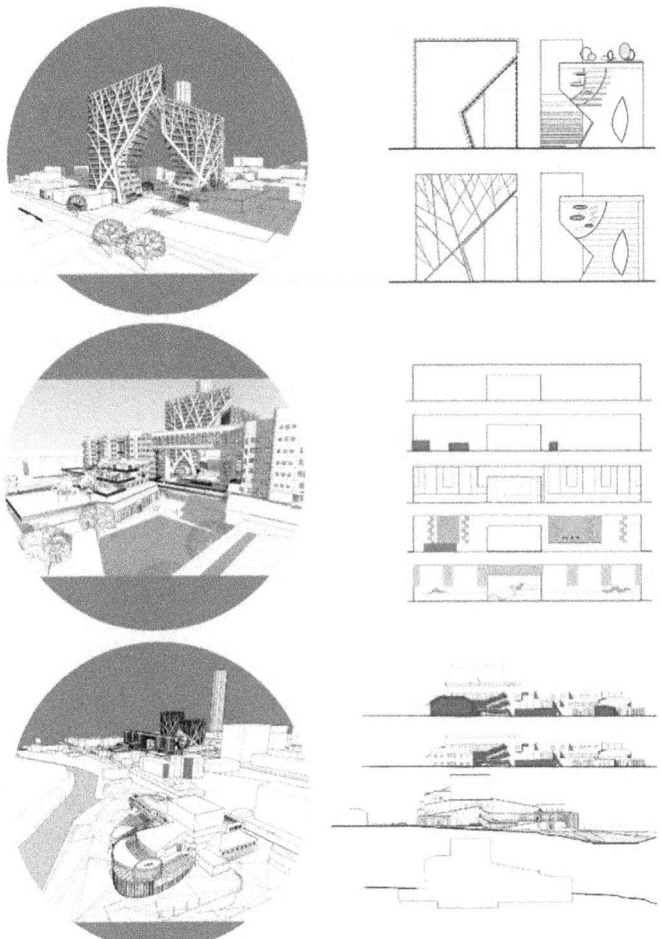

Fig. 4 The strategies embodying the principles of regionalism (top to bottom): universal building for biomedical research, apart-hotel, recreational and entertainment complex

regionalism as an approach in architectural practice and can be used in the design of biomedical clusters intended for research and medical tourism purposes.

Fig. 5 The proposed design solution for the universal building housing innovative biomedical research facility

Fig. 6 The proposed design solution for the apartment hotel

Fig. 7 The proposed design solution for the recreational and entertainment complex with a drama theater, located in the floodplain of the Iset' River

Acknowledgements The author expresses gratitude to the architect Vera Nikolaevna Kurshakova for the successful joint architectural and design activity.

References

1. Merleau-Ponty M (1996) Phenomenology of perception. Motilal Banarsidass Publisher
2. Norberg-Schulz C (1971) Existence, space and architecture. Praeger, New York
3. Yankovskaya Y (2013) Architectural theory in Russia: holding on the past or looking to the future. In: Proceedings of the Latvia University of agriculture, landscape architecture and art, vol 2. 2nd edn., pp 50–58
4. Kurshakova VN, Yankovskaya YS (2014) Architectural regionalism in the context of globalization. In: New ideas of new century: proceedings of the international scientific conference, vol 1. pp 178–183
5. Kurshakova VN, Yankovskaya YS (2015) Manifestations of regionalism in modern architecture. Modern trends in urban systems development. In: Proceedings of the international scientific conference dedicated to the 135th anniversary of the founder of ural school of architecture, pp 127–129
6. Islankina EA, Kucenko ES, Filina FN (2019) Biomedical clusters in the world: success factors and stories of the best. M.: NRU HSE
7. Yankovskaya IS, Merenkov AV (2020) "Green architecture" and sustainable modern urban development. Saint Petersburg, SPbGASU
8. Yankovskaya YS (2006) Architectural object: image and morphology. Doctoral dissertation, Moscow Institute of Architecture
9. Yankovskaya Y, Merenkov A (2013) Green Architecture: theoretical interpretation and experimental design. In: Proceedings of the Latvia University of agriculture landscape architecture, pp 77–84
10. Yankovskaya YS, Merenkov AV (2017) Image and morphology in modern theory of architecture. In: IOP conference series: materials science and engineering, vol 262, 1st edn, IOP Publishing, pp 012134
11. Yankovskaya Y, Lobanov Y, Temnov V (2018) Structural and compositional features of high-rise buildings: experimental design in Yekaterinburg. In: E3S web of conferences. vol 33, EDP Sciences, pp 01029
12. Yankovskaya Y, Datciuk T, Kondratieva L et al (2020) Structure, adaptability and security of an architectural object. In: E3S web of conferences, vol 164, EDP Sciences, pp 05007
13. Zavarichin S, Lisovskij V (2013) Conceptualism as a phenomenon of European architecture in XX Century. World Appl Sci J LNCS 23:149–152
14. Zavarikhin S, Tran NG (2016) Manifestation of "Indochinese style" in Hanoi's architecture in 1920–1950s. Archit Eng 1(3):33–40

Innovative Eco-Friendly Biocide Technologies for Improving Daily Life Safety

Stanislav Amelichkin and Alexander Tabakov

Abstract The article discusses the Silver Bullet disinfectant developed by the authors and the innovative technology of volumetric aerosol disinfection. It is shown that the effectiveness of this disinfectant significantly exceeds the leading foreign analogues, and the described disinfection technology is the most adequate for combating coronavirus infection due to the identified air-aerosol method of its spread. At the same time, the Silver Bullet disinfectant is environmentally friendly, since it was developed on the basis of hydrogen peroxide and complex silver compounds with a prolonged bacteriostatic effect.

1 Introduction

Preventing the spread of the new coronavirus infection COVID-19 is currently one of the most important global tasks. Of particular importance in the system of measures to counteract the pandemic is desinfection of the objects being contacted by people in their daily lives.

The studies published in reputable medical journals such as BMJ, The Lancet, JAMA, among others, were followed by publications concluding that the dominant pathway for the spread of coronavirus infection is by aerosol transmission [1–5]. Scientists tend to believe that coronavirus is transmitted to a greater extent by aerosol transmission, not solely by airborne transmission as was previously believed.

Coronavirus is found in the ventilation systems of hospitals, and the only way it can get there is by aerosol transmission. In particular, at the Uppsala University Hospital in Sweden, the corona virus was detected in the air filters located far from patients who could be a source of droplets [1].

S. Amelichkin
Head of Research Laboratory "Innovative Water and Air Treatment Technologies", Head of Water Supply, Sanitation and Hydraulics Department, RPA EKhA-MAG, LLC, Emperor Alexander I St. Petersburg State Transport University, Saint Petersburg, Russia

A. Tabakov (✉)
Saint Petersburg State University of Architecture and Civil Engineering, Saint Petersburg, Russia
e-mail: tabakov@mail.ru

In South Korea, residents of an apartment building got who infected with coronavirus without having contact with each other. Scientists believe the virus could have been transmitted through general ventilation. This case has been covered in a study published in the International Journal of Infectious Diseases. Experts came to the conclusion that, even though the study has investigated one specific case, it proves that such a risk exists. According to experts, the threat is currently underestimated. During a pandemic, many people may have to stay indoors to avoid interpersonal contact. But some may be exposed to viral infection when inhaled due to inadequate ventilation system, the study states [5].

The finding that coronavirus can spread by aerosol transmission requires new approaches, methods and disinfectants that would ensure the effectiveness of preventive measures aimed at combating the spread by aerosol transmission.

In recent years, the following mandatory requirements have been established for chemical disinfectants: (1) safe for humans and animals; (2) friendly to materials and equipment; (3) no toxic or other harmful substances that can penetrate surroundings or the environment under treatment. The main trend, as can be seen from the requirements, is to ensure safety while achieving the target level of decontamination effect.

Not all known disinfectants meet these requirements. When disinfecting water, air and surfaces, the active substances of many disinfectants—phenols, aldehydes, halogens, etc.—form mutagenic and carcinogenic compounds when interacting with the components of the objects under treatment.

On the disinfecting products market, there are disinfectants that do meet the above criteria, one being Sanosil Super 25 (Switzerland), which is widely used globally in many industries, agriculture, medicine, energy, housing and communal services, etc. At the same time, it should be recognized that no matter how effective and safe these products can be, they are not enough to achieve the target level of decontamination measures or to ensure normal competition in the market.

2   Materials and Methods

The empirical basis of the study encompasses the results of disinfection measures to counteract the spread of the coronavirus infection and other dangerous diseases; materials of medical and sociological studies' information on disinfectants manufacturers activity; reports on projects implemented in the field of vital functions safety; other information sources.

The study made use of the following scientific methods: (1) chemical experiment; (2) chemical analysis; (3) chemical synthesis; (4) measurement; (5) comparison; (6) modeling; (7) forecasting; (8) expert evaluation; (9) literary search; (10) description.

3 Results

A disinfectant has been developed, tested and launched on the Russian market that meets all mandatory requirements to modern disinfection products, Silver Bullet. This disinfecting product is a collaborative effort of the research team at Emperor Alexander I St. Petersburg State Transport University's Department of Water Supply, Sanitation and Hydraulics, and the team of Research and Production Association "Ekha-Mag" (St. Petersburg). In 2013, Silver Bullet disinfectant underwent a series of tests by Federal State-Financed Research Institution—State Center for Applied Microbiology and Biotechnology Research under the RF RosPotrebnadzor (Federal State Agency for Health and Consumer Rights), and was registered by the Customs Union under the trademark Silver Bullet.

Silver Bullet is an environmentally friendly disinfectant. It uses as basic components hydrogen peroxide and complex silver compounds with a prolonged bacteriostatic effect. Silver Bullet is highly effective against gram-positive and gram-negative bacteria, fungi, yeast, viruses, amoebas, protozoa, enveloped and non-enveloped viruses and even anthrax spores, which allows it to be used in the fight against particularly dangerous infections. As a biocide, Silver Bullet boasts an enhanced bactericidal activity allowing it to destroys the biofilms affecting the thermal conductivity coefficient of heat exchangers, while also ensuring high disinfection efficiency and a significant reduction in the operating costs. Silver Bullet is applied by method of wiping, irrigating or aerosoling with ultra-small particles.

The active substances of Silver Bullet are identical to those of the aforementioned Sanosil Super 25 (Switzerland), but Silver Bullet is substantially more effective and cheaper. The comparative study of hydrogen peroxide-based disinfectants is presented in Table 1 and involved the analysis of the official instructions for use:

Table 1 Hydrogen peroxide-based disinfectants: performance analysis

Application modes (disinfection targets)	Hydrogen peroxide	Sanosil super 25 (Switzerland)	Silver bullet (Russia)	Outperformance by silver bullet	
				Silver bullet versus hydrogen peroxide	Silver bullet versus sanosil super 25
Bacteria	10%–60 min	4%–60 min 6%–30 min	0.1%–60 min 0.25%–30 min	100-fold	40-fold
Tuberculosis	15%–60 min	8%-60 min	0.5%–60 min 1%–30 min	30-fold	16-fold
Viruses	12%–60 min 15%–30 min	6%–60 min	0.5%–30 min 0.25%–60 min	48-fold	24-fold
Spore-forming cells (mold, fungi)	n/a	8%–60 min	0.5%–60 min	16-fold	–
Anthrax spores	n/a	n/a	2%–60 min	–	–

(1) Instruction No.01/2013 for use of Silver Bullet disinfectant for pre-sterilization cleaning, disinfection and sterilization (St. Petersburg, 2013); (2) Instruction No.2 for use of disinfecting agent Sanosil Super 25 (Moscow, 2007); (3) Instruction No.01/12 for use of disinfectant 6% Hydrogen Peroxide (Moscow, 2012).

The comparative analysis of Sanosil Super 25, Silver Bullet and Hydrogen Peroxide has found that: (1) Silver Bullet outperforms hydrogen peroxide across all disinfection modes by an average of 50 times; (2) Silver Bullet outperforms Sanosil Super 25 by an average of more than 20 times; (3) the use of Silver Bullet can significantly reduce costs due to its lower con-centration. The 6% Sanosil Super 25 (50 ml/m3, 60 min) has been found to produce the effect completely identical to that of "cold fog" 0.5% Silver Bullet (60 min) to disinfect the air and the surfaces containing bacterial infections.

Considering that the coronavirus infection has been found to spread by aerosol transmission, Research and Production Association "Ekha-Mag" has developed MAG-SB set, an innovative technology for volumetric aerosol disinfection. The set includes a mobile aerosol generator (MAG) that generates Silver Bullet biocide aerosol mist for disinfecting indoor air and surfaces to counteract the spread of the coronavirus infection by airborne aerosol.

4   Discussion

The technology uses gas-like diffusion to spread the 1–5-microns aerosol particles (small size particles have been found for ensure greater efficiency as studies show [6] to fill the entire indoor premise with the aerosol mist of hydrogen peroxide-based, silver ion-catalyzed biocide Silver Bullet. Silver Bullet is suitable for disinfecting rooms as large as 10,000 cubic meters.

The use of this technology is an environmentally friendly process as its agent is biode-gradable (it completely decomposes into oxygen and water and does not require flushing or removal by any other method) and does not have a corrosive effect on equipment.

Unlike other methods of treatment, such as ultraviolet radiation, the technology of volu-metric aerosol disinfection MAG-SB allows to achieve three-dimensional penetration on analogy with gas; the aerosol mist can reach the surfaces that are hard-to-reach that are not in view, for example, surfaces behind radiators, suspended ceilings, furniture, etc. This property allows to have ventilation systems disinfected volumetrically, within short timeline and high-ly effectively.

For the purposes of express testing of whether the aerosol mist has filled the entire volume of the surfaces in the room being disinfected, Silver Bullet indicator strips are used. They are a tool to ensure proper monitoring of the disinfection measures for completeness and quality.

The technology described above is being used regularly for final volumetric disin-fection of the premises and ventilation systems in the classrooms and the dorms of Emperor Alexander I St. Petersburg State Transport University, and has been successfully tested on the ventilation and air conditioning systems in the Federal State-Financed Institution—V. A. Almazov Na-tional Medical Research Center, at train stations, passenger port facilities "Saint Petersburg Water Front", and many other facilities. It has been clinically proven that once disinfected, the surfaces do not develop any further growth of bacterial spores, viruses, fungi and mold.

The results of the conducted studies proving the effectiveness of Silver Bullet disinfectant and volumetric aerosol disinfection technology have been presented and discussed at the following research-to-practice conferences: (1) international conference "Practical Balneology, Environmental Hygiene, Human Ecology: Inter-disciplinary Approach" (Kislovodsk, Russia; January 14, 2020); (2) international research-to-practice conference "New Developments in Water Supply, Sanitation, Hydraulics and Water Resource Protection" (St. Petersburg, Russia, November 14, 2020); (3) international research-to-practice conference "Innovation-Driven Medical Facility Design, Construction and Infrastructure Support" (St. Petersburg, Russia, May 19–21, 2021); (4) reporting meeting of "Water—Medicine—Ecology" interna-tional asso-ciation, timed to its 25th anniversary (Verbilki. Taldom District, Moscow Region, Russia; July 9–11, 2021).

The research and medical community has found the results of the study reliable and worthy of close attention. The developer team are recommended to pursue further research in the field of environmentally friendly biocides and innovative technolo-gies for volumetric aerosol disinfection, and to intensify their efforts towards wider implementation of the approved means and methods.

5 Conclusion

Work is currently in process to further analyze the properties of Silver Bullet biocide by exploring its mechanism and effectiveness against various forms of pathogenic microorganisms in water, air, and different kinds surfaces. Also, research and devel-opment projects are in progress, one aiming to provide the disinfecting apparatus with automated dosing option depending on parameters of objects being disinfected (for example, when they develop biofilms).

The progress and results of these projects will be published in the near future.

References

1. Greenhalgh T, Jimenez JL, Prather KA et al (2021) Ten scientific reasons in support of airborne transmission of SARS-CoV-2. Lancet 397(10285):1603–1605. https://doi.org/10.1016/S0140-6736(21)00869-2
2. Pan M, Lednicky JA, Wu CY (2019) Collection, particle sizing and detection of airborne viruses. J Appl Microbiol 127(6):1596–1611

3. Tang JW, Bahnfleth WP, Bluyssen PM et al (2021) Dismantling myths on the airborne transmission of severe acute respiratory syndrome coronavirus (SARS-CoV-2). J Hosp Infect 110:89–96. https://doi.org/10.1016/j.jhin.2020.12.022

4. Eichler N, Thornley C, Swadi T et al (2021) Transmission of severe acute respiratory syndrome coronavirus 2 during border quarantine and air travel, New Zealand (Aotearoa). Emerg Infect Dis 27(5):1274. https://doi.org/10.3201/eid2705.210514

5. Hwang SE, Chang JH, Oh B et al (2021) Possible aerosol transmission of COVID-19 associated with an outbreak in an apartment in Seoul, South Korea, 2020. Int J Infect Dis 104:73–76. https://doi.org/10.1016/j.ijid.2020.12.035

6. Fennelly KP (2020) Particle sizes of infectious aerosols: implications for infection control. Lancet Respir Med 8(9):914–924. https://doi.org/10.1016/S2213-2600(20)30323-4

Infrastructure Support for Innovative Medicine: Current State, Challenges, and Prospects

Alexander Tabakov and Dmitry Popkov

Abstract The paper provides an overview of the current state of, and trends in, the high-tech infrastructures for supporting for innovative medicine in the context of informatization, digitalization, computerization and robotic automation. The paper further discusses the key challenges of infrastructures for supporting innovative medicine and proposes ways for their solution and prospects of development.

1 Introduction

The main task before medicine today is to increase life expectancy and improve quality of life for all the population groups by increasing the availability of high-quality medical care, which largely relies on high medical technologies. Achieving this task is possible by setting the medical industry on the path of innovative development.

It is gratifying to note that the awareness, and in a certain sense, inevitability of such development path has been recognized by the expert communities. It should also be recognized that medicine as a system of scientific knowledge and practical activities remains open to innovation and investing in intelligent tools, methods and technologies.

A. Tabakov (✉)
Saint Petersburg State University of Architecture and Civil Engineering, Saint Petersburg, Russia
e-mail: tabakov@mail.ru

D. Popkov
Baltic Interregional Advocate Collegium of St. Petersburg, St. Petersburg, Russia

D. Ivanov et al. (eds.), *Proceedings of ECSF 2021*, Lecture Notes in Civil Engineering 257, https://doi.org/10.1007/978-3-030-99877-6_42

351

2 Materials and Methods

The empirical basis for this study encompasses statistical data, previous sociological and economic studies, innovative healthcare centers performance data, reports on projects delivered in the field of innovative medicine, and other information sources.

The study involved the use of the following scientific methods: (1) analysis; (2) synthesis; (3) description; (4) sociological methods (document study, interviewing, overt observation); (5) statistical method; (6) modeling; (7) forecasting; (8) expert evaluation method; (9) literature search.

3 Results

Recently, the volume of high-tech healthcare services has been increasing, and so has the use of novel complex, unique and resource-intensive treatment methods with scientifically proven effectiveness, including cellular technologies, robotic technologies, information technologies and genetic engineering methods, developed on the frontlines of the medical research and related research and development areas.

To the foreground come the medical and biological sciences with significant potential for innovation in healthcare services—neurocognitive technologies, cellular and tissue engineering, genetic engineering (genomic and postgenomic technologies, molecular genetics, genomics, DNA markers, DNA microarrays), nuclear medicine and radiopharmaceuticals, pharmacogenetics, intelligent biocompatible materials, biobanking, personalized surgery, reprogramming of vital systems such as immune system, experimental microsurgery, etc. Molecular biology and medicine are increasingly seen as mutually integrated fields [2].

The medical industry has been penetrated by digitalization, informatization, computerization and robotic automation [4]. Designed to enhance healthcare system's performance, digital technologies accompany the process of patient management and follow-up in all its stages from diagnostics to treatment to rehabilitation. Artificial intelligence is now a common tool in diagnosing diseases such as cancer, providing high-speed and accurate identification [3, 4]. More and more medical centers are using big data as a way to improve performance of medical specialists [1]. Medical technologies based on brain-computer interface (hand exoskeleton conjugated to upper limb, neuro-communicators, neurotrainers based on 3D models of the body) are gaining wider use.

3D modeling systems are being developed for biocompatible medical implants for reconstruction of, and defect replacement in, organs and tissues. Medical practice is beginning to use 3D bio-printing—a technology for creating volumetric cellular models with viable, fully functioning cells. This technology plays an important role in the cultivation of organs and the development of innovative materials, primarily biomaterials, that are used as used for printing three-dimensional objects. 3D-printed tissues, medicines (and prospectively entire human organs) have potential to be used as replacements for "natural" human organs with properties and characteristics superior to those of natural organs.

The trends described above indicate not only the expansion in the range of quality medical services, but also the emergence of new, innovation-driven medicine with focus on prevention and personalized treatment that are based on in-depth understanding of human biology and all its levels starting from molecular. Innovative medicine can be said to have acquired the status of a strategically important industry.

At the same time, the normal functioning and further development of innovative medicine would be impossible without the appropriate infrastructure support. Theoretical studies into healthcare facilities' performance show that any expansion in nomenclature, availability and quality of medical services is largely contingent on infrastructural support. This dependence manifests itself most clearly in those healthcare facilities that are overburdened with demand for their services. The capacity for meeting this demand is terms of timeliness, volume and quality of services depends on the material, technical, informational, financial, economic and human resources available to each given department. With this in mind, the style of managing the infrastructures for the innovative medicine is becoming increasingly proactive, involving forecasting (including strategic forecasting) of the infrastructure needs and performance planning (including strategic) based on the projected infrastructure needs.

When identifying the development strategies for innovative medicine infrastructure, which is a highly dynamic infrastructure, the focus should be laid on the life cycles of healthcare facilities and technologies. Updates (modernization) to previous technology or emergence of a new one require the existing infrastructures to be adapted accordingly. Consequently, there should be a balance between technological processes and their supporting infrastructures. Defects (or lack or absence) in the supporting infrastructures deteriorate the quality of medical services, leading, in some cases, to the closure of healthcare providers. On the other hand, economically unjustified over-investing in the infrastructure may only prove unproductive but also lead to negative consequences by disadvantaging the under-invested areas of medical activity and burdening the current technological processes with additional expenditures. Here, the indicator of adequate investment performance is the total cost of infrastructure maintenance and losses due to lack or complete absence of the infrastructure. In conditions of limited resource availability, rational and sustainable use is important. This universal rule applies also to the infrastructures supporting innovative medicine.

As a system for ensuring healthcare performance, healthcare infrastructure includes logistical, technological, economic, personnel and information-related subsystems. The management of infrastructures for innovative medicine as a system consisting of the above subsystems, consists in ensuring their unity and effective interaction to achieve the goal of ensuring the normal functioning and progressive development of the high-tech industry under consideration. When seen as a set of resources to be distributed among the above subsystems and to provide conditions for their normal operation and development, the infrastructure for innovative medicine operates resources such as material and technical (buildings, structures, equipment, engineering systems, transport, energy and water resources, etc.), information and human resources. In terms of its functions, the infrastructure for innovative medicine classifies into material and processing infrastructure, that serves the main operational procedures; social infrastructure, i.e. provision of sufficient medical and service personnel with due qualification level; logistics (transport and storage) infrastructure to ensure uninterrupted supplies, transportation and storage at minimal cost; communication infrastructure, designed to provide reliable communication between all stakeholders in infrastructure support; research infrastructure that ensures the performance of fundamental and applied scientific research in promising areas of innovative medicine. The above classification of the infrastructures for innovative medicine is not exhaustive: the phenomenon under study is actually much more complex than their classificatory descriptions. These categories are conditionally and are mutually integrated.

Understanding the importance of adequate infrastructure for supporting innovative medicine, the expert communities fully realize that no remarkable idea can be implemented to enhance a medical service without due materials, technical support, personnel, and information resources. Know-how alone is not enough; an economically sustainable basis is needed. And innovative medicine, as the medicine involving high-tech medical care, appears to be especially vulnerable to any lack of its supporting infrastructure, as was shown above. The importance of adequate infrastructure in innovative medicine cannot be overestimated.

It is gratifying to note that the efforts to enhance the infrastructure support for innovative medicine have done some noticeable progress. Despite the difficulties, primarily of economic nature, this area is actively developing.

There emerge medical technology centers (medical technoparks)—the promising venues with high concentration of innovative medical and pharmaceutical industries that are working to promote development of medical IT technologies by creating prototypes of medical instrumentation, etc. Medical clusters are being formed, bringing together all sorts of medical facilities and establishing research and training institutions as their "anchor elements". New samples of medical and pharmaceutical equipment are being developed and introduced into production. Specialized medical facilities are being built—modern diagnostic centers, outpatient clinics, rehabilitation centers, cryobanks, ecological spaces, etc.

At the same time, there are many unresolved problems in the field of infrastructure support for innovative medicine.

Given the priority of specialized healthcare buildings, structures and engineering systems in medical care sector, one important area, and also source of challenges, in the activities related to the development of infrastructures is the management of their lifecycles and performance levels at the stage of design, construction and operation. This management involves, in particular, participation in the development of projects in order to ensure that their design and planning solutions correspond to their functional purpose; identification of optimal configuration of engineering networks and systems; purchase and installation of equipment and other material and technical means essential to the normal functioning of modern medical complexes; policy development, organization and implementation of all types of infrastructure maintenance of buildings, structures, engineering networks, systems equipment used in the delivery of medical services; re-design and redevelopment of premises in accordance with the changing technological processes, as well as technical, sanitary, ergonomic and other requirements and standards applicable at national and international levels to medical organizations; creation of an environment favorable for the provision of medical services and meeting the current sanitary and hygienic standards and consumers' expectations; comprehensive safety assurance; cleaning of premises and waste disposal; landscaping of adjacent territories for a positive therapeutic effect, etc. One way to optimize the processes of design, construction and operation of the above-mentioned facilities is offered by Building Information Modeling (BIM). BIM can serve as basis for decision-making throughout the entire life cycle of a facility, from initial design concept to engineering to construction to operation to demolition.

The above challenges are certainly not the only ones in the design, construction and operation of innovative healthcare facilities and their supporting infrastructures. Many other issues remain relevant.

Thus, the principles of public-private partnership and insurance medicine require re-interpretating from the perspective of balance between public and private interests, and possible limits for combining gratis and fee-based services in medical, sanitary, preventive, rehabilitation and pharmaceutical sectors. In the healthcare system, different organizational and legal forms, as well as forms of ownership need to be provided with rationales that take into account the specifics and the social significance of the given units within a healthcare facility. In this regard, it should be noted that the infrastructures to support innovative medicine still suffer from a number of problems of legal nature.

Technological backwardness remains an increasing problem in a number of developing regions including Russia. The medical industry and its high-tech services are inhibited by insufficient progress with modern (digital, nano, etc.) technologies and on the intellectual property market with its critically low level of inventive and patent activity. As a result, Russia has only a small share in exports and, conversely, high share in imports of medical products and services with high added value, including intellectual value, i.e. the value resulting mostly from internal integration of intellectual capital components. Together with technological dependence on the leading countries, this imbalance in the foreign trade structure, that manifests itself in Russia's enjoying a negligible share on the international markets (in terms of promoting high-tech products and services) are not only factors constraining the development of the domestic medical industry, but also threats to national security.

The problems we have raised are complex and multidimensional. Their solution lies in improving scientific, technical, methodological, organizational, economic and regulatory foundations towards higher quality of the infrastructures supporting innovative medicine. This, in turn, requires targeted scientific research in the field under consideration.

As for scientific research, the promising approaches to the study of the infrastructures supporting innovative medicine involve interdisciplinary and comparative approach. In fact, since medicine combines scientific and practical developments from a variety of fields, it is seen as an industry of strategic importance.

Let us recall here that at the turn of the millennium, the productive lines of medical research are those that involve two or more sciences, the marker of the interdisciplinary interaction being intense import-export of scientific methodology and research products and the most noticeable scientific result being contribution to the formation of a novel metamethodology. The interdisciplinary approach, as stated by science theorists, has proved highly fruitful, since the most noticeable discoveries of our time were achieved through the combined effort of specialists with different background. It is not surprising that this trend is pursued even more intensely.

Interdisciplinary research is pursued also in innovative medicine, the most promising areas lying at the confluence of different fields of knowledge and represented by bio-informatics, bioengineering, molecular medicine, nanobiotechnology, medical robotics, among others [5].

The above characteristics (systemic nature and interdisciplinarity) of the modern innovative medicine have their effect on the performance of supporting infrastructures. Complex problems, which accompany also the infrastructures for innovative medicine, can be solved only by using complex means and methods, and this rule applies not only to theoretical studies, but also practical organization of measures designed to solve the problems identified.

In this regard, the interaction of specialists with different backgrounds (physicians, pharmacists, programmers, designers, builders, experts, lawyers, economists, etc.) is welcome. With multiple approaches to solving the challenges of infrastructures supporting innovative medicine, this interaction can create a synergetic effect and bring these infrastructures to a qualitatively new level.

4   Discussion

The results of this study into the current state, trends, problems and prospects the infrastructures supporting innovative medicine have been reported and discussed at the international research-to practice conference "Design, Construction and Infrastructure of Innovative Healthcare" (St. Petersburg, Russia, May 19–21, 2021).

The scientific and medical community have found the results reliable and worthy of attention. The authors are recommended to continue scientific research in this field.

5 Conclusions

1. At the present stage of the development of medicine, its innovative nature becomes clearly evident, manifesting itself in the increasing volume of high-tech medical care and the introduction of information, digital, molecular and other high-tech technologies.
2. The current trends in healthcare sector are towards new, innovation-driven medicine with focus on prevention and personalized treatment that are based on in-depth understanding of human biology and all its levels starting from molecular.
3. In innovative medicine, the optimal performance and progress (towards expanded nomenclature, availability and quality of medical services) is largely contingent on the availability of adequate infrastructure.
4. As a system for ensuring adequate performance of healthcare facilities, infrastructure comprises mutually integrated subsystems of logistics, technologies, economy, human resources and personnel.
5. Both innovative medicine and its infrastructure are characterized by a complex, inter-sectoral, systemic nature. These systems are complex, open and dynamically developing. Similarly, the problems of infrastructural support for innovative medicine are multidimensional, requiring a range of solutions to scientific, technical, methodological, organizational, economic and regulatory issues.
6. The ideal infrastructure support for innovative medicine is the one that is sustainable, adaptive and allows to meets the current needs of the industry and give it a new impetus.
7. As a highly complex system, the infrastructure for innovative medicine should be investigated and managed using means and methods of a systemic nature. This explains high relevance of systemic interdisciplinary research in the field of infrastructural support for innovative medicine, the most productive studies lying at the confluence of multiple sciences.

References

1. Anom BY (2020) Ethics of Big Data and artificial intelligence in medicine. Ethics Med Pub Health 15:1–11
2. Cambon-Thomsen A, Ducournau P, Gourraud PA et al (2003) Biobanks for genomics and genomics for biobanks. Comp Funct Genomics 4(6):628–634. https://doi.org/10.1002/cfg.333
3. Galimova RM, Buzaev IV, Ramilevich KA et al (2019) Artificial intelligence—developments in medicine in the last two years. Chronic Dis Transl Med 5(1):64–68. https://doi.org/10.1016/j.cdtm.2018.11.004
4. Li JPO, Liu H, Ting DS et al (2021) Digital technology, tele-medicine and artificial intelligence in ophthalmology: a global perspective. Prog Retin Eye Res 82:100900. https://doi.org/10.1016/j.preteyeres.2020.100900

5. Ravid K, Faux R, Corkey B et al (2013) Building interdisciplinary biomedical research using novel collaboratives. Acad Med J Assoc Am Med Coll 88(2):179–184. https://doi.org/10.1097/ACM.0b013e31827c0f79

Ingram Content Group UK Ltd.
Milton Keynes UK
UKHW020726060623
422954UK00007B/538